Techniques and Concepts of High-Energy Physics

NATO Science Series

A Series presenting the results of scientific meetings supported under the NATO Science Programme.

The Series is published by IOS Press, Amsterdam, and Kluwer Academic Publishers in conjunction with the NATO Scientific Affairs Division

Sub-Series

I. Life and Behavioural Sciences	IOS Press
II. Mathematics, Physics and Chemistry	Kluwer Academic Publishers
III. Computer and Systems Science	IOS Press
IV. Earth and Environmental Sciences	Kluwer Academic Publishers
V. Science and Technology Policy	IOS Press

The NATO Science Series continues the series of books published formerly as the NATO ASI Series.

The NATO Science Programme offers support for collaboration in civil science between scientists of countries of the Euro-Atlantic Partnership Council. The types of scientific meeting generally supported are "Advanced Study Institutes" and "Advanced Research Workshops", although other types of meeting are supported from time to time. The NATO Science Series collects together the results of these meetings. The meetings are co-organized bij scientists from NATO countries and scientists from NATO's Partner countries – countries of the CIS and Central and Eastern Europe.

Advanced Study Institutes are high-level tutorial courses offering in-depth study of latest advances in a field.
Advanced Research Workshops are expert meetings aimed at critical assessment of a field, and identification of directions for future action.

As a consequence of the restructuring of the NATO Science Programme in 1999, the NATO Science Series has been re-organised and there are currently five sub-series as noted above. Please consult the following web sites for information on previous volumes published in the Series, as well as details of earlier sub-series.

http://www.nato.int/science
http://www.wkap.nl
http://www.iospress.nl
http://www.wtv-books.de/nato-pco.htm

Series C: Mathematical and Physical Sciences – Vol. 566

Techniques and Concepts of High-Energy Physics

edited by

Harrison B. Prosper

Florida State University,
Tallahassee, Florida, U.S.A.

and

Michael Danilov

ITEP, Moscow, Russia

Kluwer Academic Publishers

Dordrecht / Boston / London

Published in cooperation with NATO Scientific Affairs Division

Proceedings of the NATO Advanced Study Institute on
Techniques and Concepts of High-Energy Physics
St. Croix, Virgin Islands, U.S.A.
15–26 June 2000

A C.I.P. Catalogue record for this book is available from the Library of Congress.

ISBN 1-4020-0157-6 (HB)
ISBN 1-4020-0158-4 (PB)

Published by Kluwer Academic Publishers,
P.O. Box 17, 3300 AA Dordrecht, The Netherlands.

Sold and distributed in North, Central and South America
by Kluwer Academic Publishers,
101 Philip Drive, Norwell, MA 02061, U.S.A.

In all other countries, sold and distributed
by Kluwer Academic Publishers,
P.O. Box 322, 3300 AH Dordrecht, The Netherlands.

Printed on acid-free paper

Printed in the Netherlands.

Table of Contents

Preface

The eleventh Advanced Study Institute (ASI) on Techniques and Concepts of High Energy Physics marks the transition from an extraordinary century of science to one that will surely bring wonders we can scarcely imagine. It also marks a transition from its founder, the inimitable Tom Ferbel, to its new directors. We are honored to have been asked to continue the venerable tradition that Tom established. The school is his distinctive creation, and will always bear his mark.

The 2000 meeting was held at the Hotel on the Cay in St. Croix. It is an ideal location: sufficiently secluded to inspire a vigorous but informal intellectual atmosphere, yet close enough to the main island to afford opportunities to mingle with the locals and partake of their hospitality. Altogether 76 physicists both young, and not so young, participated from 18 countries. For the first time, this meeting attracted a substantial number of students from Eastern Europe, all of whom were warmly welcomed. The bulk of the financial support for the meeting was provided by the Scientific Affairs Division of the North Atlantic Treaty Organization (NATO). The ASI was co-sponsored by the U.S. Department of Energy (DOE), by the Fermi National Accelerator Laboratory (Fermilab), by the U.S. National Science Foundation (NSF), the University of Rochester, Florida State University (FSU) and the Institute for Theoretical and Experimental Physics (ITEP, Moscow).

As is the tradition, the scientific program was designed for advanced graduate students and recent PhD recipients in experimental particle physics. The present volume covers topics that update and complement those published (by Plenum and Kluwer) for the first ten ASIs. The material in this volume should be of interest to a wide audience of physicists.

We wish to thank, first and foremost, Tom Ferbel for showing us the ropes. Given the number of details to be borne in mind we are simply amazed that Tom did this for twenty years! We thank the students and lecturers who made this meeting a success, both on and off the lecture floor. To see John Schwarz dance with the next generation was quite a treat! We are grateful to the lecturers for their hard work in preparing both their lectures and their manuscripts for the Proceedings. Without such dedication this ASI would not work. The members of our Advisory Committee (whose names are listed in the back of the volume) provided excellent advice for which we thank them heartily. We thank Barbara Smalska for organizing the delightful wine tasting evening, Tuba Conka-Nurdan for organizing the excellent student presentations, and Zandy-Marie Hillis of the United States National Park Service for her wonderful description of the marine life and geology of St. Croix.

We thank Michael Witherell for making the resources of Fermilab available to us. One of us (HBP) wishes to thank the FSU High Energy Physics group for supporting this effort and Pat Rapp for support from the DOE. We could not have succeeded without the tremendous help provided by Connie Jones (Rochester) and Kathy Mork (FSU). Ken Ford (FSU) did an excellent job designing the poster for the School and we thank him for it.

We owe special thanks to Marion Hazlewood for her efficiency and hospitality, and that of her staff at the Hotel on the Cay, as well as Janene Betterton for her exemplary work at the King Christian Hotel. We thank Hurchell Greenaway and his staff at the Harbormaster who worked cordially and hard to keep us well fed and entertained. Finally, our thanks go to Dr. Fausto Pedrazzini for his strong support and the NATO Division of Scientific Affairs for their cooperation and confidence in what we have tried to do.

HARRISON B. PROSPER

MICHAEL DANILOV

THE STANDARD MODEL: 30 YEARS OF GLORY

Jacques Lefrançois
Laboratoire de l'Accélérateur Linéaire
IN2P3-CNRS et Université PARIS-SUD
Centre Scientifique d'Orsay - Bât. 200 - B.P. 34
91898 ORSAY Cedex (France)
lefranco@lal.in2p3.fr

Abstract In these three lectures, I will try to give a flavour of the achievements of the past 30 years, which saw the birth, and the detailed confirmation of the Standard Model (SM).

1. INTRODUCTION

I have to start with three disclaimers:

- Most of the subject is now history, but I am not an historian, I hope the main core will be exact but I may err on some details.
- Being an experimentalist, I will mostly focus on the main experimental results, giving, when appropriate, some emphasis to the role of well conceived apparatus. This does not mean that I minimise the key role of the theorist, it only reflects the fact that the same story told by a theorist would correspond to another interesting series of lectures.
- Finally, in some cases, I may show a bias coming from the fact that I have seen from closer what has happened in Europe.

The three lectures will cover:
QCD from the discovery of deep inelastic scattering at SLAC to the gluon discovery at Petra.
Weak interaction and the quarks and lepton families, from the neutral current discovery to the members of the third family.
Finally, I will spend some time on LEP and SLC results since they are (or have been) ideal machines for detailed standard model studies. However, since these last results are well covered in many schools and conferences it has been greatly shortened in the written version.

H.B. Prosper and M. Danilov (eds.), Techniques and Concepts of High-Energy Physics, 1–49.

2. QCD

2.1. DEEP INELASTIC AT SLAC

Up to the 1967 SLAC discovery, electron scattering was used to measure elastic and inelastic nuclear and nucleon form factors. In the language of the time the nucleon form factors (which revealed the charge distribution inside the nucleon) corresponded to the probability of hitting the meson cloud around the nucleon and keeping the nucleon intact or transforming it into an excited state (N*). In modern language they represent the probability to hit a quark inside the nucleon and nevertheless keeping the nucleon intact or transforming it into an excited 3 quark state.

The experimental breakthrough in 1967 was the construction of a high intensity 20 GeV electron linear accelerator. The apparatus was a spectrometer rather classical (for linac apparatus). The accent was put on reliability and very powerful online computer control (which was new for the time). The first result in 1968 (fig.1) showed clearly that something new was happening and that the Q^2 dependence of the cross-section was quite different from previous elastic and inelastic results and similar to what would be expected to result of scattering on point-like objects.

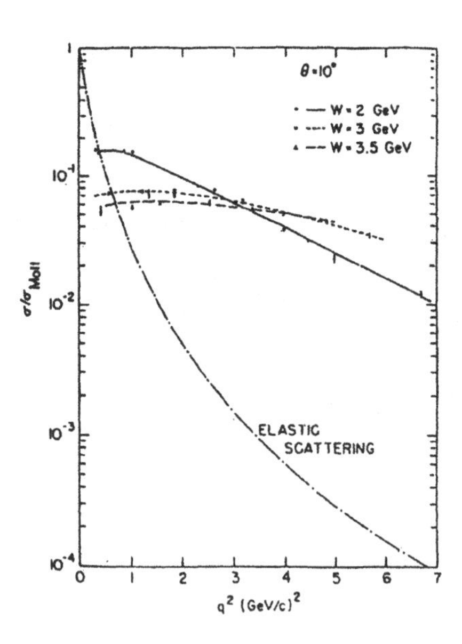

Figure 1 The elastic and inelastic cross-section as function of Q^2 showing the absence of form factor at large excitation energy.

The main variables are defined below:

$$Q^2 = 2EE'(1 - cos\theta) = 4EE' sin^2\theta/2,$$

$$\nu = E - E' \quad \omega = \frac{2M\nu}{Q^2} \quad x = \frac{Q^2}{2M\nu},$$

and the cross-section, in the most general case is

$$\frac{d^2\sigma}{d\Omega dE'} = \sigma_{Mott}(W2 + 2W1tg^2\theta/2),$$

where $\sigma_{M_{ott}}$ is a point-like cross-section.

In 1969 Bjorken predicted scaling of the W structure functions i.e. that, as Q^2 and ν went to large values 2MW1(ν,Q) and νW2(ν,Q) would become only functions of the scaling variable ω or x. Scaling was experimentally confirmed as shown in figure 2.

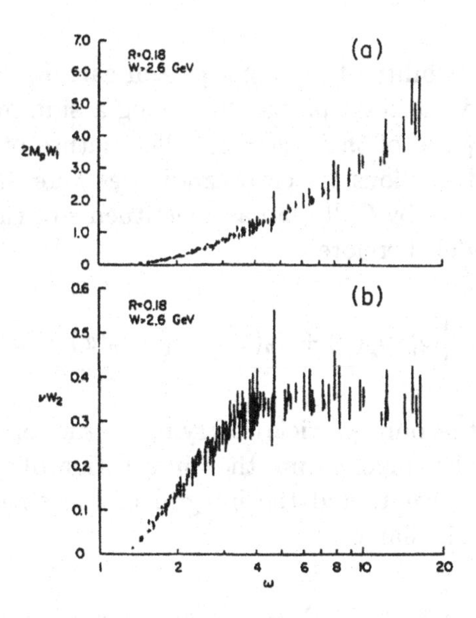

Figure 2 Structure functions at various Q^2 as function of w, showing scaling behaviour.

Actually, already in 1966, using current algebra and sum rules Bjorken had derived the following formula

$$\lim_{E\to\infty} \left[\frac{d\sigma}{dQ^2}ep + \frac{d\sigma}{dQ^2}en\right] \geq \frac{2\pi\alpha^2}{Q^4}.$$

This implied clearly a cross-section varying with Q^2 in the way expected in the case of scattering from point-like objects.

However it went unnoticed by the experimentalists preparing the SLAC experiment and the result shown in figure 1 came as a surprise.

More generally the theoretical framework was badly understood by experimentalists until about 1969. In that year Feynman invented his very intuitive parton model to explain the deep inelastic results. The parton were point-like constituents inside the nucleon; in the frame where the nucleon momentum is very large, the collision is instantaneous and the partons are independent. If the nucleon has a momentum P the partons have momentum xP. The kinematics and dynamics is then a very simple electron parton elastic scattering and the following formula is derived

$$\nu W_2(\nu, Q^2) = F_2(x) = \sum_1^N Q_i^2 \times f_i(x)$$

$f_i(x)$ being the probability of finding a parton carrying a fraction x of the momentum and the total probability being a sum over all partons weighted by the square of their charges. The nature of these partons was not immediately obvious but clear candidates were the u and d type quarks invented earlier by Gell-Man as constituents of the nucleon.

In the quark model therefore:

$$F_2^p(x) = x\left[Q_u^2(u_p(x) + \bar{u}_p(x)) + Q_d^2(d_p(x) + \bar{d}_p(x))\right]$$

If the quarks are the only particles carrying a fraction of a deuteron (or other I = 0 nuclei) momentum, then integration of the momentum distribution should give 1, and the integral of the structure function should give the quark charge:

$$\left(Q_u^2 + Q_d^2\right)/2 = 5/18 = .28$$

In case of the 2/3, 1/3 charge assignment the integral should be 0.28. Experimentally a value of 0.14 ± 0.005 was found: either the quark interpretation of partons was wrong or something else (chargeless) was carrying 50% of the nucleon momentum. (We of course now know that it is the gluons). More information was needed to solve this riddle. It was provided by neutrino experiments.

2.2. NEUTRINO SCATTERING RESULTS (1972-1974)

In this case the experimental breakthrough was the existence of a high intensity neutrino beam at the CERN 28 GeV PS and the construction of a large heavy liquid bubble chamber Gargamelle.

The neutrino beam was made possible by the invention around 1966 by Simon Van der Meer of the neutrino horn which was used to focus pions over a wide momentum and angular band. The decay of the pions giving rise to the neutrino beam. The improvement of an order of magnitude in neutrino flux can be seen in figure 3.

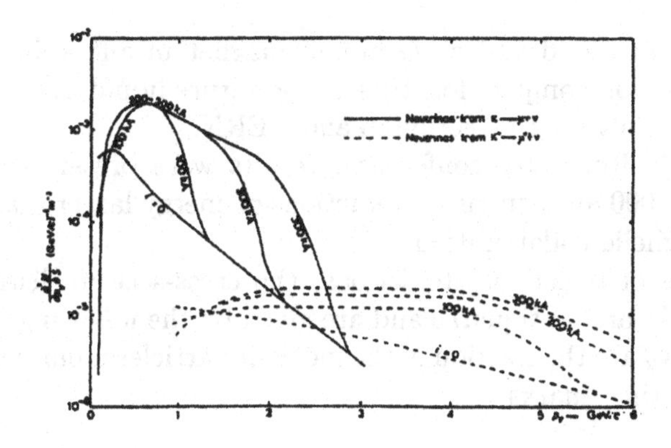

Figure 3 The neutrino yield, as function of energy for various values of the horn current.

The decision to construct Gargamelle (fig. 4) was an extremely effective one for doing medium energy neutrino physics.

Gargamelle's dimensions were 4.5 m in length and 1.5 m in diameter; it could be filled with freon of density 1.2 to 1.5 g/cm^3, a clear mass advantage compared to liquid hydrogen bubble chambers. The short radiation and interaction length of freon together with the bubble chamber resolution meant that neutrino reactions could be followed in detail which was a decisive advantage for few GeV neutrino physics when compared to massive sampling calorimeter instruments which will be more suited for high energy experiments.

Figure 4 **The Gargamelle bubble chamber.**

Finally warm liquid and a warm coil magnet meant a simpler and faster construction compared to that of the future liquid hydrogen bubble chambers built later at Fermilab and CERN.

At the 1972 Rochester conference, results were presented on 1000 neutrino and 1000 antineutrino interactions of energy larger than 1 GeV, by the Gargamelle collaboration.

In the case of ν and $\bar{\nu}$ interactions, the cross-section distinguishes between quarks and antiquarks and are given by the following formulas, where X_1 an X_2 are the fraction of the incident particles momenta carried by the interacting quarks:

$$\frac{d^2\sigma^{\nu,\bar{\nu}}}{dxdy} = \frac{G^2ME}{\pi}\left[(1-y)F_2(x) + \frac{y^2}{2} \times 2xF_1(x) \pm y\left(1-\frac{y}{2}\right)xF_3\right],$$

$$2xF_1 \approx F_2,$$

$$\sigma_\nu + \sigma_{\bar{\nu}} = \frac{G^2ME}{\pi}\frac{4}{3}\int_0^1 F_2 dx,$$

$$\int_0^1 F_2 dx = \int x\left[u + \bar{u} + d + \bar{d}\,\right] dx = 0.47 \pm 0.07.$$

Results confirmed that 50% of the nucleon's momentum is transported by particles without charge and therefore without electromagnetic and weak interactions (the gluons). Comparison between electron nucleon and neutrino nucleon interactions confirmed the fractional charge of quarks.

Furthermore from the difference $\nu - \bar{\nu}$ the xF_3 function was obtained; and it could be shown that valence quarks dominate the momentum distribution:

$$B = \frac{\int_0^1 xF_3}{\int_0^1 F_2} = \frac{\int u + d - \bar{u} - \bar{d}}{\int u + d + \bar{u} + \bar{d}} = 0.9 \pm .1$$

also the integral over F_3 gives the number of valence quarks. At the 1974 Rochester conference Gargamelle announced : $\int_0^1 F_3 = 3.2 \pm .6$ a striking confirmation of the standard model.

By 1973 the QCD theory had been proposed and strong interaction was finally understood as a force between coloured quarks mediated by an octet of coloured gluons. As a consequence of the theory a certain number of phenomena could be predicted and calculated in perturbation theory.

2.3. R(E^+E^-)

The prediction of a modification of R(e^+e^-) was one of the earliest predicted consequences of the colour concept.

If the production of hadrons in e^+e^- annihilation at high energy is considered to result from the creation of quark antiquark pairs the predicted ratio of the parton production cross-section to the μ pair cross-section will be higher by a factor 3 because of the 3 colour possibilities.

$$R = \frac{\sigma_{had}}{\sigma_{\mu^+\mu^-}} = 3 \times \sum_1^{n_f} Q_i^2 + n_{\ell \neq e,\mu}$$

The first term is the contribution of $q\bar{q}$ pairs and the second one is the contribution of a new lepton family (of charge one and without colour factor) which has to be added since it was at that time indistinguishable from hadronic events. The first results in 1973 were obtained at 4-5 GeV at the CEA storage ring (built by Harvard-MIT). They were inconclusive because of the ignorance of the charmed quark and τ lepton existence (the τ pairs were observed as hadrons since essentially everything not identified as Bhabba events or μ pair events were called hadrons). As a result the cross-section was observed to be higher than prediction (fig. 5).

Even in 1975 after the discovery of the c quark the SLAC MARKI results (fig. 6) were thought to be in disagreement with predictions. The observed R value at high energy was about 5.0 while the predicted

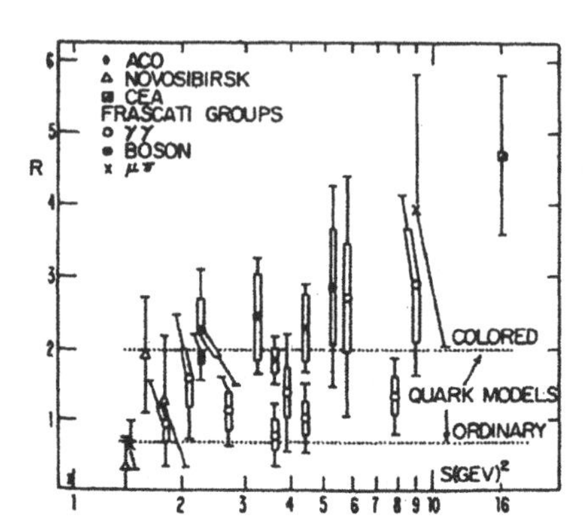

Figure 5 The R value as function of s, the square of the center of mass energy.

value was 3.33. This fact was one of the reasons of the slow acceptance of QCD. It took the confirmation of the τ lepton existence in 1975-1977 and the calculation of the first order QCD correction $(1. + \alpha s/\pi)$ to change gradually the predicted value to about 4.6.

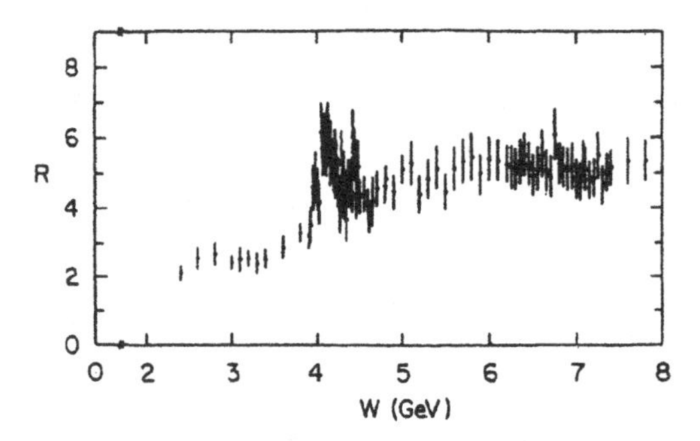

Figure 6 R values obtained by MARKI as function of the center of mass energy.

2.4. SCALING VIOLATION

By analogy with QED, QCD radiative corrections were expected in deep inelastic cross-section of electrons muons or neutrinos (fig.7).

As a result, the parton structure functions were predicted to become Q^2 dependent. There were some early but inconclusive sign of this phenomenon in 1972 SLAC data but which could not be disentangled from threshold effects at low Q^2. By the 1975 lepton photon conference ev-

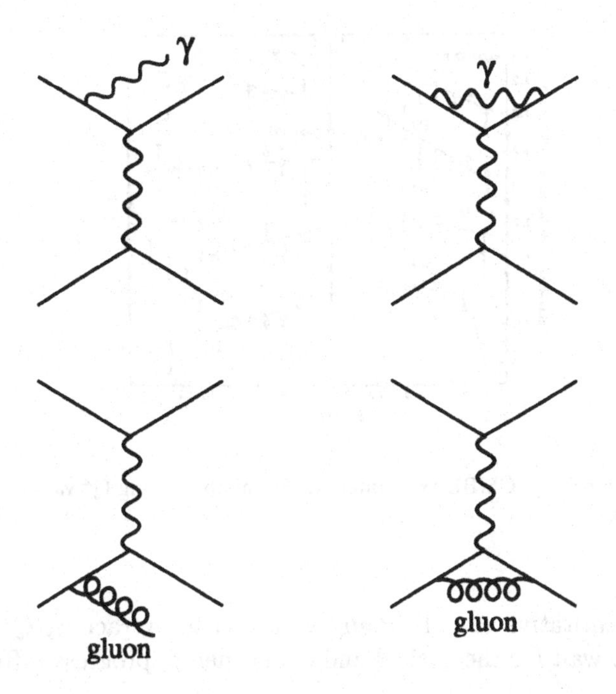

Figure 7 QED and QCD radiative corrections to deep inelastic cross-section.

idence of scaling violation were presented from the SLAC experiment (fig. 8) and from a FERMILAB experiment on muon iron scattering at high energy done by the Cornell-Michigan state-Berkeley-La Jolla collaboration (fig. 9).

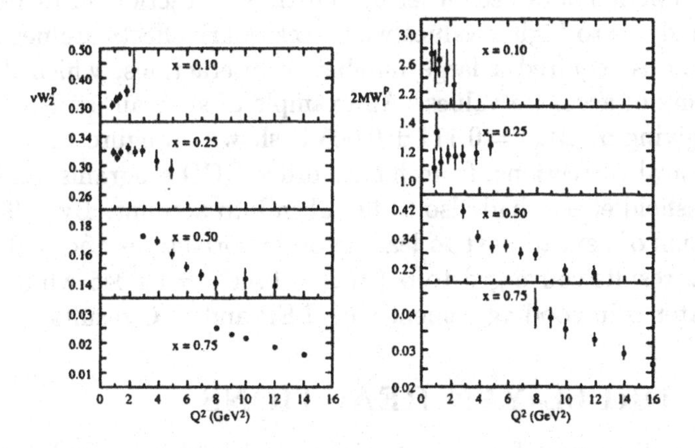

Figure 8 SLAC results on Q^2 variation of structure functions.

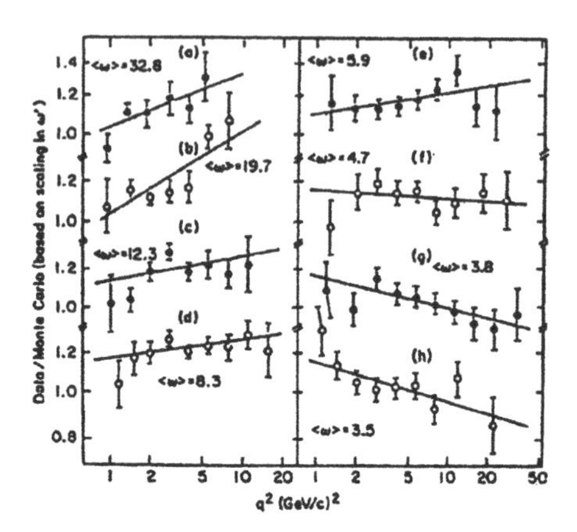

Figure 9 Result of the CMBL experiment at Fermilab showing Q^2 variation of structure functions.

For a quantitative use of scaling violation to extract $\alpha_s(Q^2)$, it was necessary to wait for theoretical and experimental progress (after 1977).

By 1977 the DGLAP equations were invented which gave a quantitative prescription of structure function evolution.

For increasing Q^2 the valence quark structure function is depleted by gluon emission feeding in the gluon structure function whose evolution in turn modifies the sea quark structure functions. Because of the ignorance of the gluon structure function, the evolution studies used to extract accurate values of α_s had to be restricted to the valence quark structure function obtained either by neutrino interaction, or by restricting the analyses to high x values where sea quark effects are negligible. Both methods required a large number of interactions, which delayed the securing of accurate values. An example of such an analysis (done in 1992) giving $\alpha_s(Mz) = 0.111 \pm 0.003$ is shown on figure 10.

Theoretical corrections, from higher order QCD diagrams, and from $1/Q^2$ threshold effects had also to be taken into account. By 1999 with the inclusion of next to next to leading order corrections and high accuracy data, results converged to $\alpha_s(Mz) = 0.1172 \pm 0.0045$ which as we will see later is in good agreement with LEP and SLC results.

2.5. DRELL-YAN REACTIONS

After the discovery of deep inelastic scattering, it was rapidly realised that there should exist a corresponding effect in hadron collisions where

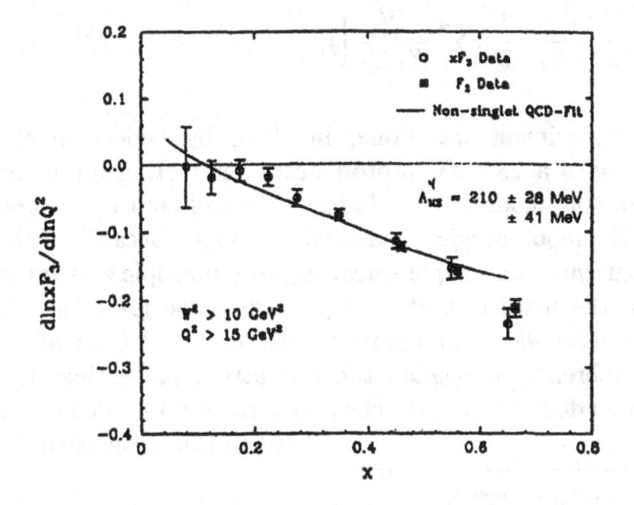

Figure 10 Extraction of Λ_{QCD} from the shape of the valence quark structure function of high x values by the CCFR neutrino experiment at Fermilab.

quark and antiquark would annihilate to create μ or electron pairs: the so called Drell-Yan effect (1970) (Fig. 11).

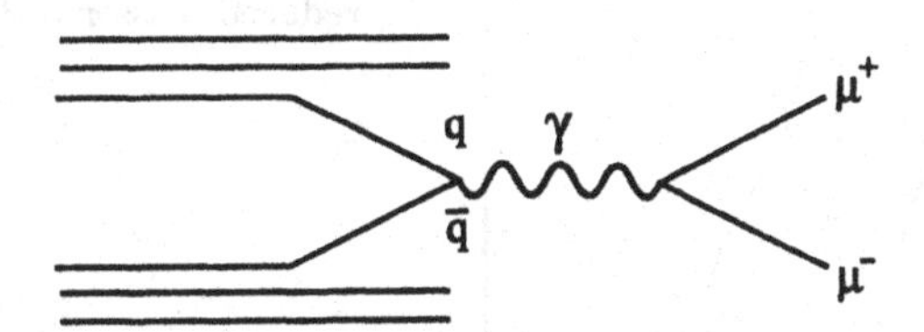

Figure 11 Feynman diagram of the Drell-Yan effect.

Application of QCD meant that the predicted cross-section was smaller by a factor 3 taking into account that annihilation can only happen if the colour of the quark and antiquark match each other. As in deep inelastic scattering, the transverse momentum of the partons are neglected in first order, kinematics and cross-section are then described by the following formulas, where X_1 and X_2 are the fraction of the incident particles momenta carried by the interacting quarks:

$$M^2_{\mu\mu} = X_1 X_2 s,$$

$$X_{\mu\mu} = \frac{2P*_L}{\sqrt{s}} = X_1 - X_2,$$

$$\frac{d^2\sigma}{dX_1dX_2} = \frac{4\pi\alpha^2}{3sX_1X_2}\times\frac{1}{3}\times\sum_i\frac{Q_i^2}{X_1X_2}\left[f_i^{h1}(X_1)f_{\bar{i}}^{h2}(X_2)+f_{\bar{i}}^{h1}(X_1)f_i^{h2}(X_2)\right].$$

The first experiment was done, in 1970, by Lederman et al., at Brookhaven, with a 28 GeV proton beam interacting on an iron target and producing muon pairs. Hadrons were filtered by an iron beam dump and the muon energies measured by range. Because of the crude energy measurement and angle smearing from multiple scattering in the beam dump, the mass resolution was very coarse (> 15%). Because of this the results were impossible to interpret and even at the 1974 Rochester conference in London the rapporteur presenting the results (Fig. 12) concluded: “it is fairly clear that theory doesn’t have much to say ured in this experiment”.

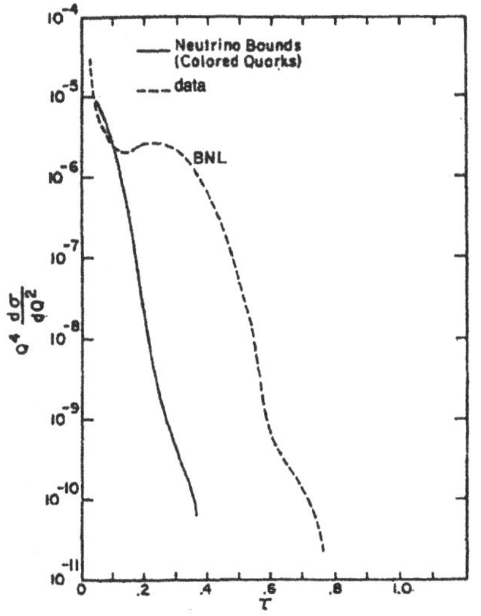

Figure 12 The first Drell-Yan cross-section measured at BNL by Lederman et al. as function of the reduced mass $\tau = M/\sqrt{s}$.

Of course we now know that the huge discrepancy was due to the production of the J/ψ resonance followed by its decay to a muon pair.

The experimental situation started to improve around 1977 when the CFS collaboration at Fermilab did a more accurate experiment both because of the higher beam energy and because of the much better apparatus (fig. 13) which will be described when the upsilon discovery will be discussed.

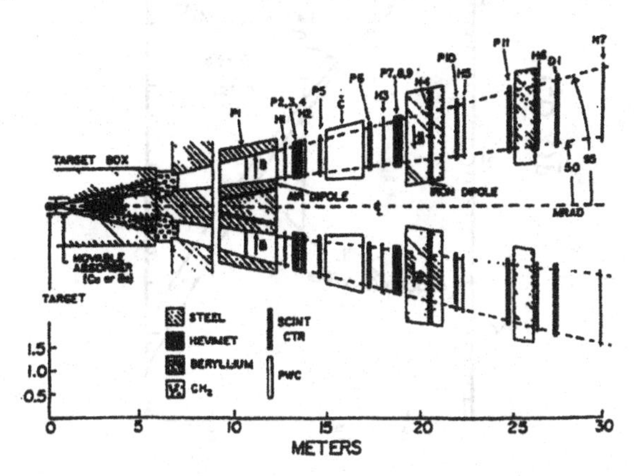

Figure 13 The two arm spectrometer used to measure Drell-Yan cross-section at Fermilab.

By the 1978 Rochester conference in Tokyo good results were presented by the CFS collaboration (and by 3 ISR experiments with smaller number of events). The results (Fig. 14) seemed to be in agreement with the Drell-Yan prediction, however the prediction depended on the antiquark structure function extracted from neutrino scattering data with rather large statistical and systematical errors.

However by 1979 it was realised that the tree level prediction of Drell-Yan cross-section should receive a large QCD correction; it was the so-called K factor calculation. The basis of the argument is that when the vertex correction of QCD are calculated at negative Q^2 (deep inelastic) or positive Q^2 (Drell-Yan) the correction differ by an extra factor of $(\ln(-1))^2$ or π^2, in the case of Drell-Yan. The final first order Drell-Yan QCD correction was therefore of order 0.6 and it was conjectured that the series $1 + 0.6$ calculated to higher order could be exponentiated and result in a K factor correction of $\exp(0.6)=1.8$ to the predicted cross-section.

Results on proton proton and pion proton cross-section gave, in 1979, early indication of the presence of the K factor; however the clearest result was presented in 1980 measurements, by the NA3 collaboration at CERN, of antiproton (and proton) Drell-Yan cross-section on nuclei. The subtracted $(\bar{p} - p)$ N cross-section corresponds to pure valence

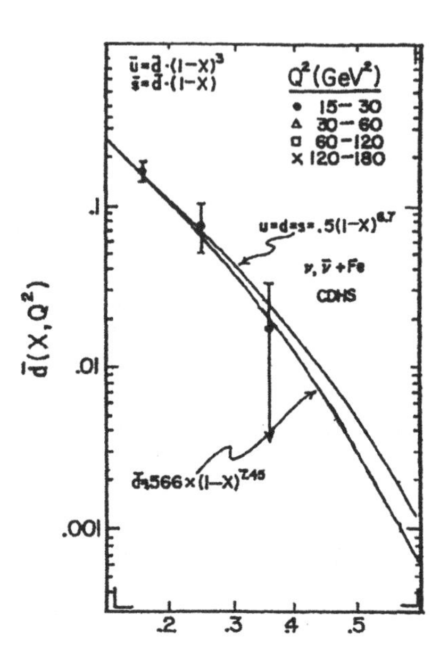

Figure 14 Nucleon six quark structure function extracted from Drell-Yan cross-section (data points) compared to fit of neutrino results.

quark interaction and can therefore be predicted accurately. As shown in figure 15 the measured and predicted structure functions agree very well; the measured K factor is 2.3 ± 0.4 in good agreement with the large predicted QCD correction.

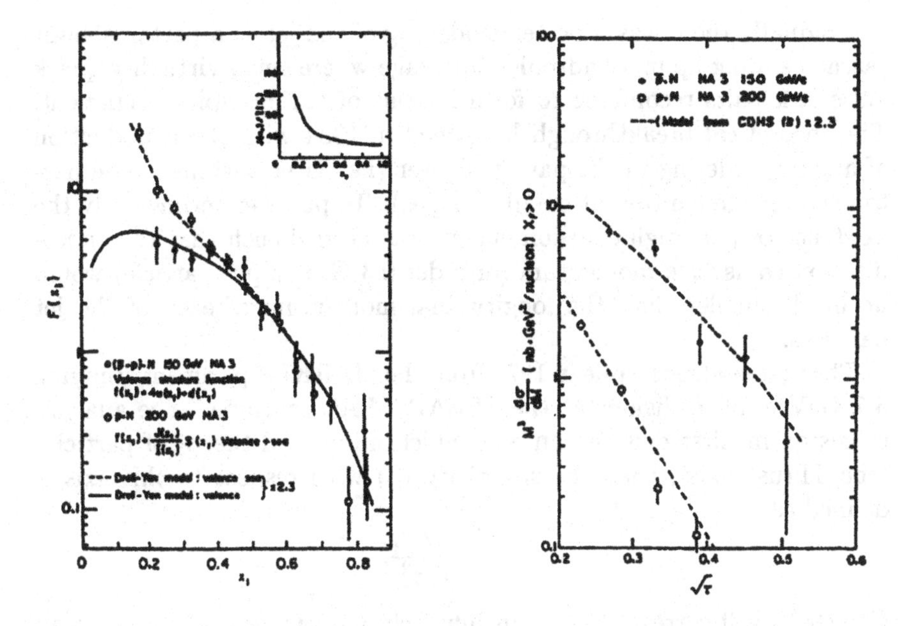

Figure 15 NA3 results on $\bar{P}N$ and PN Drell-Yan measurements compared to fit of nucleon valence quark cross-section.

2.6. OBSERVATION OF JETS

Almost immediately after the interpretation of deep inelastic scattering as lepton quark collisions, the question of the fate of the recoiling quark was raised. Already, in the 1970-1972 period, it was predicted that this recoiling quark would give birth to a jet by some sort of outside inside cascade creating $q\bar{q}$ pairs and then mesons (fig. 16).

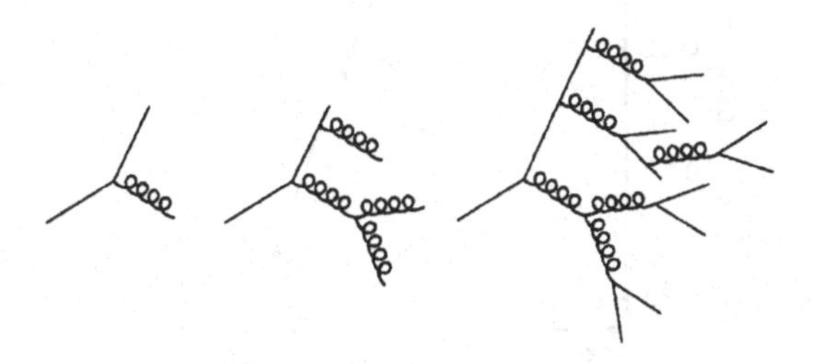

Figure 16 Parton shower cascade.

Gradually there was a better modelling of the jet as a parton shower cascade ending up in a hadronisation stage where small virtuality quark antiquark pairs recombine to form mesons of various spin assignment. The theoretical breakthrough happened in 1984 with the introduction of angular ordering in the parton shower (i.e. later partons in the parton shower are emitted at smaller angles). To prove experimentally the existence of jets, high collision energy was needed such that the hadronisation transverse momentum (of order 0.3 GeV/c per particle) would be much smaller than the longitudinal momentum of each of the jet particles.

The first evidence came in 1975 from the MARKI experiment studying 6-7 GeV e^+e^- collisions at the SPEAR/SLAC machine. The analysis consisted in, first defining an axis which minimised the p_t of particles (the Thrust axis), then the sphericity (S) with respect to this axis is defined as

$$S = \frac{3\sum P_{Ti}^2}{2\sum \vec{p}_i^{\,2}}.$$

Clearly S = 0 corresponds to an infinitely narrow jet and S = 1 to an isotropic event. Contrary to what was expected for a phase space model of particle production, it was found that the mean sphericity decreased with collision energy (fig. 17) as expected in a jet model.

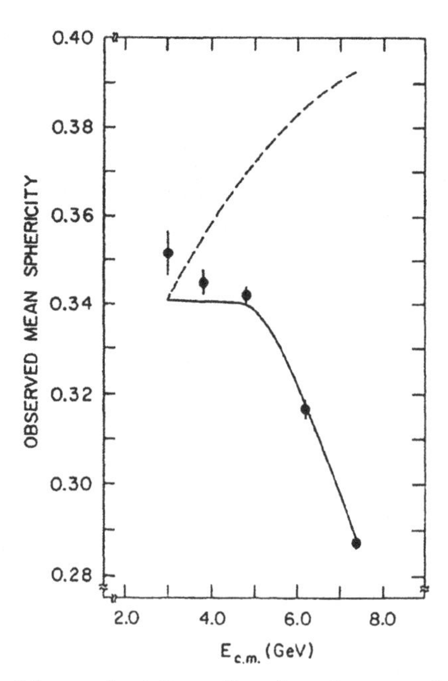

Figure 17 Mean sphericity as function of center of mass energy.

And at high energy, the S distribution of events was characteristics of a jet model (fig. 18).

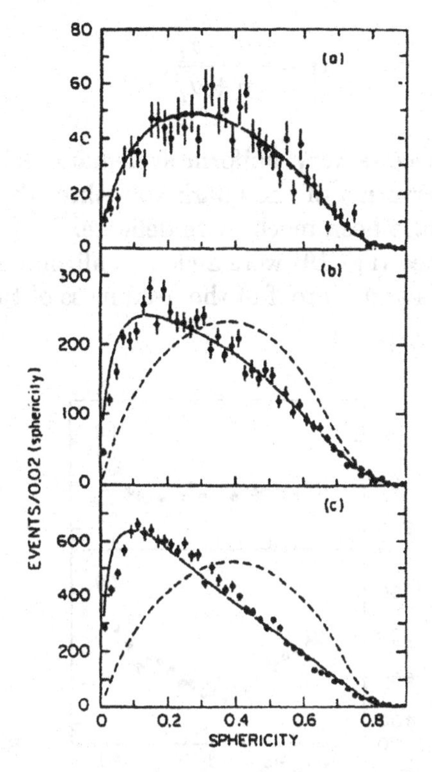

Figure 18 Event distribution as function of sphericity value for 3 center of mass energy. (a) at 3 GeV, (b) at 6 GeV, (c) at 7.4 GeV. The dashed line is a phase space model prediction, while the full line represents the jet model prediction.

One bonus of the jet observations was obtained with polarised e^+e^- beams. It had been predicted by Sokolov and Ternov, and observed before 1970 that in storage rings e^+ and e^- get gradually polarised by the emission of synchrotron radiation (except at some energies where depolarisation resonances exist in the machine). The beam polarisation P affects the azimuthal angular distribution of the annihilation product

$$e^+e^- \to \mu^+\mu^- \to 1 + cos^2\theta + P^2 sin^2\theta cos2\varphi.$$

If quark have spin 1/2, a similar formula was expected:

$$e^+e^- \to q\bar{q} \to 1 + cos^2\theta + P^2 \alpha sin^2\theta cos2\varphi,$$

where θ and ϕ of the quarks is identified to θ and ϕ of the thrust axis. P was obtained from $\mu\mu$ events $P^2 = 0.47 \pm 0.05$

$$\alpha = \frac{\sigma_T - \sigma_L}{\sigma_T + \sigma_L}.$$

Since the ϕ acceptance is very uniform systematic effects were negligible, while the observation of the quark spin through the polar angle distribution would have been much more delicate.

The results obtained (fig. 19) were a clear confirmation of the spin of the produced partons and a proof of the usefulness of the jet concept.

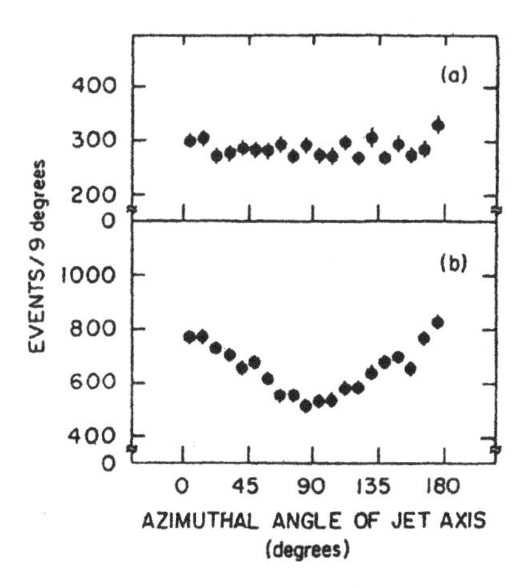

Figure 19 Jet angle orientation as function of azimuthal angle a) at 6.2 GeV where the beam polarisation is 0. b) at 7.4 GeV where beam polarisation exists.

Jets in hadronic collisions where much harder to observe. One expected a Pt distribution given by $\frac{d\sigma}{dp_t^2} \propto \frac{1}{P_t^4}$ in analogy with $\frac{d\sigma}{dQ^2} \propto \frac{1}{Q^4}$. However in almost all experiments a trigger was performed on a single high P_t particle which biased completely the jet observation. For the single particle a $\frac{1}{p_t^8}$ was observed which resulted from a complicated mixture of parton collision dynamics and parton fragmentation to a single particle. Even in 1980, (after 3 jet events had been observed in e^+e^- collisions at PETRA !), there existed no convincing and unbiased observation of jets in hadronic collisions. Either the energy was too low for isotropic experiments (NA5 experiment at the CERN SPS and similar experiments at FERMILAB) or apparatus were not isotropic enough (ISR).

The experimental breakthrough was obtained at the start of the CERN $p\bar{p}$ collider ($\sqrt{s} = 600$ GeV). At the 1982 Rochester conference in Paris, the UA2 experiment showed very convincing results. As we will see later the UA2 apparatus design was not as good as the UA1 to observe the W boson, but it was built in an almost ideal way to observe jets (fig. 20,21).

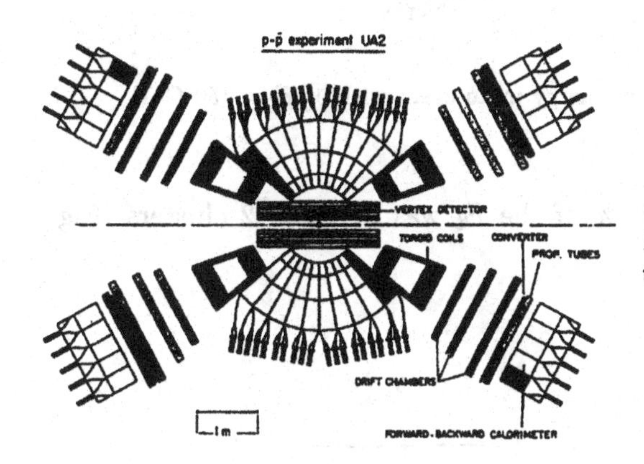

Figure 20 Polar view of the UA2 apparatus.

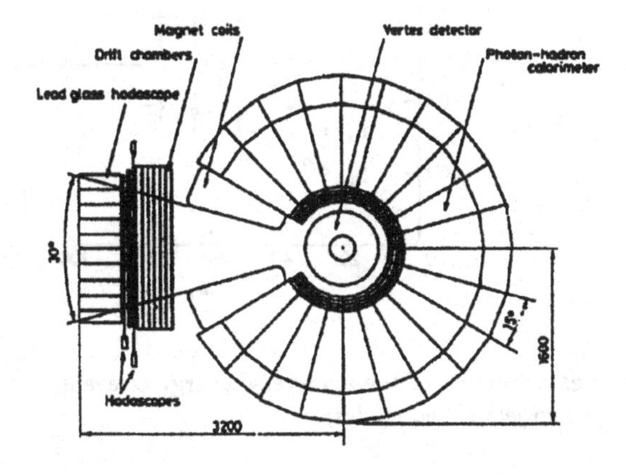

Figure 21 Azimuthal view of the UA2 apparatus.

It consisted essentially in cells of electromagnetic and hadronic calorimeter. The granularity was small for the time, 260 cells, (of $\Delta\theta \times \Delta\phi = 10^\circ \times 15^\circ$) even if it seems large compared to modern calorimeter. The trigger was an unbiased one looking for high E_t summed over all cells. It was immediately found that most of the E_t was concentrated in clusters of typically 5 cells (Fig. 22).

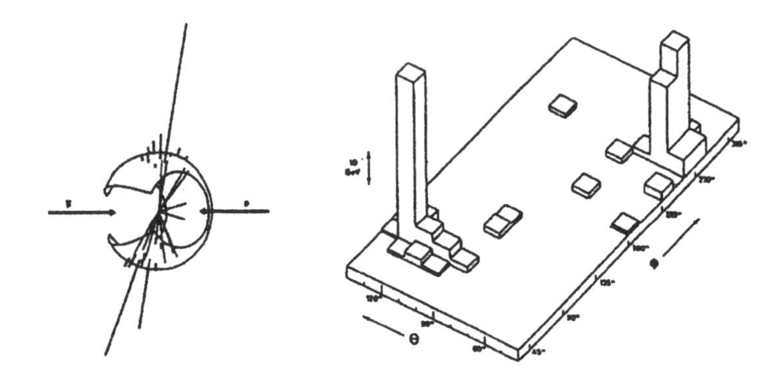

Figure 22 Angular distribution of energy in an event with $\sum E_t = 150$ GeV.

At $E_t > 80$ GeV close to 80% of the E_t was found in 2 clusters (Fig. 23).

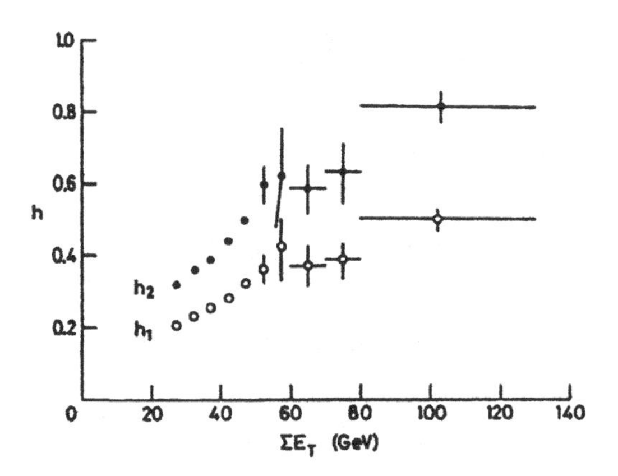

Figure 23 Fraction of the transverse energy of events measured in one (opened circles) or two jets (filled circles).

2.7. GLUON JET OBSERVATION

The idea that gluon bremsstrahlung existed, in analogy to photon radiation from charged particles, dates from the invention of QCD.

In 1976 there was a remarkable phenomenological paper by J. Ellis, M.K. Gaillard and G. Ross which pointed out that gluon radiation, in

e^+e^- hadronic collisions, would cause some 2-jet events to be "fat" on one side (contrary to pair creation of new particles which was expected to lead to events with two "fat" jets) and it was pointed out that eventually 3 jet events would occur. (Fig. 24).

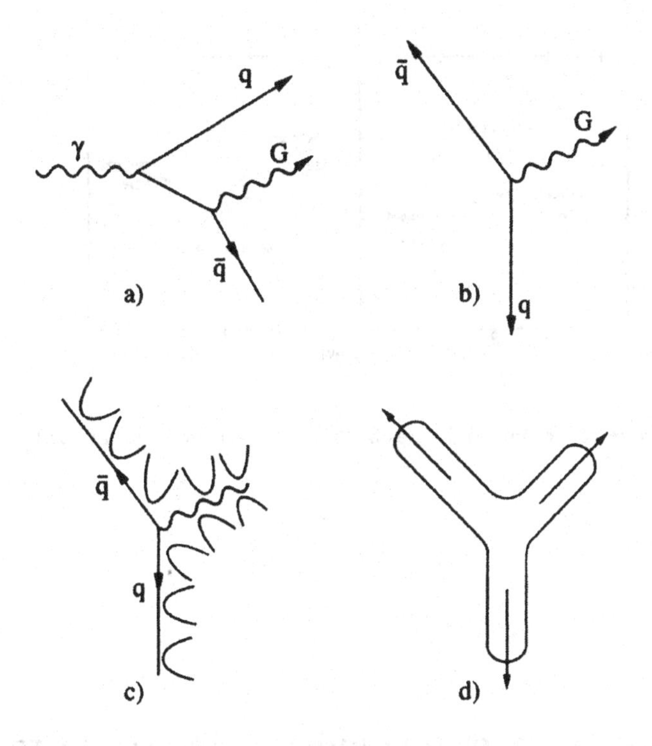

Figure 24 Schematics of gluon radiation as explained in the Ellis Gaillard Ross article.

Nevertheless the energy had to be high enough and the necessary analysis tools had to be developed by the experimentalist.

The Petra machine reached the necessary energies (27 GeV) in 1979. On the analysis side the idea was to choose, for each event, a plane which minimised the event transverse momentum perpendicular to this plane (Pt-out). At the 1979 June EPS meeting in Geneva Tasso showed a few of such planar events which possessed 3-jets structure (fig. 25) together with strikingly different Pt-in and Pt-out distributions. By the time of the electron-photon conference in Fermilab in August 1979 all four Petra experiments showed convincing distributions and the evidence for the gluon discovery was accepted.

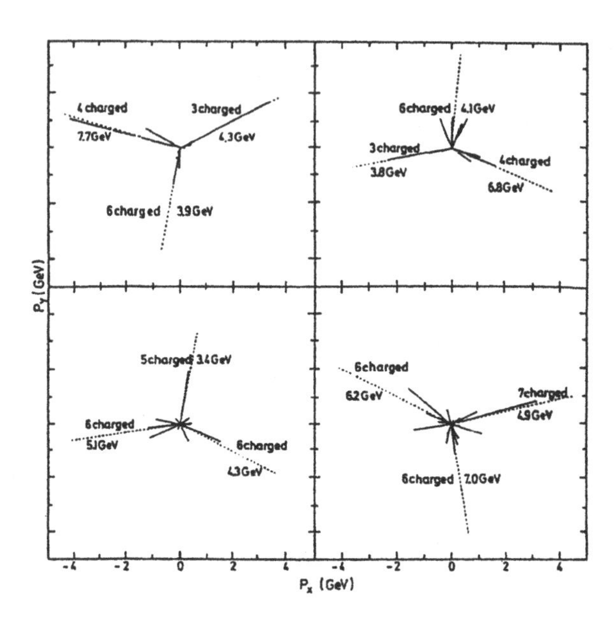

Figure 25 From candidates 3 jet events seen by Tasso in 1979.

3. WEAK INTERACTION AND QUARK AND LEPTON FAMILIES

3.1. NEUTRAL CURRENT DISCOVERY (1973-1974)

The discovery of neutral currents (NC) by the Gargamelle experiment was one of the key discovery of the past 30 years. It allowed one to confirm that the Glashow Salam Weinberg model was the correct electroweak theory. At that time, in the early seventies, other alternatives were proposed to deal with the diverging high energy behavior of the Fermi pointlike weak interaction, (for example the existence of heavy leptons or of diagonal currents).

Since the late 50's it was generally assumed (for example by Schwinger in 1957)that the pointlike nature of weak interactions was the result of the exchange of a very heavy charged particle ($W^{\pm}$), in contrast to the long-ranged QED force caused by the exchange of a massless photon.

In the late sixties the Glashow Salam Weinberg model was presented, which predicted the existence of a neutral partner of the $W^{\pm}$, the Z^0 particle, and hence the existence of neutral current weak interactions.

A big stumbling block for this model, was the experimental absence of $\Delta S = 1$ neutral currents, i.e., the absence or very small rates of decays like $K_L^0 \to \mu^+\mu^-$ or $K \to \pi\nu\bar{\nu}$.

In 1970, Glashow, Iliopoulos, Maiani mechanism (GIM) explained this puzzling fact by the proposing the existence of a new quark: the charm quark. Then in 1971, Veltman and t'Hooft proved that the electroweak theory was renormalizable, and therefore that higher order calculations could be performed.

Experimentally, neutral currents were predicted to be seen in two types of neutrino interactions:

of leptonic type:

$$\bar{\nu}_\mu e^- \Rightarrow \bar{\nu}_\mu e^-,$$

or

$$\nu_\mu e^- \Rightarrow \nu_\mu e^-,$$

with background

$$\nu_e n \to e^- p,$$

$$\bar{\nu}_e p \to e^+ + n,$$

$$\hookrightarrow +\gamma \to e^+ e^-,$$

or of hadronic type

$$\nu_\mu + N \Rightarrow \nu_\mu + X,$$

with background

$$n + N \Rightarrow n + X.$$

The background for the leptonic reaction was the elastic interaction on nucleons of the ν_e or $\bar{\nu}_e$ always present as a contamination in a ν_μ or $\bar{\nu}_\mu$ beam, the background from $\bar{\nu}_e$ being of course much smaller.

The background in the hadronic reaction was neutron scattering, the neutrons being produced by neutrino interaction at the end of the ν beam shielding.

As explained in the previous section, the Gargamelle bubble chamber started to take data in 1971. The fiducial mass was 4.5 tons out of a total mass of 20 tons. Gargamelle was an almost ideal apparatus for few GeV neutrino physics, it showed full details of the interaction and the reinteractions of neutrals. It was big enough (about 6 interaction length) to see attenuation of the incident neutron background, similarly muons could be identified by their absence of reinteraction in the liquid.

The first sign of neutral current was a $\bar{\nu}_\mu e^-$ scattering event observed in December 1972 (fig. 26). For the total exposure of 1.4×10^6 pictures 5 to 30 events were expected depending on the $\sin^2 \vartheta_W$ value, finally 3 events were seen.

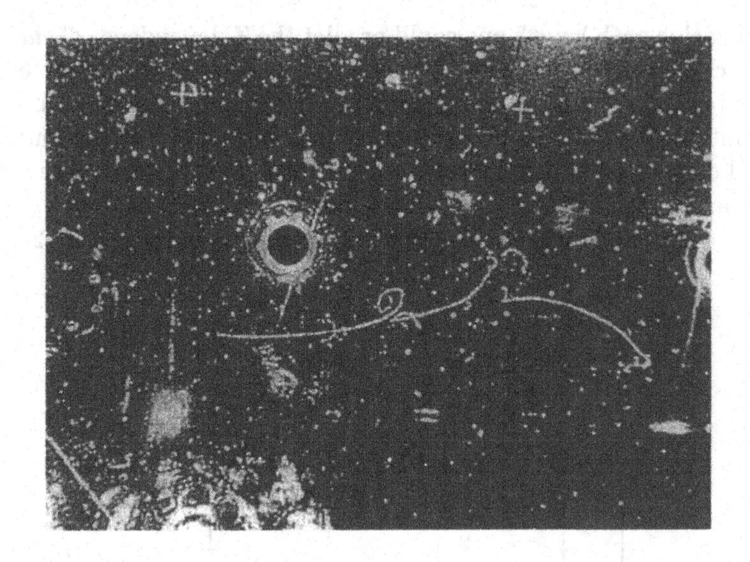

Figure 26 The first elastic $\bar{\nu}_\mu e^-$ scattering event seen in Gargamelle.

The big difficulty for the observation of hadronic neutral current events was the neutron background evaluation. As shown in figure 27 the neutron path length in the Gargamelle liquid was calibrated by looking at neutron reinteraction downstream from a neutrino interaction (the so called Associated Star).

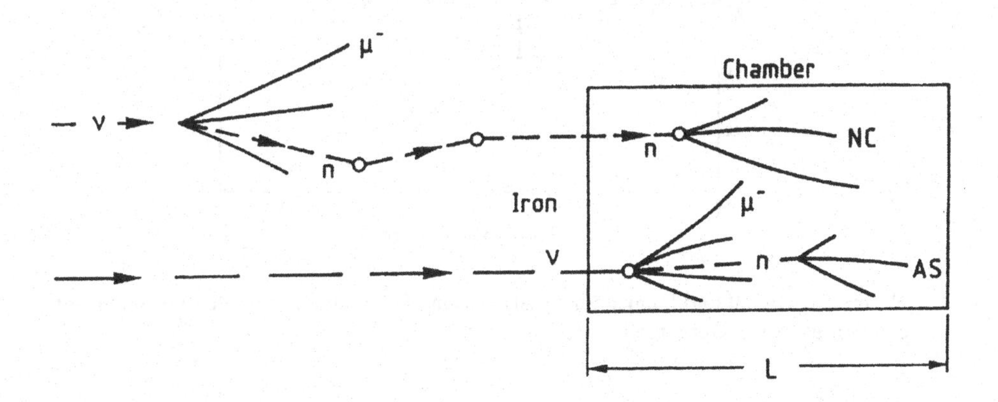

Figure 27 Schematics of neutron background events in Gargamelle and their probability and path length determination by associated stars (AS).

Knowing this path length one could predict the Z dependence of the neutron background events, while of course neutrino neutral current events were expected to be produced equally at all Z values. From the observed distribution (fig. 28), it was estimated that only 10 % of the muonless neutral current candidates were from neutron. From R the relative rate of the neutral current events to charge current events in ν_μ and $\bar{\nu}_\mu$ beam a value of $\sin^2\theta_w$ was extracted, the results are shown on figure 29.

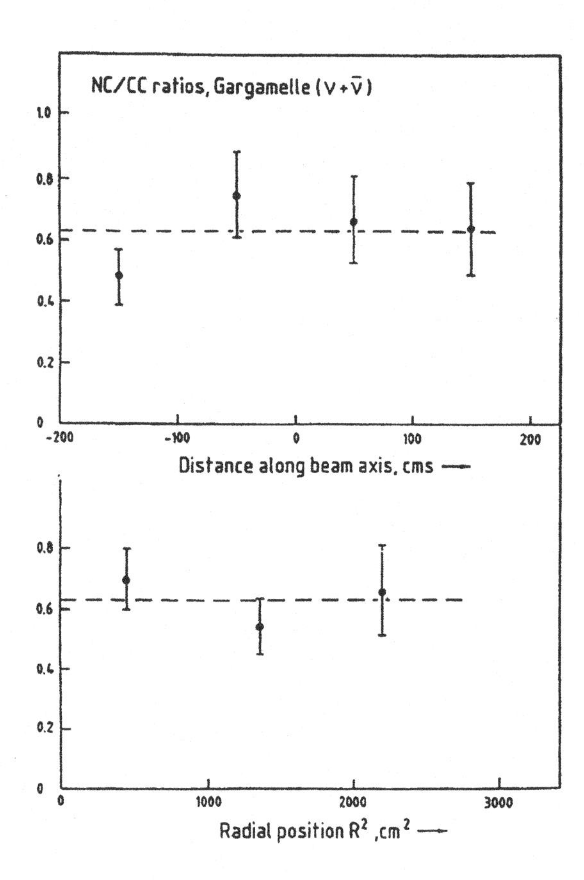

Figure 28 Radial and longitudinal distribution of the relative rate of neutral current candidates in the Gargamelle chamber.

The results were contested during 1973 by the HWPF experiment which had just started at Fermilab and which used a high energy neutrino beam and a calorimeter and magnetic spectrometer apparatus. Because of the underestimation of hadron penetration in the calorimeter, which faked muons, a fraction of the NC events were identified as

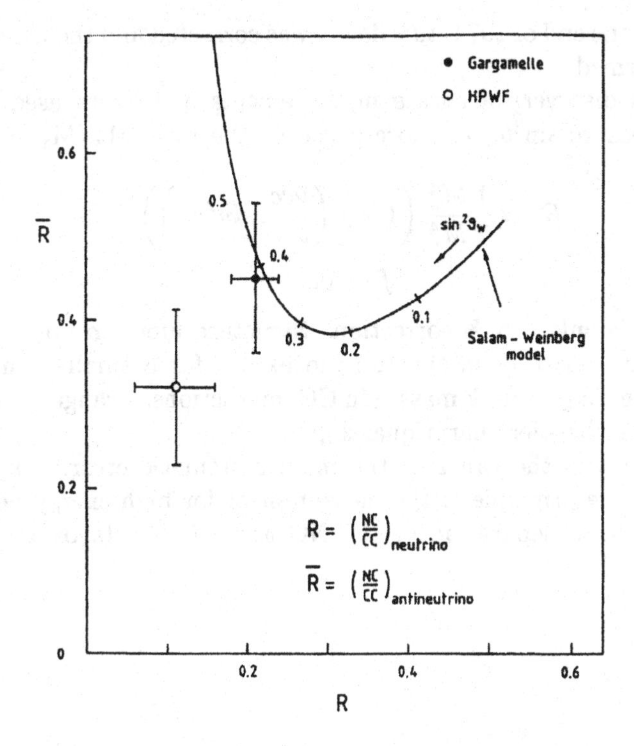

Figure 29 The 1974 Gargamelle and HPWF results on $\sin^2\theta_w$ obtained from neutral current relative rate in ν and $\bar{\nu}$ beams.

charged current (CC) by HWPF, and after background subtraction no NC events were left.

However, by the end of 1974, this defect was corrected and the Gargamelle results confirmed.

After this discovery accurate measurements of R were used (1975-2000) to measure $\sin^2\theta_W$ or more precisely the ratio $M_W/M_Z = \cos\theta_W$.

$$R_\nu \approx \frac{1}{2}\frac{M_W^4}{M_Z^4}\left(1 + f\left(\frac{\nu cc}{\bar{\nu} cc}, sin^2\theta_W\right)\right)$$

$$f \sim .05$$

The f term is only a 5 % correction. Structure function uncertainties cancel almost perfectly in the R ratio except for a small asymmetry caused by the charm quark mass. (In CC interactions, strange sea quarks are changed to heavier charm quarks).

This effect was the cause of the main systematic error. As shown in figure 30, the principle of the measurement for high energy neutrino interaction was a separation of the NC and CC events by the event length.

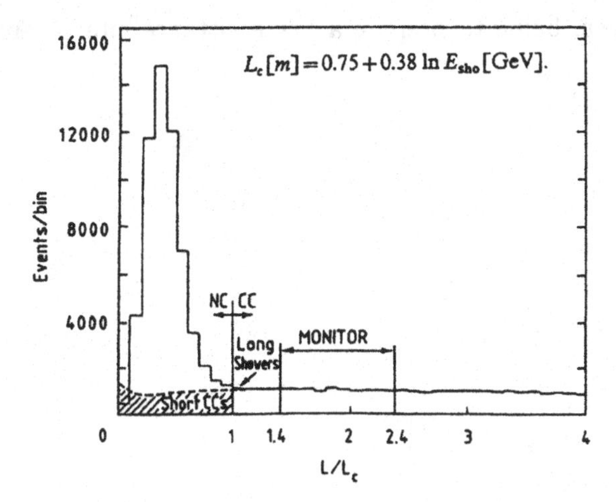

Figure 30 **Event length distribution in high energy neutrino experiments. NC events have a short length typical of hadronic shower while, for CC events, the length is dominated by muon penetration.**

R was gradually measured to 1% (CDHS and CHARM in 1987-1988) or even better CCFR (1994-1998) the derived $\sin^2\theta_w$ values were 0.231 ± 0.006 and 0.224 ± 0.004 the final error being dominated by systematics.

A breakthrough in accuracy was recently obtained by the NuTeV experiment at Fermilab. It used both ν and $\bar{\nu}$ interactions so as to depend only on valence quark interactions and therefore suppress the uncertainty of the CC events on the strange sea quarks. This last measurement gave $\sin^2\theta_w = 0.2253 \pm 0.0022$

3.2. DISCOVERY OF THE W AND Z BOSONS

After the neutral current discovery and early measurements of $\sin^2\theta_w$ the W and Z mass could be predicted to be around 80 and 90 GeV respectively, out of reach of any existing accelerator. In hadron-hadron collisions, a centre of mass energy greater than 500 GeV was needed. The first proton-proton collider, the ISR was approved for construction in 1965 and used first in 1971 but its energy was much too low.

In 1977, Cline, McIntire and Rubbia proposed to convert the Fermilab accelerator into a proton antiproton collider. The use of antiproton increased the probability of obtaining high energy quark antiquark collisions and furthermore allowed one to store both beam in a single ring, the main problem being to prepare a low emittance intense antiproton beam.

The idea of doing $p\bar{p}$ collisions had been proposed originally by Budker from Novosibirsk at the 1966 Saclay accelerator conference. In the original Fermilab proposal, as in Budker's proposal, the idea was to decrease the emittance of low energy antiprotons (cooling) by collisions with a high intensity, low emittance electron beam having the same velocity.

After the Fermilab proposal was turned down a similar proposal was presented to CERN by Rubbia and approved in 1978 and first collisions observed in 1981, an incredibly short time for such a complicated accelerator and apparatus.

3.2.1 The collider.

The machine breakthrough was the invention around 1968 by S. Van der Meer (who shared with C. Rubbia the Nobel prize for the W,Z discovery) of the principle of stochastic cooling. The principle is shown in figure 31: if a single particle position with respect to the central orbit is sensed and corrected by a kicker after (n/2 + 1/4) betatron wavelength, then clearly this particle can be placed on the central orbit.

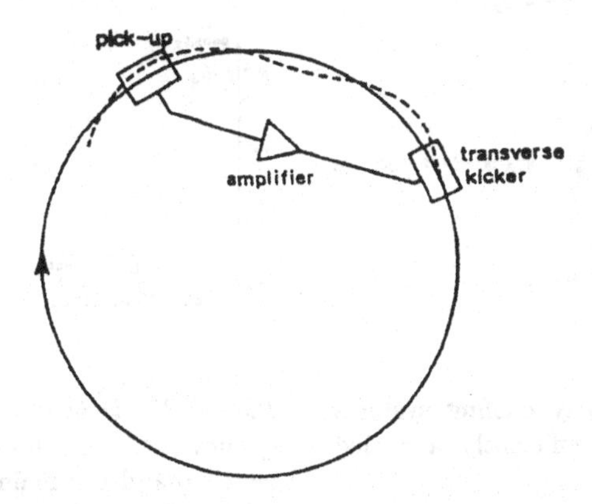

Figure 31 Schematics of the feedback needed for stochastic cooling.

If n particles are sensed, in a short time, by a high frequency system, then the mean orbit can be corrected and the beam size σ is reduced:

$$\sigma_f^2 = \sigma_i^2 - \bar{x}^2.$$

Where $\bar{x} = \sigma/\sqrt{n}$ is the average position which is corrected. The cooling continues because the sample composition changes due to the spread in revolution frequency.

Momentum or longitudinal cooling can be performed in a similar way, except that the momentum offset is detected by a shift in orbit frequency which is compensated by acceleration or deccelaration. In practice sophisticated and complicated procedures were used to cool rapidly (every 2 sec) each batch of $10^6 \bar{p}$ and cool gradually over 12 hours a full stack of $10^{11} \bar{p}$ (fig. 32 and fig. 33).

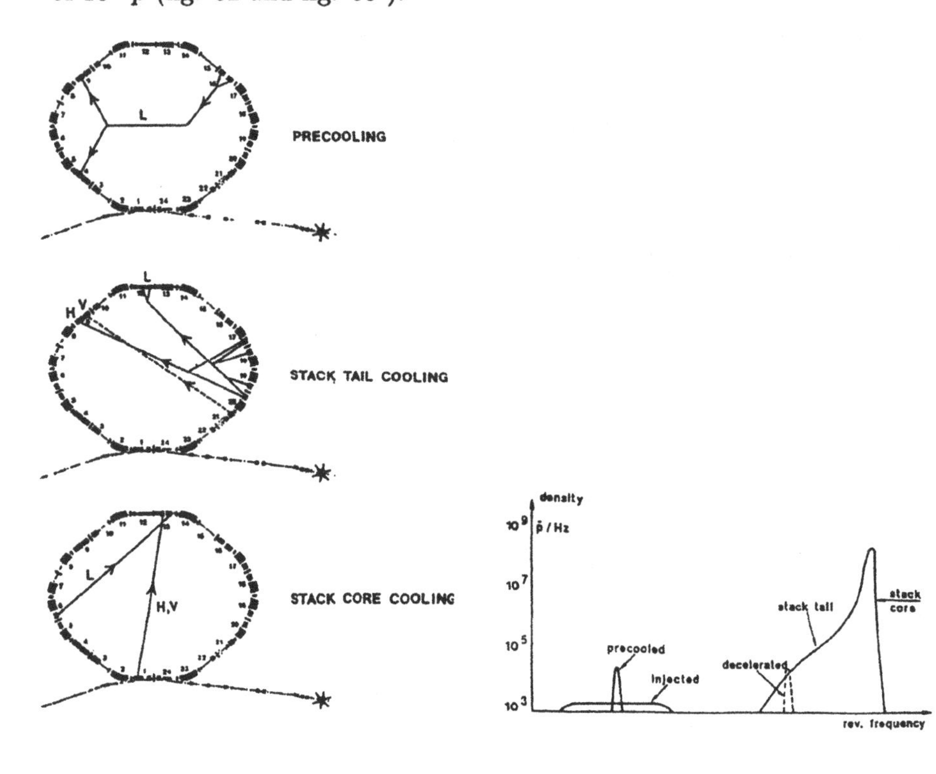

Figure 32 The many cooling systems needed to prepare efficiently a cooled stack of $10^{11} \bar{p}$.

Figure 33 Density in revolution frequency, i.e. in momentum, of the injected $\bar{p}$ and the main stack of antiprotons.

3.2.2 The experimental apparatus .

The UA2 apparatus was described in the preceding chapter, while being ideal for jet observation it was not as good as UA1 for W discovery because of incomplete solid angle, UA1 will be briefly described here.

The key observation was the following: in W$\rightarrow e\nu, \mu\nu$ decays there will be, because of the unobserved ν, 40 GeV of missing momentum in the observed particles. However in q+q̄ $\rightarrow$W the W longitudinal momentum is not known event by event, the useful signature is therefore missing transverse momentum i.e. an apparatus hermetic in azimuth and with a coverage down to small polar angle is needed. UA1 was the first of the general purpose hadron collider detectors. As can be seen in figure 34, it was a complete apparatus measuring charged particles, gammas, and neutral hadrons and identifying electrons and muons.

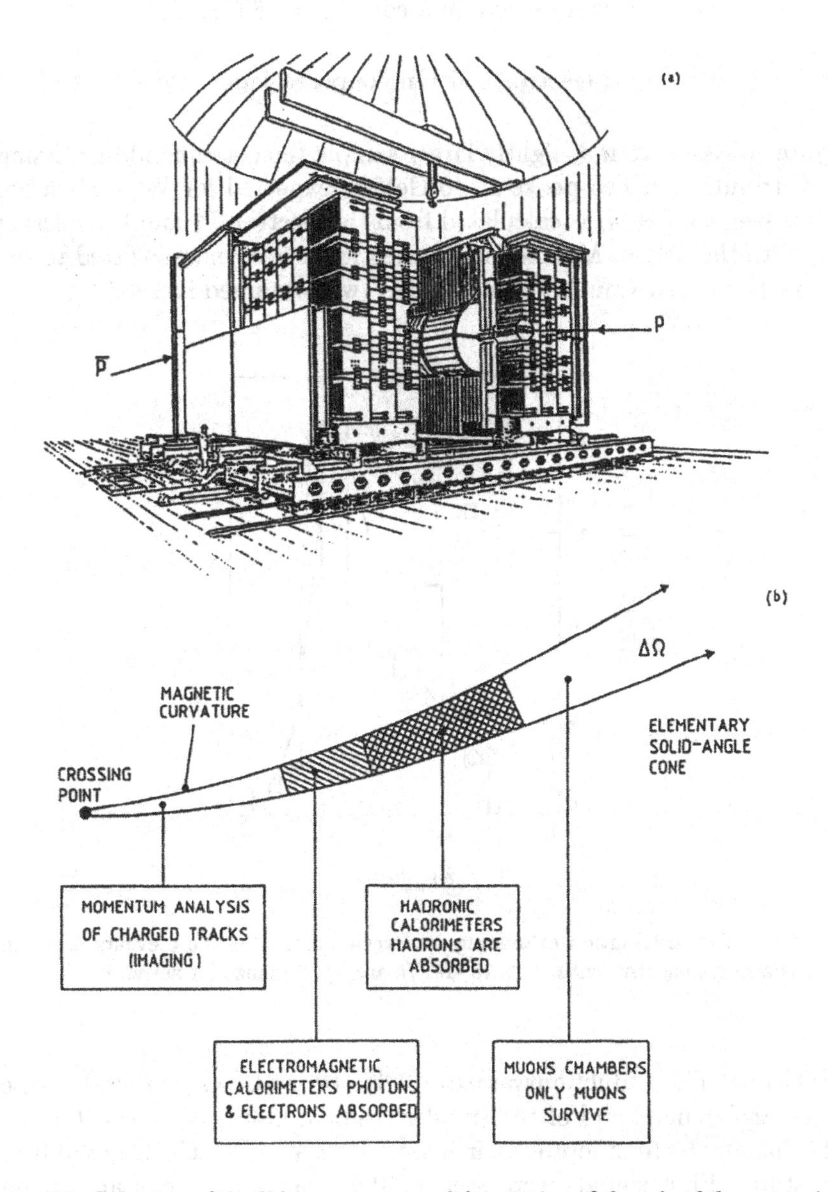

Figure 34 Side view of the UA1 apparatus and description of the role of the successive layers.

For the W discovery, the trigger selected high P_t electrons (muons were used later). The following event samples with electrons $P_t > 15$ GeV were collected:

$$291 \text{ electrons} + \text{jets in a cone} \Delta\phi = 180^\circ \pm 15^\circ,$$

$$55 \text{ electrons with no opposed jets.}$$

Figure 35 shows with a slightly larger sample the corresponding missing E_t distribution, which peaks at 40 GeV as expected for W production. A transverse mass M_t is calculated from the electron P_t and the missing E_t. With the help of Monte-Carlo correction the W mass is fitted to the M_t distribution a value of 81 ± 1.5 GeV was obtained in 1983.

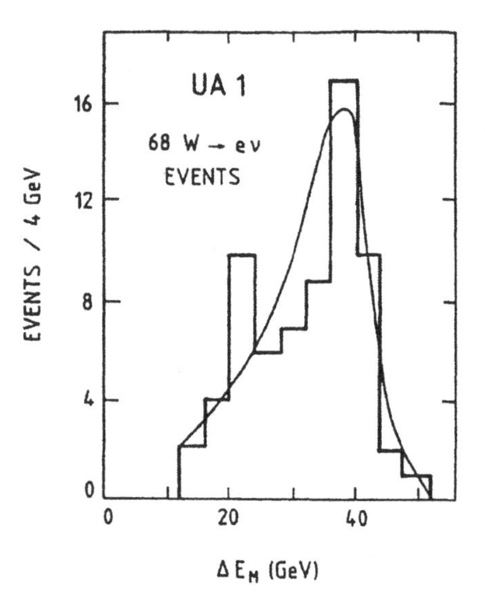

Figure 35 The distribution of missing transverse energy for those events in which there is a single electron with $P_t > 15$ GeV/c and no coplanar jet activity.

The rate of Z production was expected to be ten times smaller, because of the higher mass and of the smaller leptonic branching ratio. On the other hand, the high lepton pair mass allows an almost backgroundless signature. First signals were seen in 1994, using electrons and muon pairs in UA1 and electron pairs in UA2. A clear peak was visible, as shown in figure 36.

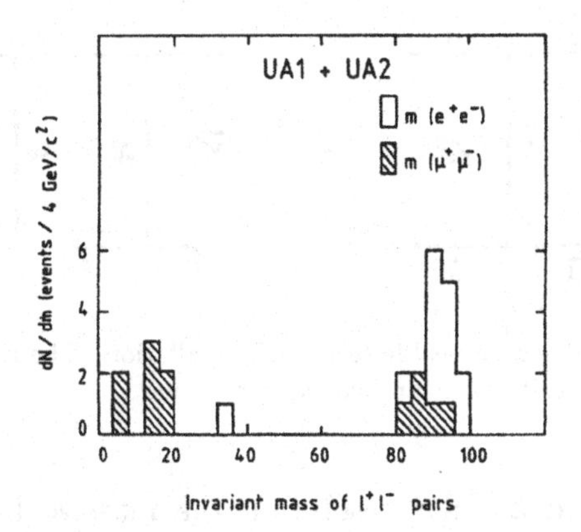

Figure 36 **Invariant mass distribution of dilepton events from UA1 and UA2 experiments. A clear Z° peak is seen at a mass of about 95 GeV/c^2.**

3.3. A NEW QUARK : CHARM (THE 1974 "NOVEMBER REVOLUTION")

The discovery of the new quark, charm, caused an earthquake-like shock in the community. At first the interpretation was not obvious and then gradually this discovery not only gave enormous confidence to the electroweak standard model but it gave an enormous boost to the idea that quarks were "real". This was based on the partly naive idea that with the discovery of a heavy quark one would obtain the equivalent to the hydrogen atom for quark physics. Retrospectively, one could say that theorists had been calling for such an object, but this was not generally realized. The GIM mechanism (discovered in 1969-1970) used a fourth quark to cancel ΔS=1 neutral currents for example in box diagrams (fig. 37). The cancellation is up to terms

$$(m_c^2 - m_u^2)/m_w^2.$$

From this the mass of the new charm quarks was predicted to be less than a few GeV. By the summer of 1974 a phenomenological article by M.K. Gaillard, B. Lee and J.L. Rosner gave recipes for charm search with detailed predictions of a narrow $c\bar{c}$ state decaying to $\mu^+\mu^-$ and e^+e^-; however these predictions were not the direct cause of the searches at Brookhaven and SLAC. The apparatus used at Brookhaven for the J discovery by S. Ting et al., was a major change compared to the previous Lederman apparatus looking at muon pairs. The main emphasis was on mass resolution and identification redundancy.

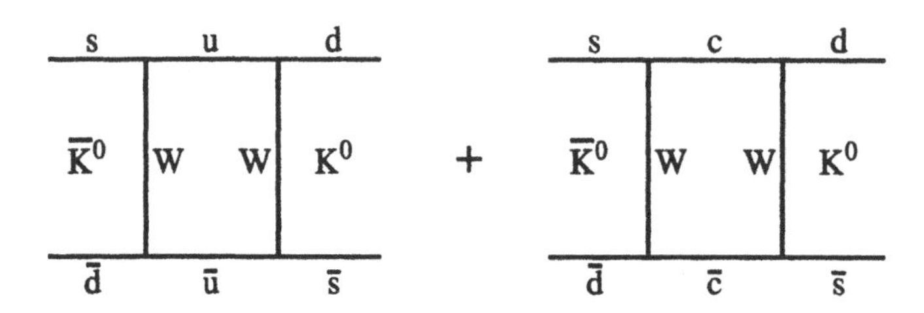

Figure 37 Box diagrams responsible for K_0 $\bar{K}_0$ oscillation. The amplitude from the two diagrams have opposite signs and cancel each other

To obtain a good angular resolution the massive hadron absorber which caused multiple scattering in the muon experiment had to be abandoned. Because of the huge background from hadron decays to muon, it was then decided to detect electron pairs instead of muons. A powerful 2 arm magnetic spectrometer was used to measure the electrons angle and momentum and the hadron background was rejected by a redundant set of Cerenkov counters and shower counters (Fig. 38).

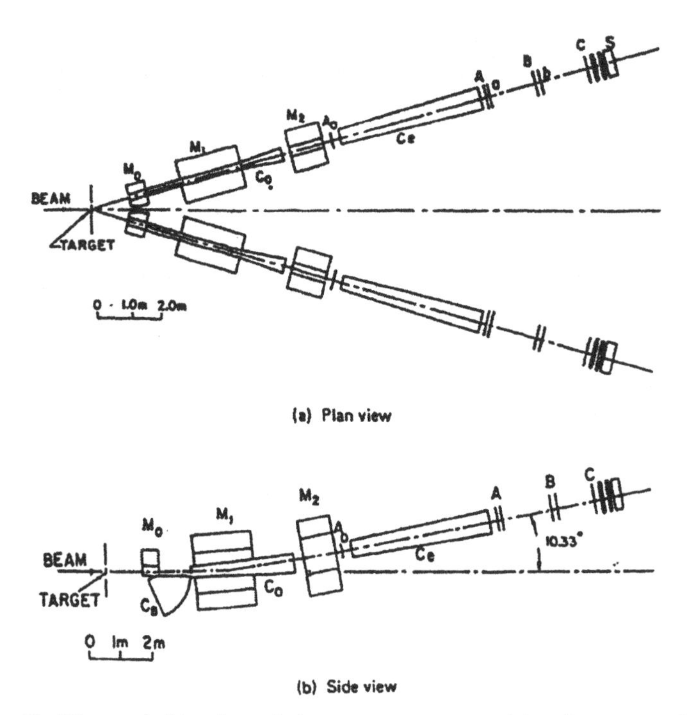

Figure 38 Plan and side view of the spectrometer used by the group of S. Ting et al. for the J discovery.

The spectrometer mass resolution was about 20 MeV a 0.6% mass resolution compared to the 16, 20 % mass resolution of the previous BNL experiment.

The mass distribution obtained is shown in figure 39 together with the result of the previous experiment which was optimised for high counting rate instead of resolution.

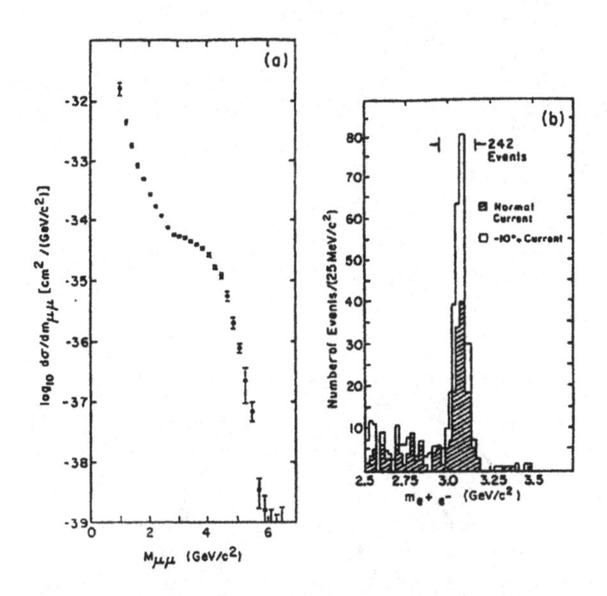

Figure 39 Dilepton mass distributions, a) in the BNL Lederman et al., experiment and b) in the S. Ting et al. experiment.

As everybody knows the $c\bar{c}$ state was discovered simultaneously at SLAC (and called the ψ). The 3-6 GeV e^+e^- colliding ring called SPEAR was probably the accelerator with the greatest number of first class discoveries in the history of our science. (Some were already mentioned in the previous chapter). These discoveries were sometime simply due to the good choice of energy but the design of the apparatus (MARKI) played an important role in the open charm and τ discoveries. The MARKI detector (fig. 40), built in 1973, can be seen as a prototype of many future e^+e^- collider detectors. It had the following main strong points: It had a full and homogeneous ϕ acceptance and a good if not excellent cosθ acceptance; it was a complete detector with charge particle tracking, gamma detection, electron and muon identification and some K,π separation using time of flight.

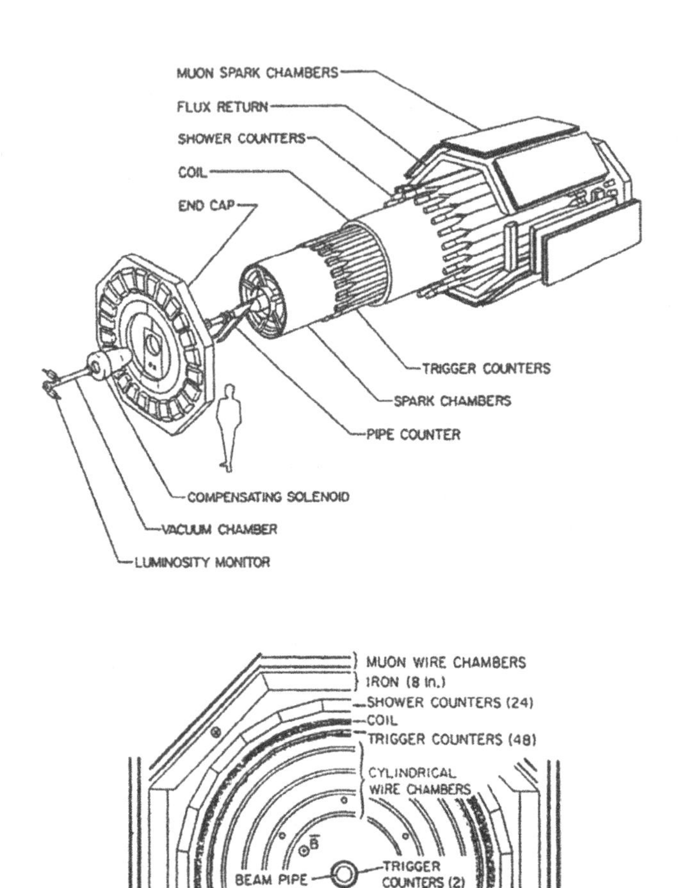

Figure 40 Views of the MARKI apparatus showing the detector layers.

The construction was done by a collaboration which resulted from a remarkable marriage of SLAC apparatus experts and Berkeley software experts (coming from the bubble chamber groups).

The ψ was discovered by a scan of the machine energy. The resonance was initially seen because the collision cross-section above the resonance is also increased due to the emission of bremshtrahlung photons by the colliding e^+e^-. When the peak cross-section was found, the increase above the non resonant region was incredibly high (a factor of more than 100 !).

Initially there was much confusion on the interpretation of the discovery and various hypothesis explored (I remember an explanation which identified the 3 GeV object with the Z° boson !).

However a few months later, detailed results on production of hadrons muons and Bhabha events were produced (fig. 41).

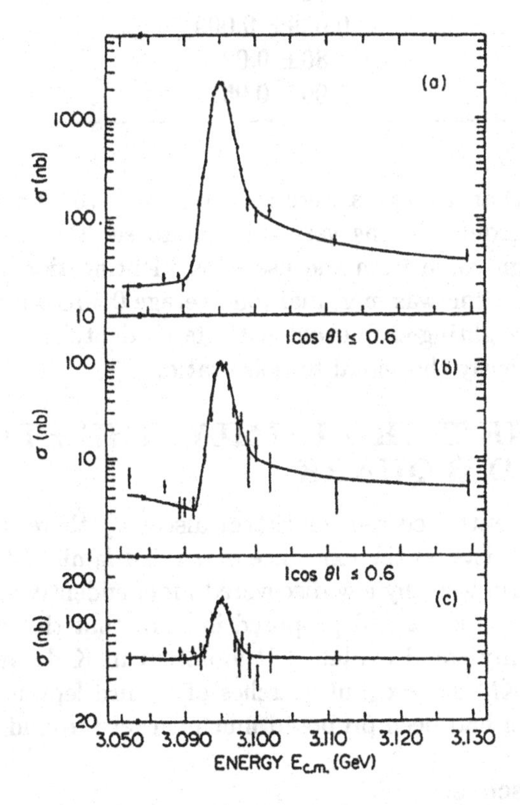

Figure 41 The ψ resonnance cross-section as function of center of mass energy in a) the hadronic charmed, b) the muon channel and c) the Bhabha channel.

From these results the branching ratios and total width of the J/ψ could be derived (Table 1), from these it was clear that, even if very narrow, the resonance was too wide to be due to a pure electromagnetic coupling (the γh branching ratio 12 KeV is narrower than the total width) and the "hidden charm" hypothesis was accepted.

Table 1 Properties of $\psi(3095)$

Mass	3095 ± 0.0004 GeV
J^{PC}	1 - -
Γ_e	4.8 ± 0.06 KeV
Γ_μ	4.8 ± 0.06 KeV
Γ_h	59 ± 14 KeV
$\Gamma_{\gamma h}$	12 ± 2 KeV
Γ	69 ± 15 KeV
Γ_e/Γ	0.069 ± 0.009
Γ_μ/Γ	0.069 ± 0.009
Γ_h/Γ	0.86 ± 0.02
Γ_μ/Γ_e	1.00 ± 0.05

However no charm mesons were seen before 1976. Because of combinatorial background it was very difficult to see the D$\rightarrow K\pi$ decay; the breakthrough came from the use of K identification using time of flight. The separation was marginal (on average 0.5 ns with a resolution of 0.5 ns) but weighting each event with its K identification probability improved sufficiently the signal to noise ratio.

3.4. THE THIRD FAMILY: THE τ LEPTON AND B QUARK

Contrary to neutral current or charm discovery there was no strong advice from theorists on the existence of a third family (it must be admitted that charm was anyhow discovered independently of the advice). Kobayashi and Maskawa had proposed in 1973 that the existence of a third quark family could explain CP violation in K decay by a phase in a 3 family CKM matrix, but searches of a third lepton heavier than muon or electron had been proposed much earlier (around 1970).

3.4.1 τ discovery.

The τ was discovered, in 1975, on the MARKI apparatus at SLAC by M. Perl et al. The key idea of the analysis was that, in analogy to $\mu \rightarrow e\nu\bar{\nu}$ decay, one expects the τ to decay both to $e\nu\bar{\nu}$ and to $\mu\nu\bar{\nu}$. Then in the reaction $e^+e^- \rightarrow \tau^+\tau^-$ it is expected that some events will be seen as acollinear $e\mu$ particles accompanied by missing momentum.

The big experimental problem of the experiment was the limited quality of the lepton identification: The μ's were identified by penetration in the iron used for magnetic flux return which was only 1.7 interaction length thick. The electrons were signed by characteristic pulses in shower counters which were however of rather poor quality. Because of this, the hadron misidentification probability was about 20% instead of the 1% or better now achieved in modern apparatus. The 1975 results are shown in table 2. The number of eμ events, when compared to the numbers of eh and μh events, could not be explained by hadron misidentification. There was therefore, in the article, the claim of a new phenomenon and the existence of a new lepton was presented carefully as a possible explanation. By 1977 the apparatus had been improved, τ decays to hadrons could be observed in eh events and the τ existence was generally accepted.

Table 2 Table of acollinear ($\Delta\theta > 20°$) events used in τ lepton discovery. All particles momenta are greater than 0.65 GeV/c.

	Total Charge = 0			Total Charge = ± 2		
Number Photons =	0	1	> 1	0	1	> 1
ee	40	111	55	0	1	0
eμ	24	8	8	0	0	3
μμ	16	15	6	0	0	0
eh	18	23	32	2	3	3
μh	15	16	31	4	0	5
hh	13	11	30	10	4	6
Sum	126	184	162	16	8	17

3.4.2 b quark discovery.

Actually, as in the case of charm, the first sign of this new quark was seen in the form of a hidden beauty $b\bar{b}$ state, the Υ, seen in its decay mode to lepton pairs (e^+e^- or $\mu^+\mu^-$). The discovery was done in 1976-1977 at Fermilab by the group of Lederman et al. It was clear by then that the secret of the discovery of narrow states was to obtain an excellent mass resolution, and, in its first attempt, the group chose to observe e^+e^- pairs, as had been done at BNL by the group of Ting et al. for the J discovery. This first run was done in 1976 in a two arm spectrometer, a mass resolution of < 70 MeV was obtained,

but, because of the open configuration of the spectrometer needed to observe electrons, the interaction rate in the target had to be limited to 5×10^9 interactions/accelerator spill. This was marginal to observe, with enough rate, a high mass resonance produced with a small cross-section. The group then switched back, in 1977, to a dimuon configuration but the breakthrough compared to the BNL dimuon experiment was to absorb the hadrons with 7 meters of Beryllium rather than with iron. This improved the radiation length to interaction length ratio by a factor 8 and consequently the angular error contribution to the mass resolution. The muons momenta were measured by powerful air-core magnet spectrometers and muon identification was checked by further absorber and momentum remeasurements in iron core magnets. A mass resolution of 200 MeV was obtained.
The apparatus is shown on figure 42. As can be seen in figure 43 a triple structure was seen corresponding to the three 1^- states $\Upsilon, \Upsilon', \Upsilon''$.

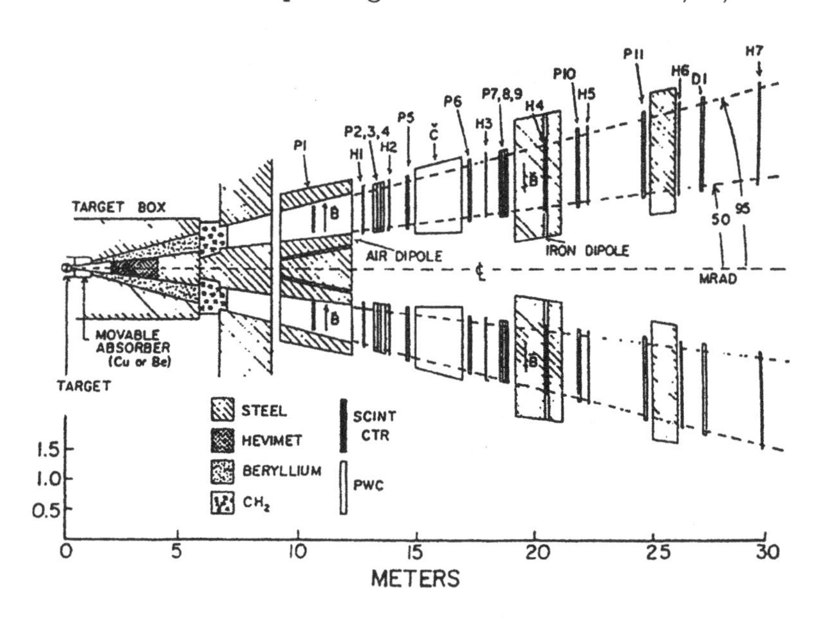

Figure 42 The dimuon spectrometer used by the group of Lederman et al. at Fermilab for the Υ discovery.

Soon after, the Υs were next observed by the DASP, PLUTO, and DESY HEIDELBERG experiments at the Doris e^+e^- collider which could be pushed to the necessary energy. The peak of the Υ and Υ' were clearly observed. As with the J/ψ when the cross-section and leptonic branching ratio were measured the leptonic partial width could be determined.

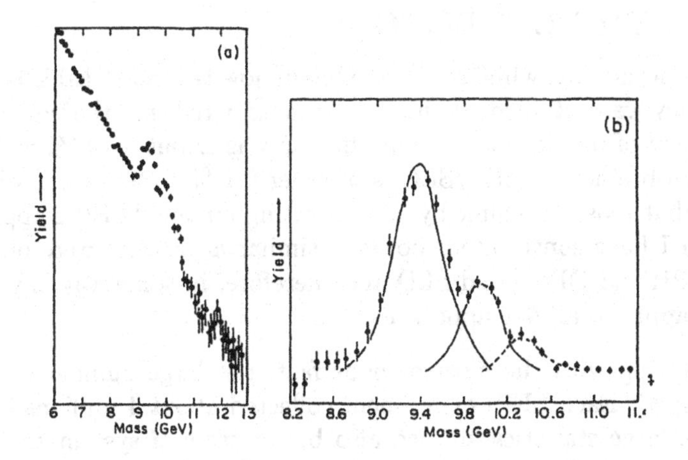

Figure 43 The dimuon mass spectrum a) observed in the experiment of Lederman et al. The spectrum after Drell-Yan background and subtraction is shown in b).

A value of Γee = 1.3 KeV was found while the equivalent value for the J/ψ was 5.4 KeV. This indicated that the b quark had a 1/3 charge (compared to the 2/3 value for the c-quark). As it was the case for charm the discovery of the B meson was much more difficult and it was only 7 years later that the Argus experiment at Doris (DESY) and the Cleo experiment at CESR (Cornell) showed evidence for the decay B$\rightarrow D$ or $D^* + \pi$.

The last quark, the top was found at Fermilab in 1995 but this is a recent story, which the attendants of this school have certainly heard on many occasions.

4. LEP AND SLC: THE IDEAL MACHINES FOR STANDARD MODEL STUDIES

The data from LEP and SLC allowed one to test the standard model with an unprecedented accuracy, an accuracy which even surpassed by factors of 3 to 4 the best expectations listed in the various workshops preparing, in the 80's, the physics before the start of the colliders. These accurate tests, as is well known, not only reinforced strongly our belief that we were using the correct theoretical framework, but also allowed us to predict points of the models (for example the Higgs mass) which could be out of present experimental reach. Since these LEP, SLC results are fairly recent and many reviews exist on this physics, in summer schools and conferences, I will be quite brief.

4.1. THE DETECTORS

I think it us worthwhile to try to explain how this remarkable progress in accuracy was obtained. It should be remembered that the typical level of accuracy of previous e^+e^- experiments was about 5-10 %, while the accuracy obtained at LEP/SLC is of order 0.1 % to 0.5 % ! I will give a somewhat biased account by concentrating on the ALEPH apparatus to which I have contributed; however similar accuracies were obtained by DELPHI L3 OPAL and SLD and therefore, in some way, my explanations apply to all five apparatus.

- Starting with the most obvious fact, the large number of events helped. Each LEPI experiment collected about 4 millions Z°, and the large statistics allowed also better tests of systematic errors. However millions of J/ψ events were collected in previous machines without comparable accuracy improvement, so large statistics is not the full answer.
- High energy helped also: Calorimeters work better at high energy, and hard to see low momentum particles play a much smaller role; finally a large fraction of Z° physics is obtained in events with two back to back jets. These events allowed powerful test of the apparatus, or model, systematics by tagging one hemisphere and studying the efficiency of jets or particles in the opposite hemisphere.

There was also an evolution to a much better apparatus. The key breakthrough, in my opinion, was the emphasis on small inefficiencies and on redundancy: If α is the fraction of events selected or measured in a redundant way and $(1-\epsilon)$ is the inefficiency of the apparatus for these selection or measurements, then it can be derived that the lost event fraction $(1-\epsilon)$ can be determined with a relative statistical error of $\sqrt{(1-\epsilon)/N\alpha}$ (N being the number of events). If $(1-\epsilon)$ is small and α large, then the systematic error from the acceptance is determined by the data and small compared to the overall statistical error. In order to implement those constraints on ϵ and α, the emphasis in ALEPH was put on an hermetic apparatus (figure 44).

A small granularity was required in order to keep high efficiency for narrow jet events or high energy τ lepton decays. The track detectors consist of 2 layers of silicon strip detectors (each with r-ϕ and Z readout) followed by an 8 layer wire chamber (the ITC) and by a TPC. The TPC gave 21 space point measurements for each track, DE/DX was measured in TPC wires but also on the space points (though with a reduced accuracy).

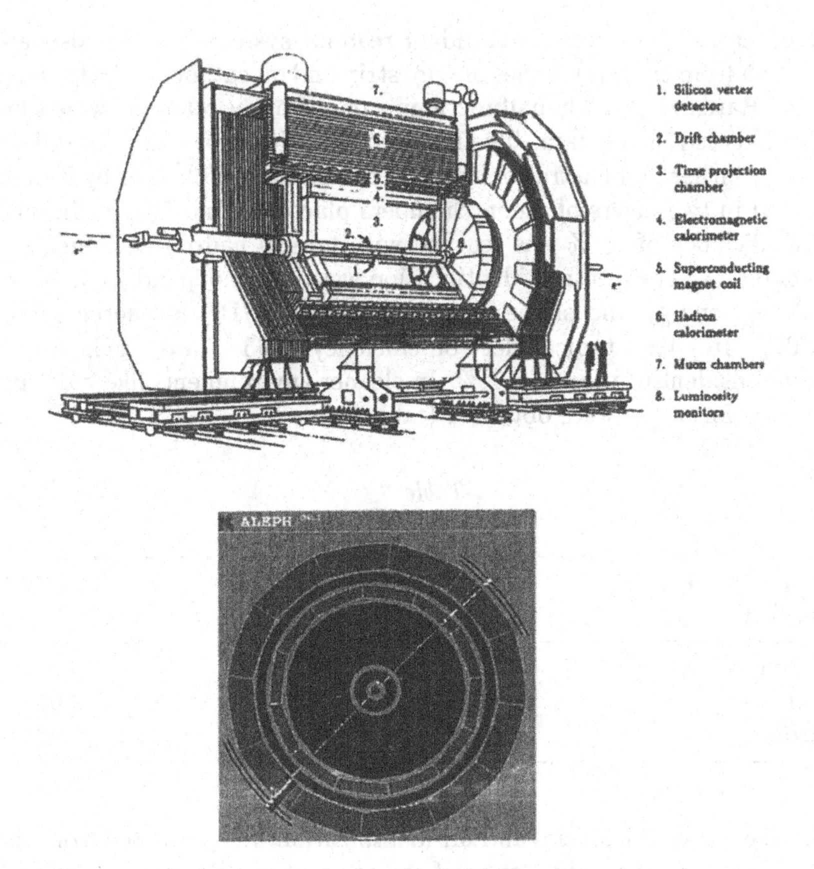

Figure 44 **Perspective view of the Aleph apparatus and r-ϕ view of a $\mu\mu$ event display.**

Because of the overlapping sector structure, there was no single track inefficiency at large angle ($|\cos\theta| < 0.8$). At smaller angle a small 1.5 % inefficiency, modulated in ϕ, was due to sector edge effect; it was calibrated to 0.03 % accuracy with redundant events. In the large angle region the track inefficiency in dense $q\bar{q}$ jet events was only 0.3 %.

The calorimeter measurements exploited also granularity and redundancy, there was a full overlap between the barrel and encap elements and, in this overlap, resolution was worsened (and calibrated) but efficiency was maintained. Holes of 3 to 5 % exist in ϕ between the calorimeter modules but holes in ECAL and HCAL were not aligned preserving efficiency for photons over all the solid angle. Both calorimeters were made of sandwiches of wire chambers and passive material (lead for ECAL, iron for HCAL).

In both cases, there was a redundant readout system: the signals were read both from the wire planes or wire strip and on cathode pads forming towers. Rare events with malfunctioning of one of the readouts were thus easily detected in the data. The muon identification was also redundant and done either by penetration through the layers of HCAL or by signals observed in two layers of muon chambers placed behind HCAL. In jets, muon efficiency of 86 % was reached with typical hadron misidentification of 0.8 %. The electron identification was done independently, either by shower energy and shape in ECAL or by DE/DX measurement in the TPC. In jets, a typical electron efficiency of 65 % was reached with hadron misidentification of 0.1 %. In cleaner environment, like τ decays better performance were obtained (table 3).

Table 3

Tau decays → identified as ↓	$e\nu\nu$	$\mu\nu\nu$	$h + n\pi^0\nu$
electron	0.995	0.000	0.006
muon	0.000	0.992	0.010
hadron	0.005	0.008	0.984

All the above efficiencies and misidentification were obtained from the data, exploiting the redundancy of the apparatus. When compared for example with the MARKI τ discovery results (table 2) one can see the identification improvement of 20-30 typical of the LEP/SLC era.

The trigger was done mostly on the existence of at least one energetic track or jet in one hemisphere, the other hemisphere offering a redundant confirmation and check. Trigger inefficiencies were therefore minute (less than 10^{-4} !).

Finally, the luminosity was measured by Bhabha events in small angle calorimeters. After an initial stage, all experiments adopted the use of high accuracy tungsten-silicon pad calorimeters which reached shower position accuracy of about 10 microns. The experimental luminosity error was thus about 0.07 % and, with the remarkable progress, obtained by the group of Jadach et al., on the evaluation of high order radiative corrections, the theoretical small angle Bhabha cross-section was known with similar accuracy. Finally the Z hadronic cross-section was evaluated with a 0.1 % accuracy, a progress of almost 2 orders of magnitude compared to statistical or systematic errors reached in previous experiments.

4.2. ELECTROWEAK RESULTS

The most important result obtained from the initial data of SLC and LEP was to establish the number of neutrino families.

From a measurement of the Z° cross-section as function of energy, it is possible to extract the total width of the Z°. If the hadronic (Γ_h) and leptonic (Γ_e Γ_μ Γ_τ) are taken from theory then the invisible width can be derived $\Gamma_{inv} = \Gamma_z - \Gamma_h - \Gamma_e - \Gamma_\mu - \Gamma_\tau$ and the number of neutrino family obtained from $N_\nu = \Gamma_{inv}/\Gamma_\nu$. However, in the years of preparation of the MARKII experiment at SLC, G. Feldman remarked that this method had a rather large statistical error for a given total integrated luminosity. He suggested instead to obtain the Z° total width from the peak hadronic cross-section,

$$\sigma_h = (12\pi\Gamma_e\Gamma_h)/M_z^2\Gamma_z^2).$$

This method because of its improved accuracy was used by all experiments for obtaining the initial value of N_ν.

In the summer of 89 MARKII at SLC obtained its first result N_ν =2.7 ± 0.7. By the end of 89, after the start of LEP in September the result from the average of ALEPH, DELPHI, L3 and OPAL was N_ν = 2.99 ± 0.13. This was a key measurement since, at that time, there was no compelling arguments (other than cosmological) against the existence of a fourth family of heavy quarks and leptons; within the usual assumption of small neutrino masses the 1989 results fixed the number of family to three. The most recent result on $N\nu$ is $N\nu = 2.9835 \pm 0.008$, this can be interpreted as a limit on unexpected decay modes of the Z° to invisible particles $\Delta\Gamma_{inv} < 2$ MeV, i.e. $\Delta\Gamma_{inv}/\Gamma_z < 0.08\%$.

The next important step in precise electroweak measurement was to obtain information on higher order electroweak corrections. Examples of such corrections induced by top quark loops are given in figure 45.

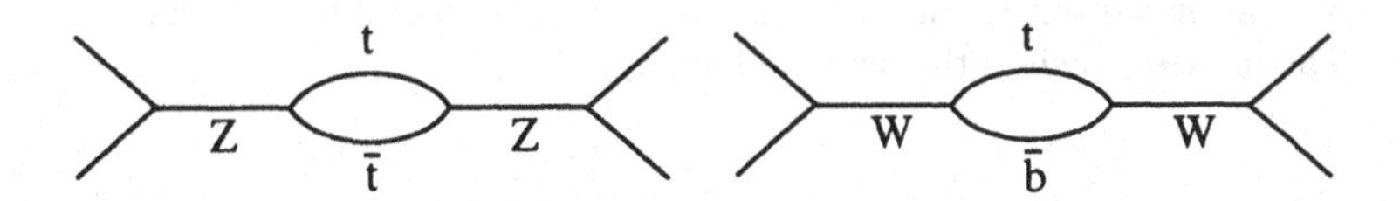

Figure 45 Top loop diagrams responsible for radiative corrections to W and Z° masses and couplings.

From measurements of G_μ, the Z° mass, and width and couplings measured from Z pole asymetries, the top quark mass could be predicted. Initially most of the accuracy came from the LEP data but later with the use of a polarized electron beam at the SLC, SLD contributed also very significantly to the evaluation of electroweak corrections.

Already the first 1989 LEP results predicted the top mass to be large (180 GeV ± 60 GeV). By the 1993 summer conference the prediction was refined to $M_{top} = 162 \pm 16$ GeV ± 20 GeV as shown on figure 46, the last error represents the uncertainty on another correction due to a Higgs boson loop, and the range of the uncertainty corresponded at the time to a range of possible values for the Higgs mass of 60 GeV to 1000 GeV.

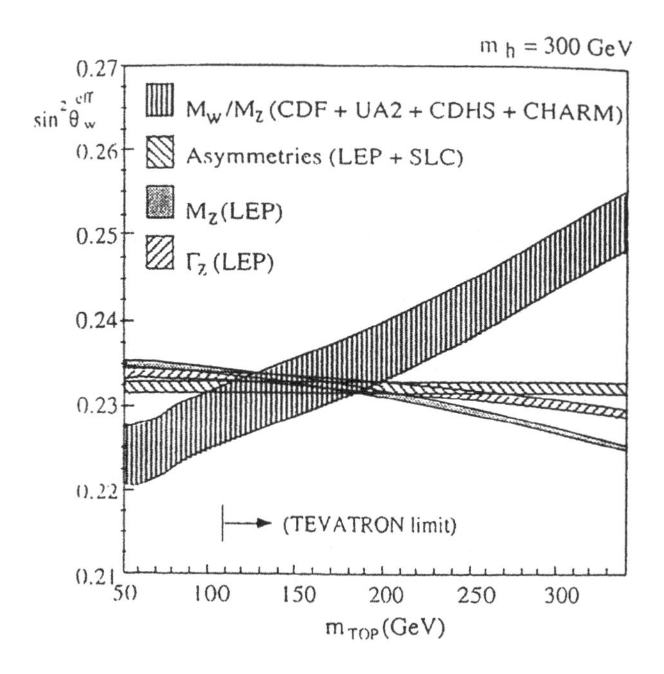

Figure 46 Calculated $\sin^2\theta_w$ values as function of M_{top} for four different experimental inputs. The intersect defines the predicted value of M_{top}.

Soon after, in 1995, the top quark discovery was announced at the Fermilab $p\bar{p}$ collider and its mass measured to be in agreement with the prediction. The most recent M_{top} measurement is $M_{top} = 174.3 \pm 5.1$ GeV and the most accurate prediction from an overall fit of all other electroweak observable, including W mass measurements at LEPII and $p\bar{p}$ colliders is $M_{top} = 167\ ^{+11}_{-8}$ GeV a stunning success for the standard model consistency. Actually inserting into the fit the experimental M_{top} value, it is possible to obtain a bound on the mass of the standard model Higgs boson. The value is inferred is less than 215 GeV at 95 % confidence level.

LEP and SLC have also contributed important and accurate results in the field of B physics and QCD but a complete review would be outside the scope of these lectures.

5. CONCLUSION

Establishing beyond doubt the main features of the standard model has been a long process. Clearly this process was often guided by remarkable theoretical insights, but decisive experimental discoveries played a key role. Progress has been more slow, difficult, and painful than is remembered by our forgiving memory.

The main ingredient of the model the Higgs boson is still missing, and signs of effects beyond the standard model have only recently been seen in ν oscillations; so the coming 2000-2010 period promises to be very interesting.

BREMSSTRAHLUNG

On the perturbative road to production of hadrons

Yuri L. Dokshitzer

LPT, Université Paris-Sud, Orsay, France

and PNPI, St. Petersburg, Russia

yuri@th.u-psud.fr

Keywords: Perturbative QCD, hard processes, multiparticle production

Abstract The QCD picture of multiparticle production in hard interactions is reviewed. We illustrate the basic properties of high energy bremsstrahlung with practical examples such as hadroproduction of J/ψ, photon radiation in $W^{\pm}$ decays, hadronic accompanyment in Higgs production. Special emphasis is given to coherence effects which determine the structure of particle multiplication in QCD jets (humpbacked plateau) and the pattern of inter-jet hadron flows (QCD string/drag effects).

1. INTRODUCTION

I planned these lectures as a stress-lifting rather than fact-finding mission. My aim is to convince you that the most simple calculations, backed by a physical intuition, are capable of delivering many amusing verifiable consequences. You will find some scattered formulae in the text. They are all simple. Trust me. They are included not to scare you away from reading but for the benefit of those (ideal) readers who would like to see how things actually come about.

We shall play some selected themes: the concept of QCD partons, physics of bremsstrahlung, conservation of color current, QCD coherence, basics of multiparticle production in hard interactions. If, having worked through these themes you will start hearing a symphony with your inner ear, then the aim of the course will have been reached.

One confession is due before we start: you will find no bibliography attached. This is because all the statements you will meet below fall into two cathegories: they are either well-known or unpublished.

H.B. Prosper and M. Danilov (eds.), Techniques and Concepts of High-Energy Physics, 51–96.

2. SMALL COUPLING, LARGE LOGARITHMS AND EVOLUTION

QCD is a *quantum* theory; have no doubt about it. This very statement seems to make the problem of describing parton systems involving $n > n_0$ gluons and quarks (with the actual value of $n_0 \sim 1$ depending on your computer) look hopeless: solving such a problem would call for sorting out and calculating $\mathcal{O}((n!)^2)$ Feynman diagrams.

Why should we worry about multiparton systems in the first place? Is it not true that the squared matrix element in the $n^{\rm th}$ order of perturbation theory is proportional to $(\alpha_s/\pi)^n \lesssim (0.1)^n$ and, thus, vanishingly small for large n? The answer to this (as to many other questions, according to the celebrated Hegel's dialectic wisdom) is: "Yes and No". Indeed,

Yes, it is very small, if we talk about a "multijet" configuration of 10 energetic quarks and gluons with large angles between them;

No, it is of order *unity*, if we address the *total* probability of having 8 extra gluons (and quarks) in addition to, say, a $q\bar{q}$ pair produced in e^+e^- annihilation at LEP.

Allowing small relative angles between partons in a process with a large hardness Q^2 results in a logarithmic enhancement of the emission probability:

$$\alpha_s \Longrightarrow \alpha_s \frac{d\Theta^2}{\Theta^2} \to \alpha_s \log Q^2 . \tag{2.1a}$$

As a result, the total probability of one parton (E) turning into two ($E_1 \sim E_2 \sim \frac{1}{2}E$) may become of order 1, in spite of the smallness of the characteristic coupling, $\alpha_s(Q^2) \propto 1/\log Q^2$. A typical example of such a "collinear" enhancement — the splitting process $g \to q\bar{q}$.

Moreover, when we consider the *gluon* offspring, another — "soft" — enhancement enters the game, which is due to the fact that the gluon bremsstrahlung tends to populate the region of *relatively* small energies ($E \simeq E_1 \gg E_2 \equiv \omega$):

$$\alpha_s \Longrightarrow \alpha_s \frac{d\omega}{\omega} \frac{d\Theta^2}{\Theta^2} \to \alpha_s \log^2 Q^2 . \tag{2.1b}$$

Thus the true perturbative "expansion parameter" responsible for parton multiplication via $q \to qg$ and $g \to gg$ may actually become *much larger* that 1!

In such circumstances we cannot trust the expansion in $\alpha_s \ll 1$ unless the logarithmically enhanced contributions (2.1) are taken full care of in all orders.

Fortunately, in spite of the complexity of high order Feynman diagrams, such a programme can be carried out. There is a physical reason for that: large contributions (2.1) originate from a specific region of phase space, which can be viewed as a sequence of parton decays strongly ordered in fluctuation times. Given such a separation in time, successive parton splittings become independent, so that the emerging picture is essentially classical. This is how the parton cascades described by the classical equations of parton balance (evolution equations) come about.

2.1. LOGARITHM IS NOT A FUNCTION

The very fact that the all-order logarithmic asymptotes can be written down in a closed form and, more than that, that they *a posteriori* prove to be quite simple, follows from the statement that constitutes a field-theoretical "article of faith":

Logarithm is not a function[1] *but a signal of simple underlying physics.*

In our context this simplicity has to do with the **classical nature** of

- *soft* enhancement of bremsstrahlung amplitudes ("infrared" singularities) and
- *collinear* enhancement of basic $1 \to 2$ parton splitting amplitudes (or "mass" singularities).

As a result, the leading logarithmic asymptotes can be found without performing laborious calculations. It suffices to invoke an intuitively clear picture of parton cascades described in probabilistic fashion in terms of sequential independent elementary parton branchings.

Whether the main enhanced terms are *Double Logarithmic* (DL) as in (2.1b) or *Single Logarithmic* (SL), (2.1a), depends on the nature of the problem under focus. The very distinction between DL and SL asymptotic regimes is often elusive. To illustrate the latter statement one may recall the QCD analysis of structure functions describing deep inelastic scattering (DIS), the subject to be discussed in detail later on in this lecture.

DIS (Q^2 large, x moderate) is a typical SL problem, with the following perturbative (PT) expansion:

$$D^{(n)}(x, Q^2) \;=\; C_n(x) \cdot \frac{1}{n!}\left[\frac{\alpha_s}{\pi}\ln\frac{Q^2}{\mu^2}\right]^n + \text{less sing. terms;} \qquad (2.2)$$

[1] ascribed to L.D.Landau

$$D(x,Q^2) \;=\; \sum_{n=0}^{\infty} D^{(n)}\,.$$

Here x is the Bjorken variable and μ^2 the finite initial virtuality of the target parton A (quark, gluon).

In general, $|C_n(x)| \sim 1$. However, in the quasi-elastic limit of $x \to 1$, when the invariant mass of the produced parton system becomes relatively small, $W^2 = Q^2(1-x)/x \ll Q^2$, the expansion coefficients in (2.2) take the form

$$\begin{aligned} C_n &\propto [C_F \ln(1-x)]^n\,; \\ D(x,Q^2) &\propto (1-x)^{-1} \exp\left\{ \frac{C_F \alpha_s}{\pi} \ln(1-x) \ln\frac{Q^2}{\mu^2} \right\}, \end{aligned} \qquad (2.3a)$$

and the problem turns out to be DL. Another important example of such a permutation has to do with the opposite limit of numerically small x. In this region the dominant contribution comes from sea-quark pairs copiously produced via gluon cascades, and the answer again exhibits the DL asymptote:

$$\begin{aligned} C_n(x) &\propto \frac{[N_c \ln x^{-1}]^n}{(n-1)!}\,; \\ D(x,Q^2) &\propto x^{-1} I_1 \left(2\sqrt{\frac{N_c \alpha_s}{\pi} \ln x^{-1} \ln\frac{Q^2}{\mu^2}} \right), \end{aligned} \qquad (2.3b)$$

with I_1 the modified Bessel function.

2.2. PUZZLE OF DIS AND QCD PARTONS

Let us invoke the deep inelastic lepton-hadron scattering — a classical example of a hard process and the standard QCD laboratory for carrying out the perturbative resummation programme.

Here, the momentum q with a large space-like virtuality $Q^2 = |q^2|$ is transferred from an incident electron (muon, neutrino) to the target proton, which then breaks up into the final multiparton $\to$ multihadron system. Introducing an invariant energy $s = 2(Pq)$ between the exchange photon (Z^0, $W^{\pm}$) q and the proton with 4-momentum P ($P^2 = M_p^2$), one writes the invariant mass of the produced hadron system which measures *inelasticity* of the process as

$$W^2 \equiv (q+P)^2 - M_p^2 = q^2 + 2(Pq) = s(1-x)\,, \qquad x \equiv \frac{Q^2}{2(Pq)} \le 1\,.$$

The cross section of the process depends on two variables: the *hardness* Q^2 and Bjorken x. For the case of *elastic* lepton-proton scattering one has $x \equiv 1$ and it is natural to write the cross section as

$$\frac{d\sigma_{el}}{dQ^2\,[dx]} = \frac{d\sigma_{\text{Ruth}}}{dQ^2} \cdot f_{el}^2(Q^2) \cdot [\delta(1-x)]\,. \tag{2.4a}$$

Here $\sigma_{\text{Ruth}} \propto \alpha^2/Q^4$ is the standard Rutherford cross section for e.m. scattering off a point charge and f_{el} stands for the elastic proton form factor.

For inclusive *inelastic* cross section one can write an analogous expression by introducing an inelastic proton "form factor" which now depends on both the momentum transfer Q^2 and the inelasticity parameter x:

$$\frac{d\sigma_{in}}{dQ^2\,dx} = \frac{d\sigma_{\text{Ruth}}}{dQ^2} \cdot f_{in}^2(x, Q^2)\,. \tag{2.4b}$$

What kind of Q^2-behavior of the form factors (2.4) could one expect in the Bjorken limit $Q^2 \to \infty$? Quantum mechanics tells us how the Q^2-behavior of the electromagnetic form factor can be related to the charge distribution inside a proton:

$$f_{el}(Q^2) = \int d^3r\,\rho(\vec{r})\,\exp\left\{i\vec{Q}\vec{r}\right\}.$$

For a point charge $\rho(\vec{r}) = \delta^3(\vec{r})$, it is obvious that $f \equiv 1$. On the contrary, for a smooth charge distribution $f(Q^2)$ falls with increasing Q^2, the faster the smoother ρ is. Experimentally, the elastic *e-p* cross section does decrease with Q^2 *much faster* that the Rutherford one ($f_{el}(Q^2)$ decays as a large power of Q^2). Does this imply that $\rho(\vec{r})$ is indeed regular so that there is no well-localized — point-charge inside a proton? If it were the case, the *inelastic* form factor would decay as well in the Bjorken limit: a tiny photon with the characteristic size $\sim 1/Q \to 0$ would penetrate through a "smooth" proton like a knife through butter, inducing neither elastic nor inelastic interactions.

However, as was first observed at SLAC in the late sixties, for a fixed x, f_{in}^2 stays practically constant with Q^2, that is, the inelastic cross section (with a given inelasticity) is similar to the Rutherford cross section (Bjorken scaling). It looks *as if* there was a point-like scattering in the guts of it, but in a rather strange way: it results in inelastic break-up dominating over the elastic channel. Quite a paradoxical picture emerged; Feynman-Bjorken partons came to the rescue.

Imagine that it is not the proton itself that is a point-charge-bearer, but some other guys (quark-partons) inside it. If those constituents were *tightly* bound to each other, the elastic channel would be bigger than, or comparable with, the inelastic one: an excitation of the parton that takes an impact would be transferred, with the help of rigid links between partons, to the proton as a whole, leading to the elastic scattering or to the formation of a quasi-elastic finite-mass system ($N\pi$, $\Delta\pi$ or so), $1-x \ll 1$.

To match the experimental pattern $f_{el}^2(Q^2) \ll f_{in}^2(Q^2)$ one has instead to view the parton ensemble as a *loosely* bound system of quasi-free particles. Only under these circumstances does knocking off one of the partons inevitably lead to deep inelastic breakup, with a negligible chance of reshuffling the excitation among partons.

The parton model, forged to explain the DIS phenomenon, was intrinsically paradoxical by itself. In sixties and seventies, there was no other way of discussing particle interactions but in the field-theoretical framework, where it remains nowadays. But all reliable (renormalizable, 4-dimensional) quantum field theories (QFTs) known by then had one feature in common: an effective interaction strength (the running coupling $g^2(Q^2)$) *increasing* with the scale of the hard process Q^2. Actually, this feature was widely believed to be a general law of nature, and for a good reason[2]. At the same time, it would be preferable to have it the other way around so as to be in accord with the parton model, which needs parton-parton interaction to *weaken* at small distances (large Q^2).

Only with the advent of non-Abelian QFTs (and QCD among them) exhibiting an anti-intuitive asymptotic-freedom behavior of the coupling, the concept of partons was to become more than a mere phenomenological model.

2.3. QCD DIS MINUTES

Typical QCD graphs for DIS amplitudes are shown in Fig. 1.

For moderate x-values (say, $x > 0.1$) the process is dominated by lepton scattering off a valence quark in the proton. The scattering cross section has a standard energy behavior $\sigma \propto x^{-2(J_{\rm ex}-1)}$, where $J_{\rm ex}$ is the spin of the exchanged particle in the t-channel. It is the quark with $J_{\rm ex}=\frac{1}{2}$ in the left picture of Fig. 1, so that the valence contribution to the cross section decreases at small x as $\sigma \propto x$.

For high-energy scattering, $x \ll 1$, the Bethe-Heitler mechanism takes over which corresponds to the t-channel *gluon* exchange: $J_{\rm ex}=1$, $\sigma \propto x^0$ =const (modulo logarithms).

In the Leading Logarithmic Approximation (LLA) one insists on picking up, for each new parton taken into consideration, a logarithmic en-

[2]relation between screening and unitarity

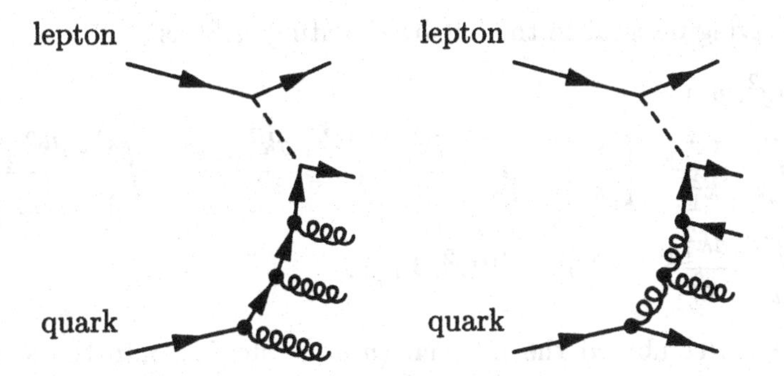

Figure 1 Valence (left) and Bethe-Heitler mechanism (right) of DIS.

hancement factor $\alpha_s \to \alpha_s \log Q^2$. In this approximation the scattering *probability* can be simply obtained by convoluting elementary probabilities of independent $1 \to 2$ parton splittings.

To cut a long story short, the appearance of the log-enhanced contributions in (2.1a) is due to the following structure

$$\begin{aligned} &\frac{1}{n!}\left[\frac{\alpha_s}{\pi}\ln\frac{Q^2}{\mu^2}\right]^n \qquad (2.5) \\ &= \left[\frac{\alpha_s}{\pi}\right]^n \int^{Q^2} \frac{dk_{\perp n}^2}{k_{\perp n}^2} \int^{k_{\perp n}^2} \frac{dk_{\perp n-1}^2}{k_{\perp n-1}^2} \cdots \int^{k_{\perp 3}^2} \frac{dk_{\perp 2}^2}{k_{\perp 2}^2} \int_{\mu^2}^{k_{\perp 2}^2} \frac{dk_{\perp 1}^2}{k_{\perp 1}^2}, \end{aligned}$$

with $k_{\perp i}^2$ the squared transverse momenta of produced partons.

To contribute to the LLA, the transverse momenta of produced partons should be strongly ordered, increasing up the "ladder": $k_{1\perp}^2 \ll \ldots \ll k_{n\perp}^2 \ll Q^2$. (At the level of Feynman *amplitudes* the ladder diagrams dominate, provided a special physical gauge is chosen for gluon fields.)

The expressions (2.2) and (2.3) exhibit an unpleasant (or wonderful, according to taste) feature: they become senseless (diverge) in the zero-quark-mass limit, $\mu \to 0$. Well, when you see a nasty thing happen beyond your reach, you can do no better than make use of it. This "mass singularity", according to (2.5), occurs in the lower limit of the $k_\perp$ integration of the very first (and only!) parton branch. Let us drag this misbehaving integral to the left by rewriting (2.5) as

$$
\begin{aligned}
&(2.5)^{[n]}(Q^2,\mu^2)\\
&= \frac{\alpha_s}{\pi}\int_{\mu^2}^{Q^2}\frac{dk_{\perp 1}^2}{k_{\perp 1}^2}\cdot\left[\frac{\alpha_s}{\pi}\right]^{n-1}\int^{Q^2}\frac{dk_{\perp n}^2}{k_{\perp n}^2}\int^{k_{\perp n}^2}\frac{dk_{\perp n-1}^2}{k_{\perp n-1}^2}\cdots\int_{k_{\perp 1}^2}^{k_{\perp 3}^2}\frac{dk_{\perp 2}^2}{k_{\perp 2}^2}\\
&= \frac{\alpha_s}{\pi}\int_{\mu^2}^{Q^2}\frac{dk_{\perp 1}^2}{k_{\perp 1}^2}\cdot\ (2.5)^{[n-1]}(Q^2,k_{\perp 1}^2),
\end{aligned}
$$

where we have combined the internal $(n-1)$ integrals into the same expression that corresponds to the previous order in α_s-expansion and has a new lower limit $k_{\perp 1}^2$ substituted for the original μ^2. Now, we can localize the μ-dependence by evaluating the logarithmic derivative:

$$
\mu^2\frac{\partial}{\partial\mu^2}\ (2.5)^{[n]} \;=\; -\frac{\alpha_s}{\pi}\cdot\ (2.5)^{[n-1]}.
$$

This equation relates the nth order of the PT expansion to the previous one. To put this symbolic relation at work one first has to recall the satellite x-dependence.

By extracting the first step one may look upon the rest as DIS off a new "target" — the parton with transverse momentum $k_{\perp 1}^2$ and a finite fraction z of the initial longitudinal momentum P. As a result, there appears an additional integration with the probability of the first splitting, $\phi(z)$, and the differential equation for the resummed $F^{(\mathrm{LLA})}$ takes the form

$$
\mu^2\frac{\partial}{\partial\mu^2}\ F\left(x,\,Q^2,\mu^2\right) \;=\; -\int_x^1\frac{dz}{z}\,\phi(z)\,\frac{\alpha_s}{\pi}\cdot F\left(\frac{x}{z},\,Q^2,\mu^2\right). \tag{2.6a}
$$

Since a logarithm (like a stick) has two ends, differentiation over the overall hardness scale Q^2 would do the same job, the result being the *evolution equation* in a familiar form:

$$
Q^2\frac{\partial}{\partial Q^2}F(x,Q^2) = \frac{\alpha_s}{\pi}\phi(x)\otimes F(x,Q^2)\,, \tag{2.6b}
$$

where the symbol $\otimes$ stands for convolution in the x-space.

2.4. LLA PARTON EVOLUTION

2.4.1 Space-like parton evolution. The decay phase space for the *space-like* evolution determining the DIS structure functions is

$$
dw^{A\to B+C} = \frac{dk_\perp^2}{k_\perp^2}\,\frac{\alpha_s(k_\perp^2)}{2\pi}\,\frac{dz}{z}\,\Phi_A^{BC}(z) \tag{2.7a}
$$

with z the longitudinal momentum fraction carried by the parton B. The functions Φ play the role of "Hamiltonian" (x-dependent kernels) of the evolution equation (2.6) for the LLA parton distributions.

In the DIS environment the initial parton A with a negative (space-like) virtuality decays into $B[z]$ with the large space-like virtual momentum $|k_B^2| \gg |k_A^2|$ and a positive virtuality (time-like) $C[1-z]$. The parton C generates a subjet of secondary partons ($\rightarrow$ hadrons) in the final state. Since no details of the final-state structure are measured in an inclusive process, the invariant mass of the subjet C is integrated over. In the dominant integration region $k_C^2 \ll |k_B^2|$ the parton C looks *quasi-real*, compared with the hard scale of $|k_B^2|$. The same is true for the initial parton A.

Splitting can be viewed as a large momentum-transfer process of scattering (turnover) of a "real" target parton A into a "real" C on the external field mediated by high-virtuality B. At the next step of evolution it is B's turn to play a role of a next target $B \equiv A'$, "real" with respect to yet deeper probe $|k_B'^2| \gg |k_B^2|$, and so on.

Successive parton decays with step-by-step *increasing* space-like virtualities (transverse momenta) constitute the picture of parton wave-function fluctuations inside the proton. The sequence proceeds until the overall hardness scale Q^2 is reached.

2.4.2 Time-like parton cascades. A similar picture emerges for the time-like branching processes determining the internal structure of jets produced, for example, in e^+e^- annihilation. Here the flow of *hardness* is opposite to that in DIS: evolution starts from a highly virtual quark with positive virtuality, originating from the e.m. vertex, while time-like virtualities of its products ("quasi-real" with respect to predecessors; "high-virtuality" with respect to offspring) degrade.

In the time-like case the longitudinal phase space is symmetric in offspring parton energy fractions, and the differential decay probability reads

$$dw^{A \to B+C} = \frac{d\Theta^2}{\Theta^2} \frac{\alpha_s(k_\perp^2)}{2\pi} dz \, \Phi_A^{BC}(z) \,. \tag{2.7b}$$

It is important to notice that the flow of *energy* (longitudinal momentum) is governed, in the LLA, by the same functions Φ_A^{BC}: it does not matter that now, in contrast to space-like evolution, A is the "virtual" one and B and C are "real". This relation (known as the Gribov-Lipatov reciprocity) is one of many wonderful symmetries that our "parton Hamiltonian" Φ_A^{BC} obeys.

2.4.3 Apparent and hidden in parton dynamics. When studying inclusive characteristics of parton cascades, one traces a single route of successive parton splittings. Having this in mind, we can drop the label that marks partons C whose fate does not concern us, $\Phi_A^{BC}(z) \Longrightarrow \Phi_A^B(z)$ ($\equiv P_{BA}$ in the standard Altarelli-Parisi notation).

For discussion of the relations between splitting functions it is convenient to strip off color factors and introduce

$$\begin{aligned} \Phi_F^F(z) &= C_F\, V_F^F(z)\,, \qquad \Phi_F^G(z) = C_F\, V_F^G(z)\,, \\ \Phi_G^F(z) &= T_R\, V_G^F(z)\,, \qquad \Phi_G^G(z) = N_c\, V_G^G(z)\,. \end{aligned} \tag{2.8}$$

Here C_F and N_c are the familiar quark and gluon "color charges", while T_R is a scientific name for one half:

$$\mathrm{Tr}(t^a t^b) = \sum_{i,k=1}^{N_c} t^a_{ik}\, t^b_{ki} \;\equiv\; T_R\, \delta^{ab} = \tfrac{1}{2}\delta^{ab}\,.$$

The splitting functions then read

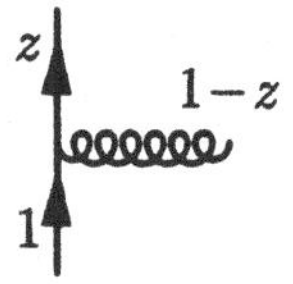

$$V_F^F(z) = \frac{1+z^2}{1-z}\,, \tag{2.9a}$$

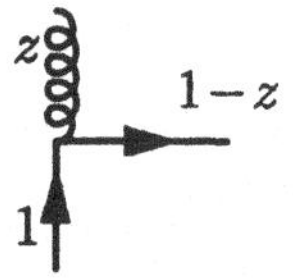

$$V_F^G(z) = \frac{1+(1-z)^2}{z}\,, \tag{2.9b}$$

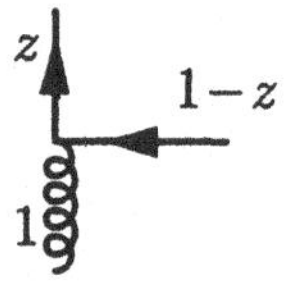

$$V_G^F(z) = z^2 + (1-z)^2\,, \tag{2.9c}$$

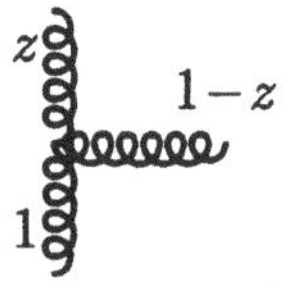

$$V_G^G(z) = 2\left[z(1-z) + \frac{1-z}{z} + \frac{z}{1-z}\right]. \tag{2.9d}$$

They have the following remarkable symmetry properties.

Parton Exchange results in an obvious relation between probabilities to find decay products with complementary momentum fractions:

$$V_A^{B(C)}(z) \;=\; V_A^{C(B)}(1-z)\,. \tag{2.10a}$$

Drell-Levy-Yan Crossing Relation emerges when one links together two splitting processes corresponding to opposite evolution "time" sequences:

$$V_B^A(z) = (-1)^{2s_A+2s_B-1}\, z\, V_A^B\left(\frac{1}{z}\right), \tag{2.10b}$$

with s_A the spin of particle A. Strictly speaking, the crossing $z \to 1/z$ relates the space-like and time-like evolution kernels, and vice versa, $V \leftrightarrow \overline{V}$. Representing (2.10b) in terms of V relies upon an identity of the LLA "Hamiltonians" mentioned above,

Gribov-Lipatov Reciprocity Relation:

$$\overline{V}(z)_{\text{time-like}} = V(z)_{\text{space-like}} \,. \tag{2.10c}$$

As we see, these relations do not leave much freedom for splitting functions. In fact, one could borrow V_F^F from QED textbooks, reconstruct V_F^G by exchanging the decay products (2.10a), and then obtain V_G^F using the crossing (2.10b). This is the way to generate all three splitting functions relevant for QED (2.9a)–(2.9c). The last gluon-gluon splitting function (2.9d) transforms into itself under both (2.10a) and (2.10b). The more surprising is the fact that the gluon self-interaction kernel actually could have been obtained "from QED" using

the Super-Symmetry Relation:

$$V_F^F(z) + V_F^G(z) = V_G^F(z) + V_G^G(z) \,. \tag{2.10d}$$

This relation exploits the existence of the supersymmetric QFT closely related to real QCD. In the SUSY-QCD, "quark" and "gluon" belong to the same (adjoint) representation of the color group, so that all color factors become identical $C_F = C_A = T_R$ (cf. (2.8)). Bearing this in mind one can spell out (2.10d) as an equality between the *total* probabilities of "quark" and "gluon" decays. The fact that it holds identically in z means that there is an infinite number of non-trivial conservation laws in this theory!

Even this is not the end of the story.

Conformal Invariance leads to a number of relations (involving derivatives) between splitting functions, the simplest of which reads

$$\left(z\frac{d}{dz} - 2\right) V_G^F(z) = \left(z\frac{d}{dz} + 1\right) V_F^G(z) \,. \tag{2.10e}$$

The general character of the symmetry properties makes them practically useful when studying subleading effects in parton distributions where one faces technically difficult calculations. For example, the supersymmetric QCD analogue had been used to choose between two contradictory calculations in the next-to-LLA (the two-loop anomalous dimensions). We illustrate the idea by another example of the most advanced next-to-next-to-leading result obtained by Gaffney and Mueller for the ratio of mean parton multiplicities in gluon and quark jets, which reads

$$\frac{C_F}{N_c}\frac{\mathcal{N}_g}{\mathcal{N}_q} \simeq 1 - \left(1 + \frac{T_f}{N_c} - 2\frac{T_f C_F}{N_c^2}\right) \cdot \left[\sqrt{\frac{\alpha_s N_c}{18\pi}} + \frac{\alpha_s N_c}{18\pi}\left(\frac{25}{8} - \frac{3T_f}{4N_c} - \frac{T_f C_F}{N_c^2}\right)\right]. \tag{2.11}$$

Here $T_f \equiv 2n_f T_R$, with $2n_f$ the number of fermions (quarks and antiquarks of n_f flavours).

The symmetry between quarks (fermions) and gluons (bosons) is hidden in QCD. It becomes manifest in another QFT which is a *supersymmetric partner* of QCD. In that theory fermions and vector bosons have the same color properties (both correspond to the adjoint representation). Equating the color factors, $N_c = C_F = T_R$, and bearing in mind another subtlety, $n_f = \frac{1}{2}$ (since the "quark" is a Majorana fermion there), it is straightforward to verify that the ratio of multiplicities in "quark" and "gluon" jets (2.11) indeed turns to unity, in all known (as well as in all unknown) orders.

2.4.4 Fluctuation Time and Evolution Times: Coherence.

Once the Hamiltonian is known, it suffices to propagate it in time to find all we want to know about the system. But was is the *time* in our context?

An attentive reader has noticed that back in (2.7) we wrote the parton splitting phase space differently: for the space-like case (2.7a) in terms of transverse momentum $k_\perp$ and for the time-like evolution (2.7b) via the decay angle Θ. Logarithmic differentials by themselves are identical, since $k_\perp^2$ and Θ^2 are proportional for fixed z. We have made this distinction to stress an important difference between a probabilistic interpretation of DIS and the e^+e^- evolution: the different *evolution times*.

To be honest, within the LLA framework it does not make much sense to argue which of evolution parameters $\ln k_{\perp i}^2$, $\ln \Theta_i^2$ or $\ln|k_i^2|$ (with k^2 the total parton virtuality) does a better job: these choices differ by *subleading* terms. A mismatch is of the order of

$$\frac{\alpha_s}{\pi}\ln^2 z\,. \tag{2.12}$$

Strictly speaking, such contributions should be treated as $\mathcal{O}(\alpha_s)$ within the LLA logic and neglected as compared to $\alpha_s \ln Q \sim 1$. These *next-to-LLA* terms become significant, however, and should be taken care of and "resummed" in all orders when numerically small values of the Bjorken x are concerned[3] such that $\frac{\alpha_s}{\pi}\ln^2 x \sim 1$.

In the DIS environment, the *transverse momentum* ordering proves to be the one that takes care of potentially disturbing corrections (2.12) in all orders, and in this sense becomes a preferable choice for constructing the probabilistic scheme for space-like parton cascades (DIS structure functions).

On the other hand, in the case of the time-like cascades — yes, you've guessed it right from (2.7b)! — it is the *relative angle* between the offspring partons which has to be kept ordered, decreasing along the evolutionary decay chain.

What is the difference between the two prescriptions, and how do they relate with the fluctuation time ordering which was claimed in the beginning of the lecture to ensure the probabilistic picture?

The time ordering is there. Simply by examining the Feynman denominators it is easy to see that the maximally enhanced contribution — collinear logarithm per each splitting, as in the DIS "ladder" (2.5) — implies the ordering in

$$t_{\text{fluct}} \simeq \frac{k_{\|}}{k_{\perp}^2}\,.$$

In the DIS kinematics the evolution goes from the proton side and, on the way towards the virtual probe Q^2, parton fluctuations are successively shorter-lived (the "probe" is faster than the lifetime of the "target"):

$$\text{time ordering} — \qquad t_i = \frac{k_{\|i}}{k_{\perp i}^2} > t_{i+1} = \frac{k_{\|i+1}}{k_{\perp i+1}^2} \tag{2.13a}$$

$$k_\perp \text{ ordering} — \qquad k_{\perp i} < k_{\perp i+1}, \tag{2.13b}$$

$$\text{mismatch} \Longrightarrow \qquad z \cdot k_{\perp i}^2 < k_{\perp i+1}^2 < k_{\perp i}^2. \tag{2.13c}$$

In the case of a time-like jet the order of events is opposite. The process starts from a large scale $s = Q^2$, and the partons of the generation $(i+1)$ live *longer* than their parent (i):

$$\text{time ordering} — \qquad t_i = \frac{k_{\|i}}{k_{\perp i}^2} < t_{i+1} = \frac{k_{\|i+1}}{k_{\perp i+1}^2} \tag{2.14a}$$

[3] The word "numerically" stands here as a warning for not confusing this kinematical region with a "parametrically" small x, such that $\alpha_s \ln 1/x \sim 1$ — the Regge region — where essentially different physics comes onto stage.

$$\text{angular ordering} — \qquad \theta_i = \frac{k_{\perp i}}{k_{\| i}} > \theta_{i+1} = \frac{k_{\perp i+1}}{k_{\| i+1}}, \tag{2.14b}$$

$$\text{mismatch} \Longrightarrow \qquad \theta_i^2 < \theta_{i+1}^2 < \frac{\theta_i^2}{z}. \tag{2.14c}$$

We see that in both situations the mismatch, (2.13c) and (2.14c), may become significant in the case of a relatively *soft* decay,

$$z \equiv \frac{k_{\| i+1}}{k_{\| i}} \ll 1,$$

that is, when soft gluon emission comes onto stage. Here we better be careful: the catch is, to be emitted *later* does not guarantee being emitted *independently*. Quantum mechanics, you know. The time-ordering (2.13a), (2.14a) proves to be too liberal in both cases. In fact, the parton multiplication in the regions (2.13c) and (2.14c) is suppressed. The sum of the contributing Feynman diagrams vanish.

Let us sort out here the DIS puzzle (2.13), and leave the time-like *angular ordering* phenomenon (2.14) to be slowly enjoyed in the next lecture.

2.4.5 Vanishing of the forward inelastic diffraction. Consider a bit of the DIS ladder — a two-step process shown by the first graph in Fig. 2. Let the second decay be *soft*, $z_2 \ll 1$. (The first one can be either soft or hard, $z_1 \lesssim 1$.) In the kinematical region (2.13c),

$$z_2 \cdot k_{1\perp}^2 \ll k_{2\perp}^2 \ll k_{1\perp}^2, \tag{2.15}$$

the *time-ordering* is still intact, which means that the momentum k_2 is transferred fast as compared with the lifetime of the first fluctuation $P \to P' + k_1$.

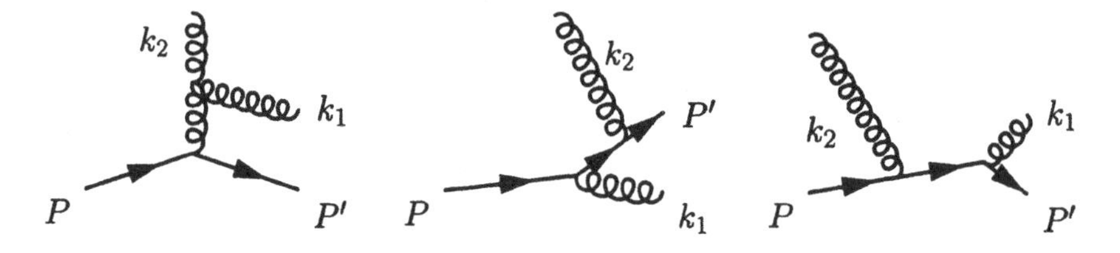

Figure 2 In the "wrong" kinematics $k_{2\perp} < k_{1\perp}$, the sum of the two space-like evolution amplitudes cancels against the final state time-like decay

Since k_2 is the softest, energy-wise, the process can be viewed as inelastic relativistic scattering $P \to P' + k_1$ in the external gluon field (k_2). The

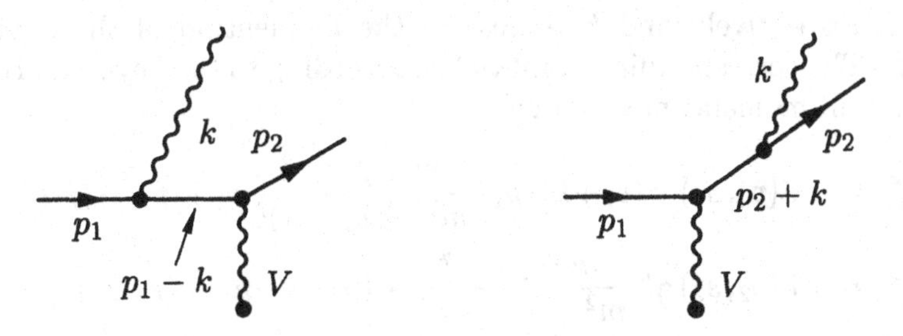

Figure 3 Photon Bremsstrahlung diagrams for scattering off an external field.

transverse size of the field is $\rho_\perp \sim k_{2\perp}^{-1}$. The characteristic size of the fluctuation $P' + k_1$, according to (2.15), is smaller: $\Delta r_{01\perp} \sim k_{1\perp}^{-1} \ll \rho_\perp$. We thus have a *compact* state propagating through the field that is smooth at distances of the order of the size of the system. In such circumstances the field cannot resolve the internal structure of the fluctuation. Components of the fluctuation, partons P' and k_1 in the first two graphs of Fig. 2, scatter *coherently*, and the total amplitude turns out to be identical and opposite in sign to that for the scattering of the initial state P (the last graph): inelastic breakup does not occur.

The cancellation between the amplitudes of Fig. 2 in the region (2.15), and thus the $k_\perp$ ordering, is a direct consequence of the conservation of current.

3. BREMSSTRAHLUNG, COHERENCE, CONSERVATION OF CURRENT

The purpose of the second lecture is to recall the basic properties of photon radiation. Once the QED bremsstrahlung is understood as an essentially coherent and, at the same time, intrinsically classical phenomenon, the physics of gluon radiation will readily follow suit. So, we shall start from the electromagnetic radiation and turn to gluons only once in this lecture, when there is something interesting to say in the specific QCD context (J/ψ production and the Low theorem).

3.1. PHOTON BREMSSTRAHLUNG

Let us consider photon bremsstrahlung induced by a charged particle (electron) which scatters off an external field (e.g., a static electromagnetic field). The derivation is included in every textbook on QED, so we confine ourselves to the essential aspects.

The lowest order Feynman diagrams for photon radiation are depicted in Fig. 3, where p_1, p_2 are the momenta of the incoming and outgoing electron respectively and k represents the momentum of the emitted photon. The corresponding amplitudes, according to the Feynman rules, are given in momentum space by

$$M_{\rm i}^{\mu} = e\,\bar{u}(p_2,s_2)\,V(p_2+k-p_1)\,\frac{m+\not{p}_1-\not{k}}{m^2-(p_1-k)^2}\,\gamma^{\mu}\,u(p_1,s_1), \quad (3.16a)$$

$$M_{\rm f}^{\mu} = e\,\bar{u}(p_2,s_2)\,\gamma^{\mu}\,\frac{m+\not{p}_2+\not{k}}{m^2-(p_2+k)^2}\,V(p_2+k-p_1)\;u(p_1,s_1). \quad (3.16b)$$

Here V stands for the basic interaction amplitude which may depend in general on the momentum transfer (for the case of scattering off the static e.m. field, $V=\gamma^0$).

First we apply the soft-photon approximation, $\omega \ll p_1^0, p_2^0$, to neglect $\not{k}$ terms in the numerators. To deal with the remaining matrix structure in the numerators of (3.16) we use the identity $\not{p}\,\gamma^{\mu} = -\gamma^{\mu}\,\not{p} + 2\,p^{\mu}$ and the Dirac equation for the on-mass-shell electrons,

$$(m+\not{p}_1)\,\gamma^{\mu}\,u(p_1) = (2p_1^{\mu} + [\,(m-\not{p}_1\,])\,u(p_1) = 2p_1^{\mu}\,u(p_1)\,,$$
$$\bar{u}(p_2)\,\gamma^{\nu}\,(m+\not{p}_2) = \bar{u}(p_2)\,([\,(m-\not{p}_2\,] + 2p_1^{\nu}) = 2p_2^{\nu}\,\bar{u}(p_2)\,.$$

Denominators for real electrons ($p_i^2 = m^2$) and the photon ($k^2 = 0$) become $m^2-(p_1-k)^2 = 2(p_1k)$ and $m^2-(p_2+k)^2 = -2(p_2k)$, so that for the total amplitude we obtain the factorized expression

$$M^{\mu} = e\,j^{\mu}\;\times\;M_{\rm el}\,. \quad (3.17a)$$

Here $M_{\rm el}$ is the Born matrix element for non-radiative (elastic) scattering,

$$M_{\rm el} = \bar{u}(p_2,s_2)\,V(p_2-p_1)\,u(p_1,s_1) \quad (3.17b)$$

(in which the photon recoil effect has been neglected, $q = p_2+k-p_1 \simeq p_2-p_1$), and j^{μ} is the *soft accompanying radiation current*

$$j^{\mu}(k) = \frac{p_1^{\mu}}{(p_1k)} - \frac{p_2^{\mu}}{(p_2k)}\,. \quad (3.17c)$$

Factorization (3.17a) is of the most general nature. The form of j^{μ} does not depend on the details of the underlying process, neither on the nature of participating charges (electron spin, in particular). The only thing which matters is the momenta and charges of incoming and outgoing particles. Generalization to an arbitrary process is straightforward and results in assembling the contributions due to all initial and final particles, weighted with their respective charges.

The soft current (3.17c) has a classical nature. It can be derived from the classical electrodynamics by considering the potential induced by change of the e.m. current due to scattering.

3.2. CLASSICAL CONSIDERATION

From classical field theory we know that it is the acceleration of a charge that causes electromagnetic radiation. Electromagnetic current participating in field formation in the course of scattering consists of two terms (we suppress the charge e for simplicity)

$$\vec{C} = \vec{C}_1 + \vec{C}_2\,, \quad \begin{cases} \vec{C}_1 = \vec{v}_1\,\delta^3(\vec{r} - \vec{v}_1\,t)\cdot\vartheta(t_0 - t)\,, \\ \vec{C}_2 = \vec{v}_2\,\delta^3(\vec{r} - \vec{v}_2\,t)\cdot\vartheta(t - t_0)\,, \end{cases} \tag{3.18a}$$

with $\vec{v}_{1(2)}$ the velocity of the initial (final) charge moving along the classical trajectory $\vec{r} = \vec{v}_i t$. By t_0 we denote the moment in time when the scattering occurs and the velocity abruptly changes. To achieve a Lorentz-covariant description one adds to (3.18a) an equation for the propagation of the charge-density D to be treated as the zero-component of the 4-vector current C^μ,

$$C_i^\mu(t,\vec{r}) \;=\; \Big(D_i(t,\vec{r}),\, \vec{C}_i(t,\vec{r})\Big) \;\equiv\; v_i^\mu\, D_i\,, \tag{3.18b}$$

with v_i^μ the 4-velocity vectors

$$v_i^\mu \;=\; (1,\, \vec{v}_i)\,.$$

The emission amplitude for a field component with 4-momentum $(\omega, \vec{k}\,)$ is proportional to the Fourier transform of the total current. For the two terms of the current we have

$$\begin{aligned} C_1^\mu(k) &= \int\limits_{-\infty}^{\infty} dt \int d^3r\, e^{ix^\nu k_\nu}\, C_1^\mu(t,\vec{r}) = v_1^\mu \int_{-\infty}^{0} d\tau\, e^{ik^0(t_0+\tau) - i(\vec{k}\cdot\vec{v}_i)\tau} \\ &= \frac{-i\ v_1^\mu\, e^{ik^0 t_0}}{k^0 - (\vec{k}\cdot\vec{v}_1)}\,, \end{aligned} \tag{3.19a}$$

$$\begin{aligned} C_2^\mu(k) &= \int\limits_{-\infty}^{\infty} dt \int d^3r\, e^{ix^\nu k_\nu}\, C_2^\mu(t,\vec{r}) = v_2^\mu \int_{0}^{+\infty} d\tau\, e^{ik^0(t_0+\tau) - i(\vec{k}\cdot\vec{v}_i)\tau} \\ &= \frac{i\ v_2^\mu\, e^{ik^0 t_0}}{k^0 - (\vec{k}\cdot\vec{v}_2)}\,. \end{aligned} \tag{3.19b}$$

The solution of the Maxwell equation for the field potential induced by the current (3.19) reads

$$\begin{aligned} A^\mu(x) &= \int \frac{d^4k}{(2\pi)^4}\, e^{-ix^\mu k_\mu}\, [-2\pi i\delta(k^2)]\cdot C^\mu(k) \\ &= \int \frac{d^3k}{2\omega(2\pi)^3}\, e^{-i\omega x^0 + i(\vec{k}\cdot\vec{x})}\cdot A^\mu(k)\,, \end{aligned} \tag{3.20}$$

where

$$A^\mu(k) = A_2^\mu(k) - A_1^\mu(k)\,;$$
$$A_i^\mu(k) = \frac{v_i^\mu \cdot e^{i\omega t_0}}{\omega(1 - v_i \cos\Theta_i)}; \quad \omega = |\vec{k}|\,, \ (\vec{k}\cdot\vec{v}_i) \equiv \omega v_i \cos\Theta_i. \tag{3.21}$$

Here Θ_i are the angles between the direction of the photon momentum and that of the corresponding (initial/final) charge.

Rewriting (3.21) in the covariant form

$$\frac{v_i^\mu}{\omega - (\vec{k}\cdot\vec{v}_i)} = \frac{E_i\, v_i^\mu}{E_i\,(\omega - (\vec{k}\cdot\vec{v}_i))} = \frac{p_i^\mu}{(p_i\, k)}\,,$$

we observe that the classical 4-vector "potential" (3.21), as expected, is identical to the quantum amplitude j^μ (3.17c), apart from an overall phase factor $\exp(i\omega t_0)$. The latter is irrelevant for calculating the observable *cross section* (see, however, section 3.4.3).

We conclude that the classical consideration gives the correct accompanying radiation pattern in the *soft-photon* limit. This is natural because in such circumstances (negligible recoil) it is legitimate to keep charges moving along their classical trajectories, which remain unperturbed in the course of sending away radiation.

3.3. SOFT RADIATION CROSS SECTION

To calculate the radiation probability we square the amplitude projected onto a photon polarization state ε_μ^λ, sum over λ and supply the photon phase space factor to write down

$$\mathrm{d}W = e^2 \sum_{\lambda=1,2} \left|\varepsilon_\mu^\lambda j^\mu\right|^2 \frac{\omega^2\, \mathrm{d}\omega\, \mathrm{d}\Omega_\gamma}{2\,\omega\,(2\pi)^3}\, \mathrm{d}W_{\mathrm{el}}\,. \tag{3.22}$$

The sum runs over two physical polarization states of the real photon, described by normalized polarization vectors orthogonal to its momentum:

$$\epsilon_\lambda^\mu(k)\cdot\epsilon_{\mu,\lambda'}^*(k) = -\delta_{\lambda\lambda'}\,, \quad \epsilon_\lambda^\mu(k)\cdot k_\mu = 0\,; \qquad \lambda, \lambda' = 1,2\,.$$

Within these conditions the polarization vectors may be chosen differently. Due to the gauge invariance such an uncertainty does not affect physical observables. Indeed, the polarization tensor may be represented as

$$\sum_{\lambda=1,2} \epsilon_\lambda^\mu \epsilon_\lambda^{*\,\nu} \;=\; -g^{\mu\nu} \;+\; \text{ tensor proportional to } k^\mu \text{ and/or } k^\nu\,. \tag{3.23}$$

The latter, however, can be dropped since the classical current (3.17c) is explicitly conserving, $(j^\mu k_\mu) = 0$. Therefore one may enjoy the gauge invariance and employ an arbitrary gauge, instead of using the physical polarizations, to calculate accompanying photon production.

The Feynman gauge being the simplest choice,

$$\sum_{\lambda=1,2} \epsilon_\lambda^\mu \epsilon_\lambda^{*\nu} \Longrightarrow -g^{\mu\nu},$$

we arrive at

$$\begin{aligned} dN \equiv \frac{dW}{dW_{\text{el}}} &= -\frac{\alpha}{4\pi^2} (j^\mu)^2 \, \omega \, d\omega \, d\Omega_\gamma \\ &\simeq \frac{\alpha}{\pi} \frac{d\omega}{\omega} \frac{d\Omega_\gamma}{2\pi} \frac{1-\cos\Theta_s}{(1-\cos\Theta_1)(1-\cos\Theta_2)} . \end{aligned} \tag{3.24}$$

The latter expression corresponds to the relativistic approximation $1 - v_1, 1 - v_2 \ll 1$:

$$-(j^\mu)^2 = \frac{2(p_1 p_2)}{(p_1 k)(p_2 k)} + \mathcal{O}\left(\frac{m^2}{p_0^2}\right) \simeq \frac{2}{\omega^2} \frac{(1 - \vec{n}_1 \cdot \vec{n}_2)}{(1 - \vec{n}_1 \cdot \vec{n})(1 - \vec{n}_2 \cdot \vec{n})};$$

it disregards the contribution of *very* small emission angles $\Theta_i^2 \lesssim (1 - v_i^2) = m^2/p_{0i}^2 \ll 1$, where the soft radiation vanishes (the so-called "Dead Cone" region).

If the photon is emitted at a small angle with respect to, say, the incoming particle, i.e. $\Theta_1 \ll \Theta_2 \simeq \Theta_s$, the radiation spectrum (3.24) simplifies to

$$dN \simeq \frac{\alpha}{\pi} \frac{\sin\Theta_1 \, d\Theta_1}{(1-\cos\Theta_1)} \frac{d\omega}{\omega} \simeq \frac{\alpha}{\pi} \frac{d\Theta_1^2}{\Theta_1^2} \frac{d\omega}{\omega} .$$

Two bremsstrahlung cones appear, centered around incoming and outgoing electron momenta. Inside these cones the radiation has a *double-logarithmic* structure, exhibiting both the *soft* ($d\omega/\omega$) and *collinear* ($d\Theta^2/\Theta^2$) enhancements.

3.3.1 Low-Barnett-Kroll wisdom. Soft factorization (3.17a) is an essence of the celebrated soft bremsstrahlung theorem, formulated by Low in 1956 for the case of scalar charged particles and later generalized by Barnett and Kroll to charged fermions. The very classical nature of soft radiation makes it universal with respect to intrinsic quantum properties of participating objects and the nature of the underlying scattering process: it is only the classical movement of electromagnetic charges that matters.

It is interesting that according to the LBK theorem both the leading $d\omega/\omega$ and the first subleading, $\propto d\omega$, pieces of the soft photon spectrum prove to be "classical".

For the sake of simplicity we shall leave aside the angular structure of the accompanying photon emission and concentrate on the energy dependence. Then, the relation between the basic cross section $\sigma^{(0)}$ and that with one additional photon with energy ω can be represented symbolically as

$$d\sigma^{(1)}(p_i,\omega) \propto \frac{\alpha}{\pi}\frac{d\omega}{\omega}\left[\left(1-\frac{\omega}{E}\right)\cdot\sigma^{(0)}(p_i)+\left(\frac{\omega}{E}\right)^2\cdot\tilde{\sigma}(p_i,\omega)\right]. \tag{3.25}$$

The first term in the right-hand side is proportional to the non-radiative cross section $\sigma^{(0)}$. The second term involves the new ω-dependent cross section $\tilde{\sigma}$ which is finite at $\omega = 0$, so that this contribution is suppressed for small photon energies as $(\omega/E)^2$.

This general structure has important consequences, the most serious of which can be formulated, in a dramatic fashion, as

3.3.2 Soft Photons don't carry quantum numbers. We are inclined to think that the photon has definite quantum numbers (negative C-parity, in particular). Imagine that the basic process is forbidden, say, by C-parity conservation. Why not to take off the veto by adding a photon to the system? Surely enough it can be done. There is, however, a price to pay: the selection rules cannot be overcome by *soft* radiation. Since the classical part of the radiative cross section in (3.25) is explicitly proportional to the non-radiative cross section $\sigma^{(0)} = 0$, only *energetic* photons (described by the $\tilde{\sigma}$ term) could do the job. The energy distribution

$$|M|^2\cdot\frac{d^3k}{\omega} \propto \omega d\omega$$

is typical for a quantum particle, where the production matrix element M is finite in the $\omega \to 0$ limit, $M = \mathcal{O}(1)$. An enhanced radiation matrix element, $M \propto \omega^{-1}$ characterizes a classical field rather than a quantum object.

So, the price one has to pay to overrule the quantum-number veto by emitting a soft photon with $\omega \ll E$ is the suppression factor

$$\left(\frac{\omega}{E}\right)^2 \ll 1\,.$$

We conclude that the photons that are capable of changing the quantum numbers of the system (be it parity, C-parity or angular momentum) cannot be *soft*. Neither can they be *collinear*, by the way, as it follows from the

3.3.3 Gribov Bremsstrahlung theorem.

This powerful generalisation of the Low theorem states that a simple factorization holds at the level of the *matrix element*, provided the photon transverse momentum with respect to the radiating charged particle is small compared to the momentum transfers characterizing the underlying scattering process:

$$M^{(1)} \propto \frac{(\vec{k}_\perp \cdot \vec{e})}{k_\perp^2} \cdot M^{(0)} + \tilde{M}. \tag{3.26}$$

Here again $\tilde{M}$ = const in the $k_\perp \to 0$ limit. This factorization holds for *hard* photons ($\omega \sim E$) as well as for soft ones.

Both the Low-Barnett-Kroll and the Gribov theorems hold in QCD as well. In particular, it is the Gribov collinear factorization that leads to the probabilistic evolution picture describing collinear QCD parton multiplication we have discussed in the first lecture.

In the QCD context, our statement that "soft photons don't carry quantum numbers" should be strengthened to even more provocative (but true)

3.3.4 Soft Gluons don't carry away no color.

Don't rush to protest. Just think it over. In more respectable terms this title could be abbreviated as the NSFL (no-soft-free-lunch) theorem.

Imagine we want to produce a heavy quark $Q\overline{Q}$ bound state ("onium") in a hadron-hadron collision. The C-even (χ_Q) mesons can be produced by fusing two quasi-real gluons (with opposite colors) from the QCD parton clouds of the colliding hadrons:

$$(g+g)_{(1)} \to Q+\overline{Q} \to \chi_Q\,. \tag{3.27}$$

In particular, radiative decays of such χ_c mesons are responsible for about 40% of the J/ψ yield. How about the remaining 60% ? To directly create a J/ψ (or ψ' — ${}^3S_1$ C-odd $c\bar{c}$ states) two gluons isn't enough. A C-odd meson can decay into, or couple to, *three* photons (like para-positronium does), a photon plus two gluons, or *three* gluons (in a color-symmetric d_{abc} state).

So, we need one more gluon to attach, for example, in the final state:

$$(g+g)_{(8)} \to \left(Q+\overline{Q}\right)_{(8)} \to J/\psi + g\,. \tag{3.28}$$

To pick up an initial gg pair in a color octet state is easier than in the singlet as in (3.27). This, however, does not help to avoid the trouble:

the perturbative cross section turns out to be too small to meet the need. It underestimates the Tevatron $p\bar{p}$ data on direct J/ψ and ψ' production by a large factor (up to 50, at large $p_\perp$).

That very same effect that makes the J/ψ so narrow a meson with the small hadronic decay width $\Gamma_{J/\psi}/M \propto \alpha_s^3(M)$, suppresses its perturbative production cross section (3.28) as well.

Since the PT approach apparently fails, it seemed natural to blame the non-perturbative (NP) physics. Why not to perturbatively form a color-octet "J/ψ" and then to get rid of color in a smooth (free of charge) non-perturbative way? To *evaporate* color does not look problematic: on the one hand, the soft glue distribution is $d\omega/\omega = \mathcal{O}(1)$, on the other hand, the coupling α_s/π in the NP domain may be of the order of unity as well. So why not?

The LBK theorem tells us that either the radiation is soft-enhanced, $\propto d\omega/\omega = \mathcal{O}(1)$, and *classical*, or hard, $\propto \omega d\omega$ and capable of changing the quantum state of the system. Therefore, to rightfully participate in the J/ψ formation as a quantum field, a NP gluon with $\omega \sim \Lambda_{\rm QCD}$ would have to bring in the suppression factor

$$\left(\frac{\Lambda_{\rm QCD}}{M_c}\right)^2 \ll 1\,.$$

The *language* of the LBK is perturbative, 'tis true. The question is, and a serious one indeed, whether the NP phenomena respect the basic dynamical features that its PT counterpart does? Or shall we rather forget about quantum mechanics, color conservation, etc. and accept an "anything goes" motto in the NP domain?

To avoid our discussion turning theological, we better address another verifiable issue namely, photoproduction of J/ψ at HERA. Here we have instead of (3.28) the fusion process of a real (photoproduction) or virtual (electroproduction) photon with a quasi-real space-like gluon from the parton cloud of the target proton:

$$\gamma^{(*)} + g \to \left(Q + \overline{Q}\right)_{(8)} \to J/\psi + g\,. \tag{3.29}$$

If the final-state gluon were soft NP junk, the J/ψ meson would carry the whole photon momentum and its distribution in Feynman z would peak at $z = 1$ as $(1-z)^{-1}$. The HERA experiments have found instead a flat*ish* (if not vanishing) z-spectrum at large z. The NSFL theorem seems to be up and running.

By the way, the conventional PT treatment of the photoproduction (3.29) is reportedly doing well. So, what is wrong with the hadroproduction then? Strictly speaking, the problem is still open. An alternative to (3.28) would be to look for the third (hard or hard*ish*) gluon in the initial state.[4]

The NSFL QCD discourse has taken us quite far from the mainstream of the introductory lecture. Let us return to the basic properties of QED bremsstrahlung and make a comparative study of

3.4. INDEPENDENT AND COHERENT RADIATION

In the Feynman gauge, the accompanying radiation factor dN in (3.24) is dominated by the *interference* between the two emitters:

$$dN \propto -\left[\frac{p_1^\mu}{(p_1k)} - \frac{p_2^\mu}{(p_2k)}\right]^2 \approx \frac{2(p_1p_2)}{(p_1k)(p_2k)}.$$

Therefore it does not provide a satisfactory answer to the question, which part of radiation is due to the initial charge and which is due to the final one?

There is a way, however, to give a reasonable answer to this question. To do that one has to sacrifice simplicity of the Feynman-gauge calculation and recall the original expression (3.22) for the cross section in terms of physical photon polarizations. It is natural to choose the so-called *radiative* (temporal) gauge based on the 3-vector potential $\vec{A}$, with the scalar component set to zero, $A_0 \equiv 0$. Our photon is then described by (real) 3-vectors orthogonal to one another and to its 3-momentum:

$$(\vec{\epsilon}_\lambda \cdot \vec{\epsilon}_{\lambda'}) = \delta_{\lambda\lambda'}, \qquad (\vec{\epsilon}_\lambda \cdot \vec{k}\,) = 0. \tag{3.30}$$

This explicitly leaves us with *two* physical polarization states. Summing over polarizations obviously results in

$$dN \propto \sum_{\lambda=1,2} \left|\vec{j}(k)\cdot\vec{e}_\lambda\right|^2 = \sum_{\alpha,\beta=1\ldots3} \vec{j}^{\,\alpha}(k)\cdot[\,\delta_{\alpha\beta} - \vec{n}_\alpha\vec{n}_\beta\,]\cdot\vec{j}^{\,\beta}(k), \tag{3.31}$$

with α, β the 3-dimensional indices. We now substitute the soft current (3.17c) in the 3-vector form, $p_i^\mu \to \vec{v}_i p_{0i}$, and make use of the relations

$$(\vec{v}_i)_\alpha \left[\delta_{\alpha\beta} - \frac{k_\alpha\, k_\beta}{\vec{k}^2}\right](\vec{v}_i)_\beta \;=\; v_i^2 \sin^2\Theta_i\,, \tag{3.32a}$$

$$(\vec{v}_1)_\alpha \left[\delta_{\alpha\beta} - \frac{k_\alpha\, k_\beta}{\vec{k}^2}\right](\vec{v}_2)_\beta \;=\; v_1v_2(\cos\Theta_{12} - \cos\Theta_1\cos\Theta_2)\,, \tag{3.32b}$$

[4]an interesting, reliable and predictive model for production of onia in the gluon field of colliding hadrons is being developed by Paul Hoyer and collaborators, see hep-ph/0004234 and references therein

to finally arrive at

$$dN = \frac{\alpha}{\pi}\left\{\mathcal{R}_1 + \mathcal{R}_2 - 2\mathcal{J}\right\}\cdot\frac{d\omega}{\omega}\frac{d\Omega}{4\pi}. \tag{3.33a}$$

Here

$$\mathcal{R}_i = \frac{v_i^2 \sin^2\Theta_i}{(1-v_i\cos\Theta_i)^2}, \quad i=1,2, \tag{3.33b}$$

$$\mathcal{J} \equiv \frac{v_1 v_2(\cos\Theta_{12} - \cos\Theta_1\cos\Theta_2)}{(1-v_1\cos\Theta_1)(1-v_2\cos\Theta_2)}. \tag{3.33c}$$

The contributions $\mathcal{R}_{1,2}$ can be looked upon as being due to *independent radiation* off initial and final charges, while the $\mathcal{J}$-term accounts for *interference* between them. The independent and interference contribution, taken together, describe the *coherent* emission. It is straightforward to verify that (3.33) is identical to the Feynman-gauge result (3.24):

$$\mathcal{R}_{\text{coher}} \equiv \mathcal{R}_{\text{indep}} - 2\mathcal{J} = -\omega^2(j^\mu)^2, \qquad \mathcal{R}_{\text{indep}} \equiv \mathcal{R}_1 + \mathcal{R}_2. \tag{3.34}$$

3.4.1 The role of interference: strict angular ordering.

In the relativistic limit we have

$$\mathcal{R}_1 \simeq \frac{\sin^2\Theta_1}{(1-\cos\Theta_1)^2} = \frac{2}{a_1} - 1, \tag{3.35a}$$

$$\mathcal{J} \simeq \frac{\cos\Theta_{12} - \cos\Theta_1\cos\Theta_2}{(1-\cos\Theta_1)(1-\cos\Theta_2)} = \frac{a_1 + a_2 - a_{12}}{a_1 a_2} - 1 \tag{3.35b}$$

where we introduced a convenient notation

$$a_1 = 1 - \vec{n}\vec{n}_1 = 1 - \cos\Theta_1, \quad a_2 = 1 - \cos\Theta_2,$$
$$a_{12} = 1 - \vec{n}_1\vec{n}_2 = 1 - \cos\Theta_s.$$

The variables a are small when the angles are small: $a \simeq \frac{1}{2}\Theta^2$.

The independent radiation has a typical logarithmic behavior up to large angles $a_1 \lesssim 1$:

$$dN_1 \propto \mathcal{R}_1 \sin\Theta d\Theta \propto \frac{da_1}{a_1}.$$

However, the interference effectively cuts off the radiation at angles exceeding the scattering angle:

$$dN \propto \mathcal{R}_{\text{coher}} \sin\Theta d\Theta = 2a_{12}\frac{da}{a_1 a_2} \propto \frac{da}{a^2} \propto \frac{d\Theta^2}{\Theta^4}, \quad a_1 \simeq a_2 \gg a_{12}.$$

To quantify this coherent effect, let us combine an independent contribution with a half of the interference one to define

$$V_1 = \mathcal{R}_1 - \mathcal{J} = \frac{2}{a_1} - \frac{a_1 + a_2 - a_{12}}{a_1 a_2} = \frac{a_{12} + a_2 - a_1}{a_1 a_2}\,,$$
$$V_2 = \mathcal{R}_2 - \mathcal{J} = \frac{2}{a_2} - \frac{a_1 + a_2 - a_{12}}{a_1 a_2} = \frac{a_{12} + a_1 - a_2}{a_1 a_2}\,; \tag{3.36a}$$
$$\mathcal{R}_{\text{coher}} = V_1 + V_2\,. \tag{3.36b}$$

The emission probability V_i can be still considered as "belonging" to the charge #i (V_1 is singular when $a_1 \to 0$, and vice versa). At the same time these are no longer *independent* probabilities, since V_1 explicitly depends on the direction of the partner-charge #2; *conditional* probabilities, so to say.

It is straightforward to verify the following remarkable property of the "conditional" distributions V: after *averaging* over the azimuthal angle of the radiated quantum, $\vec{n}$, with respect to the direction of the parent charge, $\vec{n}_1$, the probability $V_1(\vec{n}, \vec{n}_1; \vec{n}_2)$ *vanishes* outside the Θ_s-cone. Namely

$$\langle V_1 \rangle_{\text{azimuth}} \equiv \int_0^{2\pi} \frac{d\phi_{n,n_1}}{2\pi}\, V_1(\vec{n}, \vec{n}_1; \vec{n}_2) = \frac{2}{a_1} \vartheta\left(a_{12} - a_1\right)\,. \tag{3.37}$$

It is only a_2 that changes under the integral (3.37), while a_1, and obviously a_{12}, stay fixed. The result follows from the angular integral

$$\int_0^{2\pi} \frac{d\phi_{n,n_1}}{2\pi} \frac{1}{a_2} = \frac{1}{|\cos\Theta_1 - \cos\Theta_s|} = \frac{1}{|a_{12} - a_1|}\,.$$

Naturally, a similar expression for V_2 emerges after the averaging over the azimuth around $\vec{n}_2$ is performed.

We conclude that as long as the *total* (angular-integrated) emission probability is concerned, the result can be expressed as a sum of two independent bremsstrahlung cones centered around $\vec{n}_1$ and $\vec{n}_2$, both having the finite opening half-angle Θ_s.

This nice property is known as a "strict angular ordering". It is an essential part of the so-called Modified Leading Log Approximation (MLLA), which describes the internal structure of parton jets with a single-logarithmic accuracy.

3.4.2 Angular ordering on the back of envelope. What is the reason for radiation at angles exceeding the scattering angle to be suppressed? Let us try our physical intuition and consider semi-classically how the radiation process really develops.

A physical electron is a charge surrounded by its proper Coulomb field. In quantum language the Lorentz-contracted Coulomb-disk attached to a relativistic particle may be treated as consisting of photons virtually emitted and, in due time, reabsorbed by the core charge. Such virtual emission and absorption processes form a coherent state which we call a physical electron ("dressed" particle).

This coherence is partially destroyed when the charge experiences an impact. As a result, a part of intrinsic field fluctuations gets released in the form of real photon radiation: the bremsstrahlung cone in the direction of the initial momentum develops. On the other hand, the deflected charge now leaves the interaction region as a "half-dressed" object with its proper field-coat lacking some field components (eventually those that were lost at the first stage). In the process of regenerating the new Coulomb-disk adjusted to the final-momentum direction, an extra radiation takes place giving rise to the second bremsstrahlung cone.

Now we need to be more specific to find out which momentum components of the electromagnetic coat do actually take leave.

A typical time interval between emission and reabsorption of the photon k by the initial electron p_1 may be estimated as the Lorentz-dilated lifetime of the virtual intermediate electron state $(p_1 - k)$ (see the left graph in Fig.3),

$$t_{\text{fluct}} \sim \frac{E_1}{|m^2 - (p_1 - k)^2|} = \frac{E_1}{2p_1 k} \sim \frac{1}{\omega\Theta^2} \simeq \frac{\omega}{k_\perp^2} \,. \tag{3.38}$$

Here we restricted ourselves, for simplicity, to small radiation angles, $k_\perp \approx \omega\Theta \ll k_{||} \approx \omega$. The fluctuation time (3.38) may become macroscopically large for small photon energies ω and enters as a characteristic parameter in a number of QED processes. As an example, let us mention the so called Landau-Pomeranchuk effect — suppression of soft radiation off a charge that experiences multiple scattering propagating through a medium. Quanta with too large a wavelength get not enough time to be properly formed before successive scattering occurs, so that the resulting bremsstrahlung spectrum behaves as $dN \propto d\omega/\sqrt{\omega}$ instead of the standard logarithmic $d\omega/\omega$ distribution.

The characteristic time scale (3.38) responsible for this and many other radiative phenomena is often referred to as the *formation time.*

Now imagine that within this interval the core charge was kicked by some external interaction and has changed direction by some Θ_s. Whether the photon will be reabsorbed or not depends on the position of the scattered charge with respect to the point where the photon was expecting to meet it "at the end of the day". That is, we need to compare the spatial displacement of the core charge $\Delta\vec{r}$ with the characteristic size of the photon field, $\lambda_{||} \sim \omega^{-1}$, $\lambda_{\perp} \sim k_{\perp}{}^{-1}$:

$$\begin{aligned} \Delta r_{||} &\sim \left| v_{2||} - v_{1||} \right| \cdot t_{\text{fluct}} \sim \Theta_s^2 \cdot \frac{1}{\omega\Theta^2} = \left(\frac{\Theta_s}{\Theta}\right)^2 \lambda_{||} \;\Leftrightarrow\; \lambda_{||}\,; \\ \Delta r_{\perp} &\sim \quad c\Theta_s \cdot t_{\text{fluct}} \sim \Theta_s \cdot \frac{1}{\omega\Theta^2} \quad = \left(\frac{\Theta_s}{\Theta}\right) \lambda_{\perp} \;\Leftrightarrow\; \lambda_{\perp}\,. \end{aligned} \tag{3.39}$$

For large scattering angles, $\Theta_s \sim 1$, the charge displacement exceeds the photon wavelength for arbitrary Θ, so that the two full-size bremsstrahlung cones are present. For numerically small $\Theta_s \ll 1$, however, it is only photons with $\Theta \lesssim \Theta_s$ that can notice the charge being displaced and thus the coherence of the state being disturbed. Therefore only the radiation at angles smaller than the scattering angle actually emerges. The other field components have too large a wavelength and are easily reabsorbed *as if* there were no scattering at all.

So what counts is a change in the current, which is sharp enough to be noticed by the "to-be-emitted" quantum within the characteristic formation/field-fluctuation time (3.38) of the latter.

Radiation at large angles has too short a formation time to become aware of the acceleration of the charge. No scattering — no radiation.

The same argument applies to the dual process of production of two opposite charges (decay of a neutral object, vacuum pair production, etc.). The only difference is that now one has to take for $\Delta\vec{r}$ not a displacement between the initial and the final charges, but the actual distance between the produced particles (spatial size of a dipole), to be compared with the radiation wavelength.

3.4.3 Time delay and decoherence effects. Till now we were dealing with particle scattering/production as *instant* processes. Such they usually are (as compared with typical formation times). Nevertheless, let us imagine that our electron in Fig. 3 is delayed by some finite $\Delta t = \tau$ "in the V-vertex". For example, as if some metastable state was formed with characteristic lifetime $\tau = \Gamma^{-1}$.

In such a case one would have to take into consideration an extra *longitudinal* charge displacement due to a finite delay, and (3.39) would be modified as

$$\Delta r_{||} \sim \left(\frac{\Theta_s}{\Theta}\right)^2 \lambda_{||} \; + c\tau \;\Leftrightarrow\; \lambda_{||} \sim \omega^{-1}\,.$$

Now the condition $\Delta r_{||} < \lambda_{||}$ for the radiation at $\Theta > \Theta_s$ to be coherently suppressed implies an additional restriction $\tau < \omega^{-1}$. For large enough values of the delay time, $\tau \gg E^{-1}$, this new condition seriously affects radiation with comparatively large energies $\omega > \tau^{-1}$ (but still *soft* in the overall energy scale, $\omega \ll E_i$). Such photons acquire sufficiently large resolutions for coherence to be completely destroyed by the delay. Therefore they are bound to form two independent bremsstrahlung cones even for $\Theta_s = 0$.

So we would expect the accompanying radiation pattern to be that of the coherent antenna $\mathcal{R}_{\text{coher}}$ for softer radiation, $\tau^{-1} > \omega$, and, on the contrary, a sum of two independent sources $\mathcal{R}_1 + \mathcal{R}_2$ for relatively hard photons, $\tau^{-1} < \omega \ll E$. This qualitative expectation has a nice quantitative approval.

The initial and final electron currents in (3.17c) now acquire *different phases* due to the difference between the *Freeze*! and *Move*! times t_{01} and t_{02} (cf. (3.19)):

$$j^\mu \Longrightarrow j^\mu_{\text{del.}}(k) = \frac{p_1^\mu}{(p_1 k)}\, e^{i\omega t_{01}} - \frac{p_2^\mu}{(p_2 k)}\, e^{i\omega t_{02}}\,. \tag{3.40}$$

Now we should be careful when calculating the radiation probability, since the new current (3.40) is no longer conserved: $(j^\mu{}_{\text{del.}}\, k_\mu) \neq 0$. In particular, we cannot use the Feynman-gauge square of this current. The conservation could be formally rescued by adding the term describing our charge being frozen within the time interval $t_{02} - t_{01}$, namely

$$\delta j^\mu_{\text{del.}}(k) = -\frac{\delta^{0\mu}}{\omega}\left\{ e^{i\omega t_{01}} - e^{i\omega t_{02}} \right\}.$$

However, we can still use the physical polarization method which remains perfectly applicable. The relative phase enters in the interference term, so that the soft radiation pattern gets modified according to

$$\mathrm{d}N = \frac{\alpha}{\pi}\left\{ \mathcal{R}_1 + \mathcal{R}_2 - 2\,\mathcal{J}\cdot \mathrm{Re}\left[e^{i\omega(t_{01}-t_{02})} \right] \right\} \frac{\mathrm{d}\omega}{\omega}\frac{\mathrm{d}\Omega}{4\pi}\,. \tag{3.41}$$

To make our pedagogical setup more realistic, imagine that it was the formation of a meta-stable (resonant) state that caused the delay. In such a case the delay-time $\tau \equiv t_{01} - t_{02}$ is distributed according to the characteristic decay exponent

$$\left[\Gamma \int_0^\infty d\tau\, e^{-\Gamma\tau} \right].$$

Averaging (3.41) with this distribution immediately results in a simple Γ-dependent expression, namely,

$$\mathrm{d}N = \frac{\alpha}{\pi}\left\{ \mathcal{R}_1 + \mathcal{R}_2 - 2\,\mathcal{J}\cdot \chi_\Gamma(\omega) \right\} \frac{\mathrm{d}\omega}{\omega}\frac{\mathrm{d}\Omega}{4\pi}$$

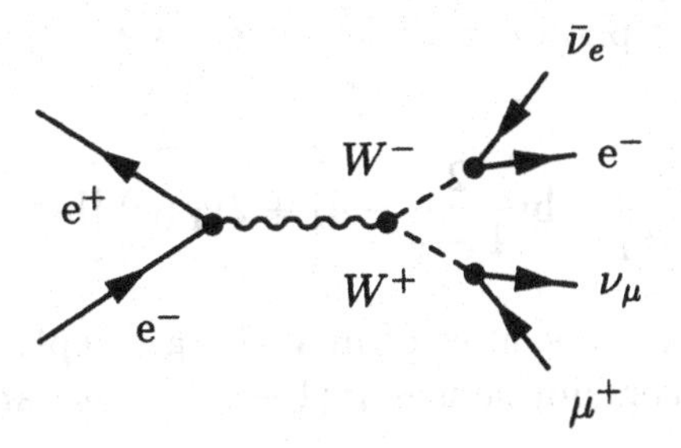

Figure 4 Leptonic decay of a W^+W^- pair as an illustration of time-dependent decoherence effects.

with the *profile factor*

$$\chi_\Gamma \equiv \mathrm{Re}\left[\Gamma\int_0^\infty d\tau\, e^{-\Gamma\tau}\cdot e^{i\omega\tau}\right] = \mathrm{Re}\left[\frac{\Gamma}{\Gamma - i\omega}\right] = \frac{\Gamma^2}{\Gamma^2+\omega^2}\,.$$

The answer can be written as a *mixture* of independent and coherent patterns with the weights depending on the ratio ω/Γ via the profile function χ_Γ,

$$\mathrm{d}N \;=\; \frac{\alpha}{\pi}\frac{\mathrm{d}\omega}{\omega}\frac{\mathrm{d}\Omega}{4\pi}\left\{\left[1-\chi_\Gamma(\omega)\right]\cdot \mathcal{R}_{\mathrm{indep}}(\vec{n}) + \chi_\Gamma(\omega)\cdot\mathcal{R}_{\mathrm{coher}}(\vec{n})\right\}\,. \quad (3.42)$$

$\chi(\omega)$ acts as a "switch": for long-wave radiation $\chi(\omega \ll \Gamma)\to 1$, the standard coherent antenna pattern appears; vice versa, for large frequencies $\chi(\omega\gg\Gamma)\to 0$, the coherence between charges is dashed away, as we expected.

Example: soft photons and the W-width. This simple phenomenon finds an intriguing and important practical implication in the dual channel. Suppose that in the e^+e^- annihilation process a pair of non-relativistic W^+W^- is produced. An intermediate boson has a finite life-time, $\Gamma\simeq 2\,\mathrm{GeV}$, and decays either leptonically or into a quark pair that produces two hadron jets at the end of the day. Thus, the W^+ and W^- decay independently of one another and produce ultra-relativistic electromagnetic currents within a characteristic time interval $|\Delta t_0|\sim\Gamma^{-1}$. The process is displayed in Fig. 4.

Therefore one meets exactly the same "delayed acceleration" scenario as applied to the final-state currents. As a result, (3.42) describes the photon radiation accompanying *leptonic* decays of non-relativistic W^+ and W^-.

Integrating over photon emission angles we derive the total photon multiplicity:

$$\omega\frac{dN}{d\omega} \propto \left[\ln\frac{2}{1-v_1} + \ln\frac{2}{1-v_2} - 4\right] + 2\chi_\Gamma(\omega)\left[\ln\frac{1-\cos\Theta_{12}}{2} + 1\right]. \tag{3.43}$$

where v_1 and v_2 are velocities of final charged leptons (e, μ or τ). The main "collinear" contributions $\sim \ln(1-v_1)^{-1} \gg 1$ are naturally ω- and Θ_{12}-independent.

A non-trivial ω-dependence of the profile function χ_Γ comes together with the functional dependence on the angle Θ_{12} between the leptons. The ω-dependent term in (3.43) *enhances* the accompanying photon multiplicity for large values of Θ_{12} and, vice versa, acts *destructively* if the angle between leptons happens to be below the critical value

$$\begin{aligned} \cos\Theta_{\text{crit}} &= 1 - 2\exp(-1) \approx 0.264\,, \\ \Theta_{\text{crit}} &\approx 1.30 \approx \frac{5\pi}{12} = 75^\circ\,. \end{aligned} \tag{3.44}$$

This suggests a programme of measuring the W-width Γ_W by studying the variation of the radiation yield with Θ_{12}.

4. BACK TO QCD

4.1. QCD SCATTERING AND CROSS-CHANNEL RADIATION

Both the qualitative arguments of the previous lecture and the quantitative analysis of the two-particle antenna pattern apply to the QCD process of gluon emission in the course of quark scattering. So two gluon-bremsstrahlung cones with the opening angles restricted by the scattering angle Θ_s would be expected to appear.

There is an important subtlety, however. In the QED case it was deflection of an electron that changed the e.m. current and caused photon radiation. In QCD there is another option, namely to "repaint" the quark. Rotation of the *color state* would affect the color current as well and, therefore, must lead to gluon radiation irrespectively of whether the quark-momentum direction has changed or not.

This is what happens when a quark scatters off a *color* field. To be specific, one may consider as an example two channels of Higgs production in hadron-hadron collisions.

At very high energies two mechanisms of Higgs production become competitive: $W^+W^- \to H$ and the gluon-gluon fusion $gg \to H$ (see Fig. 5).

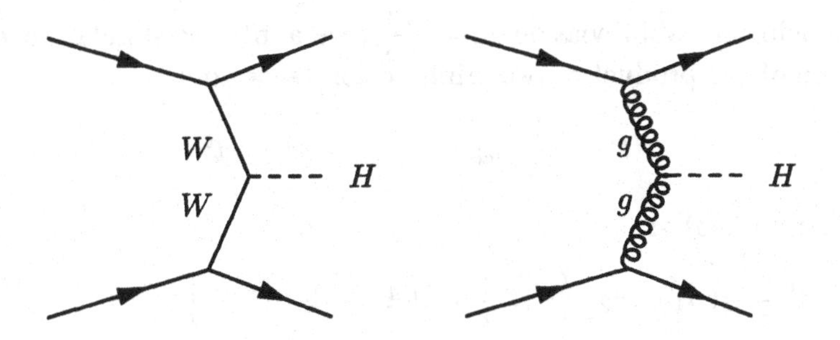

Figure 5 WW and gluon-gluon fusion graphs for Higgs production

Since the typical momentum transfer is large, of the order of the Higgs mass, $(-t) \sim M_H^2$, Higgs production is a *hard* process. Colliding quarks experience hard scattering with characteristic scattering angles $\Theta_s^2 \simeq |t|/s \sim M_H^2/s$. As far as the accompanying gluon radiation is concerned, the two subprocesses differ with respect to the nature of the "external field", which is *colorless* for the W-exchange and *colorful* for the gluon fusion.

The gluon bremsstrahlung amplitudes for the second case are shown in Fig. 6. In principle, a graph with the gluon-gluon interaction vertex should also be considered. However, in the limit $k_\perp \ll q_\perp$, with $\vec{q}_\perp \approx \vec{p}_{2\perp} - \vec{p}_{1\perp}$ the momentum transfer in the scattering process, emission off the external lines dominates (the "soft insertion rules").

The accompanying soft radiation current j^μ factors out from the Feynman amplitudes of Fig. 6, the only difference with the Abelian current (3.17c) being the order of the color generators:

$$j^\mu = \left[t^b\, t^a \left(\frac{p_1^\mu}{(p_1 k)} \right) - t^a\, t^b \left(\frac{p_2^\mu}{(p_2 k)} \right) \right] . \tag{4.45}$$

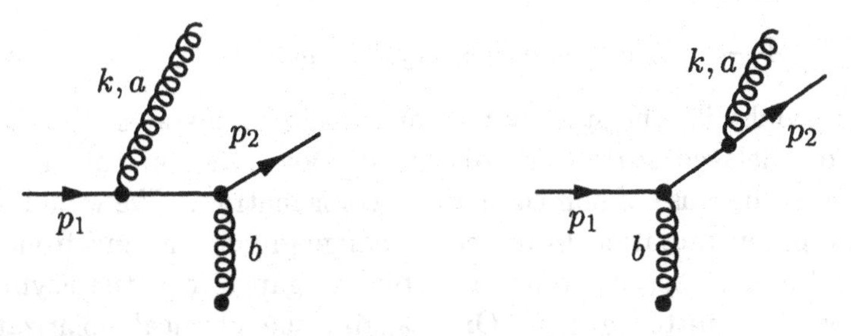

Figure 6 Gluonic Bremsstrahlung diagrams for $k_\perp \ll q_\perp$. The characters a and b denote the colors of the radiated and exchanged gluons.

Introducing the abbreviation $A_i = \frac{p_i^\mu}{(p_i k)}$, we apply the standard decomposition of the product of two triplet color generators,

$$t^a t^b = \frac{1}{2N_c}\delta_{ab} + \tfrac{1}{2}\left(d_{abc} + \mathrm{i} f_{abc}\right) t^c ,$$

to rewrite (4.45) as

$$\begin{aligned} j^\mu &= \tfrac{1}{2}(A_1 - A_2)\left\{t^b, t^a\right\} + \tfrac{1}{2}(A_1 + A_2)\left[t^b, t^a\right] \\ &= \tfrac{1}{2}(A_1 - A_2)\left(\frac{1}{N}\delta^{ab} + d^{abc} t^c\right) - \tfrac{1}{2}(A_1 + A_2)\,\mathrm{i} f^{abc} t^c . \end{aligned}$$

To find the emission *probability* we need to construct the product of the currents and sum over colors. Three color structures do not "interfere", so it suffices to evaluate the squares of the singlet, $\mathbf{8}_s$ and $\mathbf{8}_a$ structures:

$$\begin{aligned} \sum_{a,b}\left(\frac{1}{2N_c}\delta_{ab}\right)^2 &= \left(\frac{1}{2N}\right)^2 (N_c^2 - 1) = \frac{1}{2N_c}\cdot C_F ; \\ \sum_{a,b}\left(\tfrac{1}{2} d_{abc}\, t^c\right)^2 &= \frac{1}{4}\frac{N_c^2 - 4}{N_c}\,(t^c)^2 \quad = \frac{N_c^2 - 4}{4N_c}\cdot C_F ; \\ \sum_{a,b}\left(\tfrac{1}{2} i f_{abc}\, t^c\right)^2 &= \frac{1}{4} N_c\,(t^c)^2 \quad = \frac{N_c}{4}\cdot C_F . \end{aligned}$$

The common factor $C_F = \left(t^b\right)^2$ belongs to the Born (non-radiative) cross section, so that the radiation spectrum takes the form

$$\begin{aligned} \mathrm{d}N \propto \frac{1}{C_F}\sum_{\text{color}} j^\mu \cdot (j_\mu)^* &= \left(\frac{1}{2N_c} + \frac{N_c^2 - 4}{4N_c}\right)(A_1 - A_2)\cdot(A_1 - A_2) \\ &+ \frac{N_c}{4}(A_1 + A_2)\cdot(A_1 + A_2) . \end{aligned}$$

Simple algebra leads to

$$\mathrm{d}N \propto C_F\,(A_1 - A_2)\cdot(A_1 - A_2) \;+\; N_c\, A_1\cdot A_2 . \tag{4.46}$$

Dots here symbolize the sum over gluon polarization states. Similar to the case of "delayed scattering" discussed above, the current (4.45) *is not conserved* because of non-commuting color matrices. We would need to include gluon radiation from the exchange-gluon line *and* from the source, to be in a position to use an arbitrary gauge (e.g. the Feynman gauge) for the emitted gluon. Once again, the physical polarization technique (3.30) simplifies our task. To obtain the true accompanying radiation pattern (in the $k_\perp \ll q_\perp$ region) it suffices to use the projectors

(3.32) for the dots in (4.46). In particular,

$$A_1 \cdot A_2 \equiv \sum_{\lambda=1,2} \left(A_1 e^{(\lambda)}\right) \left(A_2 e^{(\lambda)}\right)^* = \mathcal{J} \qquad \{ \neq -(A_1 A_2) \text{ sic!} \} .$$

Accompanying radiation intensity finally takes the form

$$\mathrm{d}N \;\propto\; C_F\, \mathcal{R}_{\text{coher}} + N_c\, \mathcal{J} . \tag{4.47}$$

The first term proportional to the squared quark charge is responsible, as we already know, for two narrow bremsstrahlung cones around the incoming and outgoing quarks, $\Theta_1, \Theta_2 \leq \Theta_s$. On top of that an additional, purely non-Abelian, contribution shows up, which is proportional to the *gluon* charge. It is given by the interference distribution (3.33c), (3.35b),

$$\mathcal{J} = \frac{a_1 + a_2 - a_{12}}{a_1 a_2} - 1 ,$$

which remains *non-singular* in the forward regions $\Theta_1 \ll \Theta_s$ and $\Theta_2 \ll \Theta_s$. At the same time, it populates large emission angles $\Theta = \Theta_1 \approx \Theta_2 \gg \Theta_s$ where

$$\mathrm{d}N \propto \mathrm{d}\Omega\, \mathcal{J} \propto \sin\Theta\, \mathrm{d}\Theta \left(\frac{2}{a} - 1\right) \propto \frac{\mathrm{d}\Theta^2}{\Theta^2} . \tag{4.48}$$

Indeed, evaluating the azimuthal average, say, around the *incoming* quark direction we obtain

$$\int \frac{d\phi_1}{2\pi}\, \mathcal{J} = \frac{1}{a_1}\left(1 + \frac{a_1 - a_{12}}{|a_1 - a_{12}|}\right) - 1 = \frac{2}{a_1}\, \vartheta(\Theta_1 - \Theta_s) - 1 .$$

Thus we conclude that the third complementary bremsstrahlung cone emerges. It basically corresponds to radiation at angles *larger* than the scattering angle and its intensity is proportional to the color charge of the t-channel exchange.

We could have guessed without actually performing the calculation that at large angles the gluon radiation is related to the *gluon* color charge. As far as large emission angles $\Theta \gg \Theta_s$ are concerned, one may identify the directions of initial and final particles to simplify the total radiation amplitude as

$$j^\mu = T^b T^a \cdot \frac{p_1^\mu}{p_1 k} - T^a T^b \cdot \frac{p_2^\mu}{p_2 k} \approx \left(T^b T^a - T^a T^b\right) \cdot \frac{p^\mu}{pk} .$$

Recalling the general commutation relation for the $SU(N_c)$ generators,

$$\left[T^a(R)\, ,\, T^b(R)\right] = \mathrm{i} \sum_c f_{abc}\, T^c(R) , \tag{4.49}$$

we immediately obtain the factor $N_c \propto (\mathrm{i}f_{abc})^2$ as the proper color charge. Since (4.49) holds for arbitrary color representation R, we see that the accompanying gluon radiation at large angles $\Theta > \Theta_s$ does not depend on the nature of the projectile.

The bremsstrahlung gluons we are discussing transform, in the end of the day, into observable final hadrons. We are ready now to derive an interesting physical prediction from our QCD soft radiation exercise.

Translating the emission angle into (pseudo)rapidity $\eta = \ln \Theta^{-1}$, the logarithmic angular distribution (4.48) converts into the rapidity plateau. We conclude that in the case of the gluon fusion mechanism, the second in Fig. 5, the hadronic accompaniment should form a practically uniform rapidity plateau. Indeed, the hadron density in the center (small η, large c.m.s. angles) is proportional to the gluon color charge N_c, while in the "fragmentation regions" ($\eta_{\max} > |\eta| > \ln \Theta_s^{-1}$, or $\Theta < \Theta_s$) the two quark-generated bremsstrahlung cones give, roughly speaking, the density $\sim 2 \times C_F \approx N_c$.

At the same time, the WW-fusion events (the first graph in Fig. 5) should have an essentially different final state structure. Here we have a colorless exchange, and the QED-type angular ordering, $\Theta < \Theta_s$, restricts the hadronic accompaniment to the two projectile fragmentation humps as broad as $\Delta\eta = \eta_{\max} - \ln \Theta_s \simeq \ln M_H$, while the central rapidity region should be devoid of hadrons. The "rapidity gap" is expected which spans over $|\eta| < \ln(\sqrt{s}/M_H)$.

4.2. CONSERVATION OF COLOR AND QCD ANGULAR ORDERING

In physical terms universality of the generator algebra is intimately related with *conservation of color*. To illustrate this point let us consider production of a quark-gluon pair in some hard process and address the question of how this system radiates. Let p and k be the momenta of the quark and the gluon, with b the octet color index of the latter. For the sake of simplicity we concentrate on *soft* accompanying radiation, which determines the bulk of particle multiplicity inside jets, the structure of the hadronic plateau, etc. As far as emission of a soft gluon with momentum $\ell \ll k, p$ is concerned, the so-called "soft insertion rules" apply, which tell us that the Feynman diagrams dominate where ℓ is radiated off the external (real) partons — the final quark line p and the gluon k. The corresponding Feynman amplitudes are shown in Fig. 7.

Do two emission amplitudes interfere with each other? It depends on the direction of the radiated gluon $\vec{\ell}$.

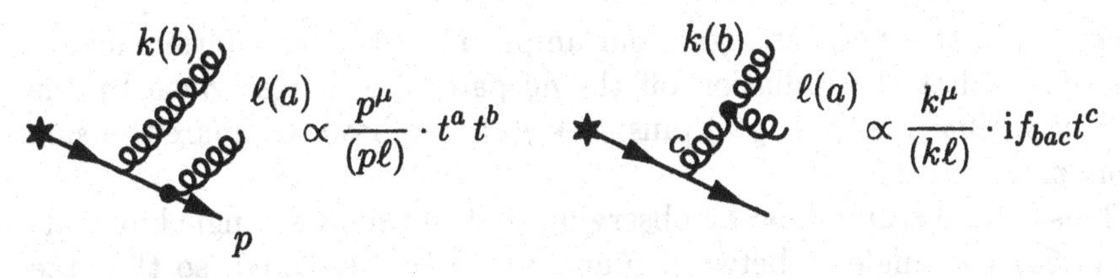

Figure 7 **Feynman diagrams for radiation of the soft gluon with momentum ℓ and color a off the qg system.**

In the first place, there are two bremsstrahlung cones centered around the directions of $\vec{p}$ and $\vec{k}$:

$$\text{quark cone:} \quad \Theta_{\vec{\ell}} \equiv \Theta_{\vec{\ell},\vec{p}} \ll \Theta \approx \Theta_{\vec{\ell},\vec{k}}\,,$$
$$\text{gluon cone:} \quad \Theta_{\vec{\ell}} \equiv \Theta_{\vec{\ell},\vec{k}} \ll \Theta \approx \Theta_{\vec{\ell},\vec{p}}\,,$$

with Θ the angle between $\vec{p}$ and $\vec{k}$ — the aperture of the qg fork. In these regions one of the two amplitudes of Fig. 7 is much larger than the other, and the interference is negligible: the gluon ℓ is radiated independently and participates in the formation of the quark and gluon sub-jets.

If Θ is sufficiently large and the gluon k sufficiently energetic (relatively hard, $k \sim p$), these two sub-jets can be distinguished in the final state. The particle density in q and g jets should be remarkably different. It should be proportional (at least asymptotically) to the probability of soft gluon radiation which, in turn, is proportional to the "squared color charge" of the jet-generating parton, quark or gluon:

$$\left(\ell\frac{\mathrm{d}n}{\mathrm{d}\ell}\right)^{g}_{\Theta_{\vec{\ell}}<\Theta} : \left(\ell\frac{\mathrm{d}n}{\mathrm{d}\ell}\right)^{q}_{\Theta_{\vec{\ell}}<\Theta} = N_c : \frac{N_c^2-1}{2N_c} = 3 : \frac{4}{3} = \frac{9}{4}\,.$$

Multijet configurations are comparatively rare: emission of an additional hard gluon $k \sim p$ at large angles $\Theta \sim 1$ constitutes a fraction $\alpha_s/\pi \lesssim 10\%$ of all events. Typically k would prefer to belong to the quark bremsstrahlung cone itself, that is to have $\Theta \ll 1$. In such circumstances the question arises about the structure of the accompanying radiation at comparatively *large* angles

$$\Theta_{\vec{\ell}} \equiv \Theta_{\vec{\ell},\vec{p}} \simeq \Theta_{\vec{\ell},\vec{k}} \gg \Theta\,. \tag{4.50}$$

If the quark and the gluon were acting as independent emitters, we would expect the particle density to increase correspondingly and to overshoot the standard quark jet density by the factor

$$\left(\ell\frac{\mathrm{d}n}{\mathrm{d}\ell}\right)^{g+q}_{\Theta_{\vec{\ell}}>\Theta} : \left(\ell\frac{\mathrm{d}n}{\mathrm{d}\ell}\right)^{q}_{\Theta_{\vec{\ell}}>\Theta} = N_c : \frac{N_c^2-1}{2N_c} + 1 = \frac{13}{4}\,. \tag{4.51}$$

However, in this angular region our amplitudes start to interfere significantly, so that the radiation off the qg pair is no longer given by the *sum of probabilities* $q \to g(\ell)$ plus $g \to g(\ell)$. We have to square the *sum of amplitudes* instead.

This can be easily done by observing that in the large-angle kinematics (4.50) the angle Θ between $\vec{p}$ and $\vec{k}$ can be neglected, so that the accompanying soft radiation factors in Fig. 7 become indistinguishable,

$$\frac{p^\mu}{(p\ell)} \simeq \frac{k^\mu}{(k\ell)}.$$

Thus the Lorentz structure of the amplitudes becomes the same and it suffices to sum the color factors:

$$t^a t^b + \mathrm{i} f_{bac} t^c = t^a t^b + \left[t^b, t^a\right] \equiv t^b t^a. \tag{4.52}$$

We conclude that the coherent sum of two amplitudes of Fig. 7 results in radiation at large angles *as if* off the initial quark, as shown in Fig. 8.

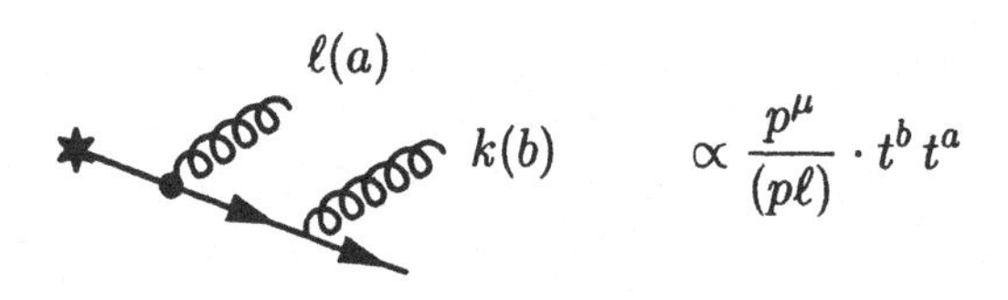

Figure 8 Soft radiation at large angles is determined by the total color charge

This means that the naive probabilistic expectation of enhanced density (4.51) fails and the particle yield is equal to that for the quark-initiated jet instead:

$$(\text{gluon+quark}) \quad \frac{9}{4} + 1 = \frac{13}{4} \qquad \Longrightarrow \qquad 1 \quad (\text{quark}).$$

It actually does not matter whether the gluon k was present at all, or whether there was instead a whole bunch of partons with small relative angles between them. Soft gluon radiation at large angles is sensitive only to the *total* color charge of the final parton system, which equals the color charge of the initial parton. This physically transparent statement holds not only for the quark as in Figs. 7, 8 but for an arbitrary object R (gluon, diquark, ..., you name it) as an initial object. In this case the matrices $t = T(3)$ should be replaced by the generators $T(R)$ corresponding to the color representation R, and (4.52) holds due to the universality of the generator algebra (4.49).

4.3. HUMPBACKED PLATEAU AND LPHD

QCD coherence is crucial for treating particle multiplication **inside** jets, as well as for hadron flows **in-between** jets.

For dessert, we are going to derive together the QCD "prediction" of the inclusive energy spectrum of relatively soft particles from QCD jets. I put the word *prediction* in quotation marks on purpose. This is a good example to illustrate the problem of filling the gap between the QCD formulae, talking quarks and gluons, and phenomena dealing, obviously, with hadrons.

Let me first make a statement:

> **It is QCD coherence that allows the prediction of the inclusive soft particle yield in jets practically from the "first principles".**

4.3.1 Solving the DIS evolution.

You have all the reasons to feel suspicious about this. Indeed, in the first lecture we stressed the similarity between the dynamics of the evolution of space-like (DIS structure functions) and time-like systems (jets). On the other hand, you are definitely aware of the fact that the DIS structure functions cannot be calculated perturbatively. There are input parton distributions for the target proton, which have to be plugged in as an initial condition for the evolution at some finite hardness scale $Q_0 = \mathcal{O}(1\,\mathrm{GeV})$. These initial distributions cannot be calculated "from first principles" nowadays but are subject to fitting. What PT QCD controls then, is the scaling violation pattern. Namely, it tells us how the parton densities change with the changing scale of the transverse-momentum probe:

$$\frac{\partial}{\partial \ln k_\perp} D(x, k_\perp) = \frac{\alpha_s(k_\perp)}{\pi} \int_x^1 \frac{\mathrm{d}z}{z}\, P(z)\, D\left(\frac{x}{z}, k_\perp\right). \tag{4.53}$$

It is convenient to present our "wavefunction" D and "Hamiltonian" P in terms of the complex moment ω, which is Mellin conjugate to the momentum fraction x:

$$D_\omega = \int_0^1 \mathrm{d}x\, x^\omega \cdot D(x)\,, \qquad D(x) = x^{-1} \int_{(\Gamma)} \frac{\mathrm{d}\omega}{2\pi \mathrm{i}}\, x^{-\omega} \cdot D_\omega\,; \tag{4.54a}$$

$$P_\omega = \int_0^1 \mathrm{d}z\, z^\omega \cdot P(z)\,, \qquad P(z) = z^{-1} \int_{(\Gamma)} \frac{\mathrm{d}\omega}{2\pi \mathrm{i}}\, z^{-\omega} \cdot P_\omega\,, \tag{4.54b}$$

where the contour Γ runs parallel to the imaginary axis, to the right from singularities of D_ω (P_ω). It is like trading the coordinate ($\ln x$) for the momentum (ω) in a Schrödinger equation.

Substituting (4.54) into (4.53) we see that the evolution equation becomes algebraic and describes propagation in "time" $\mathrm{d}t = \frac{\alpha_s}{\pi} \mathrm{d}\ln k_\perp$ of a free quantum mechanical "particle" with momentum ω and the dispersion law $E(\omega) = P_\omega$:

$$\hat{d}\, D_\omega(k_\perp) \;=\; \frac{\alpha_s(k_\perp)}{\pi} \cdot P_\omega\, D_\omega(k_\perp)\,; \qquad \hat{d} \equiv \frac{\partial}{\partial \ln k_\perp}\,. \tag{4.55}$$

To continue the analogy, our wavefunction D is in fact a multi-component object. It embodies the distributions of valence quarks, gluons and secondary sea quarks which evolve and mix according the 2×2 matrix LLA Hamiltonian (2.9).

At small x, however, the picture simplifies. Here the valence distribution is negligible, $\mathcal{O}(x)$, while the gluon and sea quark components form a system of two coupled oscillators which is easy to diagonalise. What matters is one of the two energy eigenvalues (one of the two branches of the dispersion rule) that is *singular* at $\omega = 0$. The problem becomes essentially one-dimensional. Sea quarks are driven by the gluon distribution while the latter is dominated by gluon cascades. Correspondingly, the leading energy branch is determined by gluon-gluon splitting (2.9d), with a subleading correction coming from the $g \to q(\bar{q}) \to g$ transitions,

$$P_\omega \;=\; \frac{2N_c}{\omega} - a \,+\, \mathcal{O}(\omega)\;, \quad a = \frac{11N_c}{6} + \frac{n_f}{3N_c^2}\,. \tag{4.56}$$

The solution of (4.55) is straightforward:

$$D_\omega(k_\perp) \;=\; D_\omega(Q_0)\cdot \exp\left\{ \int_{Q_0}^{k_\perp} \frac{dk}{k}\, \gamma_\omega(\alpha_s(k)) \right\}, \tag{4.57a}$$

$$\gamma_\omega(\alpha_s) \;=\; \frac{\alpha_s}{\pi}\, P_\omega\,. \tag{4.57b}$$

The structure (4.57a) is of the most general nature. It follows from *renormalizability* of the theory, and does not rely on the LLA which we used to derive it. The function $\gamma(\alpha_s)$ is known as the "anomalous dimension".[5] It can be perfected by including higher orders of the PT expansion. Actually, modern analyses of scaling violation are based on the improved next-to-LLA (two-loop) anomalous dimension, which includes α_s^2 corrections to the LLA expression (4.57b).

The structure (4.57a) of the x-moments of parton distributions (DIS structure functions) gives an example of a clever separation of PT and NP effects; in this particular case — in the form of two factors. It is

[5] This nice name is a relict of those good old days when particle and solid state physicists used to have common theory seminars. If the coupling α_s were constant (had a "fixed point"), then (4.57a) would produce the function with a non-integer (non-canonical) dimension $D(Q) \propto Q^\gamma$ (analogy — critical indices of thermodynamical functions near the phase transition point).

the ω-dependence of the input function $D_\omega(Q_0)$ ("initial parton distributions") that limits predictability of the Bjorken-x dependence of DIS cross sections.

So, how comes then that in the time-like channel the PT answer turns out to be more robust?

4.3.2 Coherent hump. We are ready to discuss the time-like case, with $D_j^h(x,Q)$ now the inclusive distribution of particles h with the energy fraction (Feynman-x) $x \ll 1$ from a jet (parton j) produced at a large hardness scale Q.

Here the general structure (4.57a) still holds. We need, however, to revisit the expression (4.57b) for the anomalous dimension because, as we have learned, the proper evolution time is now different from the case of DIS.

In the time-like jet evolution, due to Angular Ordering, the evolution equation becomes non-local in $k_\perp$ space:

$$\frac{\partial}{\partial \ln k_\perp} D(x,k_\perp) = \frac{\alpha_s(k_\perp)}{\pi} \int_x^1 \frac{dz}{z} P(z)\, D\left(\frac{x}{z},\, z\cdot k_\perp\right). \tag{4.58}$$

Indeed, successive parton splittings are ordered according to

$$\theta = \frac{k_\perp}{k_{||}} > \theta' = \frac{k'_\perp}{k'_{||}}\,.$$

Differentiating $D(k_\perp)$ over the scale of the "probe", $k_\perp$, results then in the substitution

$$k'_\perp = \frac{k'_{||}}{k_{||}}\cdot k_\perp \;\equiv\; z\cdot k_\perp$$

in the argument of the distribution of the next generation $D(k'_\perp)$.

The evolution equation (4.58) can be elegantly cracked using the Taylor-expansion trick,

$$D(z\cdot k_\perp) = \exp\left\{\ln z \frac{\partial}{\partial \ln k_\perp}\right\} D(k_\perp) \;=\; z^{\frac{\partial}{\partial \ln k_\perp}}\cdot D(k_\perp). \tag{4.59}$$

Turning as before to moment space (4.54), we observe that the solution comes out similar to that for DIS, (4.57), but for one detail. The exponent $\hat{d}$ of the additional z-factor in (4.59) combines with the Mellin moment ω to make the argument of the splitting function P a *differential operator* rather than a complex number:

$$\hat{d}\cdot D_\omega = \frac{\alpha_s}{\pi} P_{\omega+\hat{d}}\cdot D_\omega\,. \tag{4.60}$$

This leads to the differential equation

$$\left(P^{-1}_{\omega+\hat{d}}\, \hat{d} \;-\; \frac{\alpha_s}{\pi} - \left[P^{-1}_{\omega+\hat{d}}, \frac{\alpha_s}{\pi} \right] P_{\omega+\hat{d}} \right) \cdot D = 0 \,. \tag{4.61}$$

Recall that, since we are interested in the small-x region, the essential moments are small, $\omega \ll 1$.

For the sake of illustration, let us keep only the most singular piece in the "dispersion law" (4.56) and neglect the commutator term in (4.61) generating a subleading correction $\propto \hat{d}\alpha_s \sim \alpha_s^2$. In this approximation (DLA),

$$P_\omega \simeq \frac{2N_c}{\omega} \,, \tag{4.62}$$

(4.61) immediately gives a quadratic equation for the anomalous dimension,[6]

$$(\omega + \gamma_\omega)\gamma_\omega \;-\; \frac{2N_c\alpha_s}{\pi} + \mathcal{O}\left(\frac{\alpha_s^2}{\omega} \right) \;=\; 0 \,. \tag{4.63}$$

The leading anomalous dimension following from (4.63) is

$$\gamma_\omega = \frac{\omega}{2} \left(-1 + \sqrt{1 + \frac{8N_c\alpha_s}{\pi\, \omega^2}} \right) . \tag{4.64}$$

When expanded to first order in α_s, it coincides with that for the space-like evolution, $\gamma_\omega \simeq \alpha_s/\pi \cdot P_\omega$, with P given in (4.62). Such an expansion, however, fails when characteristic $\omega \sim 1/|\ln x|$ becomes as small as $\sqrt{\alpha_s}$, that is when

$$\frac{8N_c\alpha_s}{\pi} \ln^2 x \gtrsim 1 \,.$$

This inequality is an elaboration of the estimate (2.12), which we obtained from heuristic arguments in the first lecture.

Now what remains to be done is to substitute our new weird anomalous dimension into (4.57a) and perform the inverse Mellin transform to find $D(x)$. If there were no QCD parton cascading, we would expect the particle *density* $xD(x)$ to be constant (Feynman plateau). It is straightforward to derive that plugging in the DLA anomalous dimension (4.64) results in the plateau density increasing with Q and with a maximum

[6] It suffices to use the next-to-leading approximation to the splitting function (4.56) and to keep the subleading correction coming from differentiation of the running coupling in (4.61) to get the more accurate MLLA anomalous dimension γ_ω.

(hump) "midway" between the smallest and the highest parton energies, namely, at $x_{\max} \simeq \sqrt{Q_0/Q}$. The subleading MLLA effects shift the hump to smaller parton energies,

$$\ln \frac{1}{x_{\max}} = \ln \frac{Q}{Q_0}\left(\frac{1}{2} + c\cdot\sqrt{\alpha_s} + \ldots\right) \simeq 0.65\,\ln \frac{Q}{Q_0}\,,$$

with c a known analytically calculated number. Moreover, defying naive probabilistic intuition, the softest particles do not multiply at all. The density of particles (partons) with $x \sim Q_0/Q$ stays constant while that of their more energetic companions increases with the hardness of the process Q.

This is a powerful legitimate consequence of PT QCD coherence. We turn now to another, no less powerful though less legitimate, consequence.

4.3.3 Coherent damping of the Landau singularity. The time-like DLA anomalous dimension (4.64), as well as its MLLA improved version, has a curious property. Namely, in sharp contrast with DIS, it allows the momentum integral in (4.57) to be extended to very small scales. Even integrating down to $Q_0 = \Lambda$, the position of the "Landau pole" in the coupling, one gets a finite answer for the distribution (the so-called *limiting spectrum*), simply because the $\sqrt{\alpha_s(k)}$ singularity happens to be integrable!

It would have been poor taste to trust this formal integrability, since the very PT approach to the problem (selection of dominant contributions, parton evolution picture, etc.) relied on α_s being a numerically small parameter. However, the important thing is that, due to time-like coherence effects, the (still perturbative but "smallish") scales, where $\alpha_s(k) \gg \omega^2$, contribute to γ basically in a ω-*independent* way, $\gamma + \omega/2 \propto \sqrt{\alpha_s(k)} \neq f(\omega)$. This means that "smallish" momentum scales k affect only an overall *normalization* without affecting the *shape* of the x-distribution.

Since this is the role of the "smallish" scales, it is natural to expect the same for the truly small — non-perturbative — scales where the partons transform into the final hadrons. This idea has been formulated as a hypothesis of local parton-hadron duality (LPHD). Mathematically, this hypothesis reduces to the statement (guess) that the NP factor in (4.57a) has a finite $\omega \to 0$ limit:

$$D_\omega^{(h)}(Q_0) \;\to\; K^h = \text{const}, \quad \omega \to 0\,.$$

Thus, according to LPHD, the x-shape of the so-called "limiting" parton spectrum which is obtained by formally setting $Q_0 = \Lambda$ in the evolution equations, should be mathematically similar to that of the inclusive distribution of **hadrons** (h). Another essential property is that the "conversion coefficient" K^h should be a true constant independent of the hardness of the process producing the jet under consideration.

4.3.4 Brave gluon counting. The comparison of the limiting spectrum with the inclusive spectrum of all charged hadrons (dominated by $\pi^{\pm}$) was pioneered by Glen Cowan (ALEPH) and by the OPAL collaboration, and has become a standard test of analytic QCD predictions.

In Fig. 9 (DELPHI), the comparison is made of the all-charged hadron spectra at various annihilation energies Q with the so-called "distorted

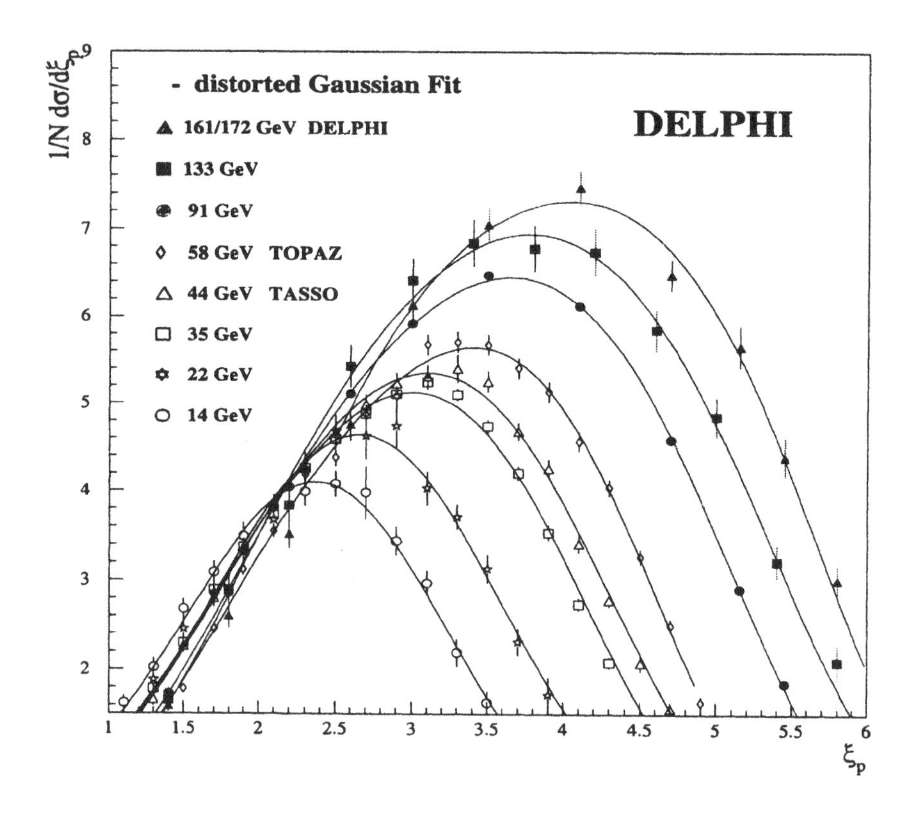

Figure 9 Inclusive distribution of charged hadrons produced in e^+e^- annihilation

Gaussian" fit which employs the first four moments (the mean, width, skewness and kurtosis) of the MLLA distribution around its maximum.

Is it nothing but one more test of QCQ? Not quite. Such close similarity is deeply puzzling, even worrysome, rather than a successful test.

Indeed, after a little exercise in translating the values of the logarithmic variable $\xi = \ln(E_{\rm jet}/p)$ in Fig. 9 into GeV you will see that the hadron momenta at the maxima are, for example, $p= \frac{1}{2}Q\cdot e^{-\xi_{\rm max}} \simeq 0.42$, 0.85 and 1.0 GeV for Q=14, 35 and 91 GeV, respectively.

Is it not surprising that the PT QCD spectrum is mirrored by that of the pions (which constitute 90% of all charged hadrons produced in jets) with momenta well below 1 GeV?!

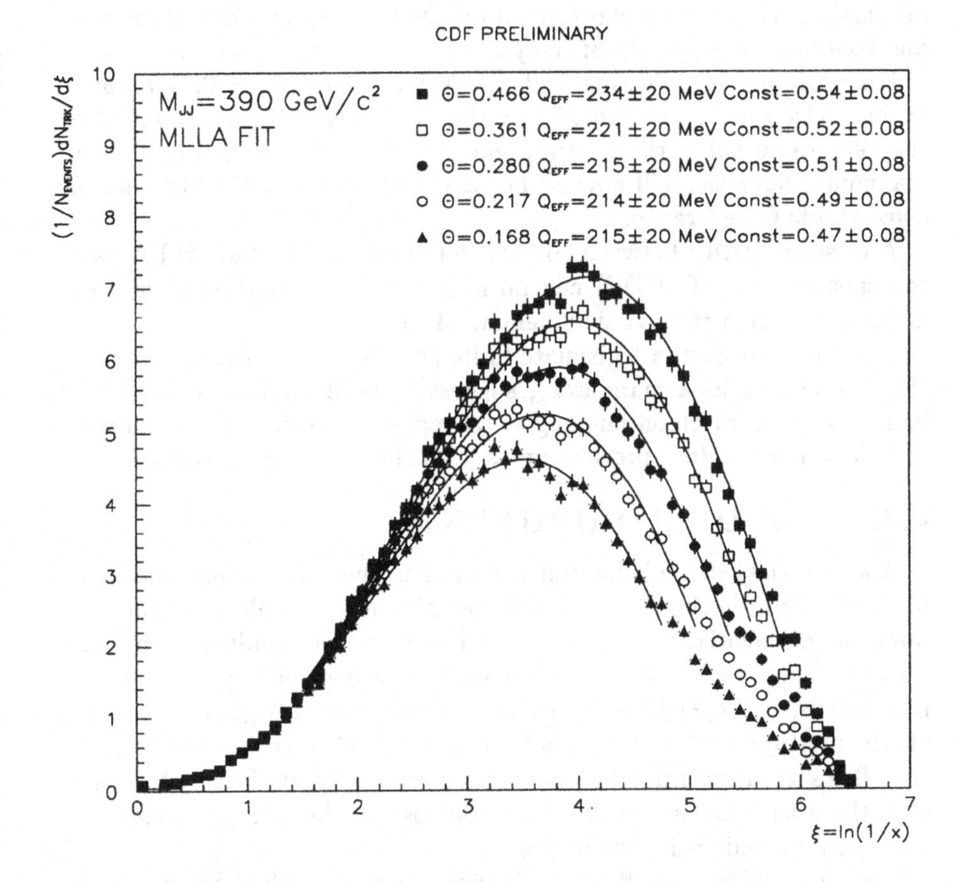

Figure 10 Inclusive energy distribution of charged hadrons in large-$p_\perp$ jets. D.Goulianos, *Proceedings 32nd Recontres de Moriond*, Les Arcs, France, March 1997.

For this very reason, the observation of the parton-hadron similarity was initially met with serious and well-grounded scepticism: it looked more natural (and was more comfortable) to blame the finite hadron mass effects for the falloff of the spectrum at large ξ (small momenta) rather than seriously believe in the applicability of the PT QCD consideration down to such disturbingly small momentum scales.

This worry has recently been answered by CDF.

Theoretically, it is not the energy of the jet but the maximal parton transverse momentum inside it, $k_{\perp\max} \simeq E_{\rm jet} \sin\frac{\Theta}{2}$, that determines the hardness scale and thus the yield and the distribution of the accompanying radiation. This means that by choosing a small opening angle Θ around the jet axis one can study relatively small hardness scales but in a cleaner environment: due to the Lorentz boost effect, eventually all particles that form a short small-Q^2 QCD hump become relativistic and concentrate at the tip of the jet.

For example, by selecting hadrons inside a cone $\Theta \simeq 0.14$ around an energetic quark jet with $E_{\rm jet} \simeq 100$ GeV (LEP-II), one should see the very same curve that corresponds to $Q = 14$ GeV in Fig. 9. Its maximum, however, will now be boosted from dubious 450 MeV into a comfortable 6 GeV range.

A close similarity between the hadron yield and the full MLLA parton spectra (Fig. 10, CDF) can no longer be considered accidental or attributed to non-relativistic kinematical effects.

The fact that even a legitimate finite smearing due to hadronization effects does not look mandatory, suggests a deep duality between the hadron and quark-gluon languages, as applied to such a global characteristic of multihadron production as an inclusive energy spectrum.

4.4. QCD RADIOPHYSICS

Another class of multihadron production phenomena that speak in favor of LPHD is the so-called inter-jet physics. It deals with particle flows in the angular regions between jets in various multi-jet configurations. These particles do not belong to any particular jet, and their production, at the PT level, is governed by coherent soft gluon radiation off the multi-jet system as a whole. Due to QCD coherence, these particle flows are insensitive to internal structure of underlying jets. The only thing that matters is the color topology of the primary system of hard partons and their kinematics.

The ratios of particle flows in different inter-jet valleys are given by parameter-free PT predictions and reveal the so-called "string" or "drag" effects. For a given kinematical jet configuration, such ratios depend only on the number of colors (N_c).

For example, the ratio of the multiplicity flow between a quark (antiquark) and a gluon to that in the $q\bar{q}$ valley in symmetric ("Mercedes") three-jet $q\bar{q}g$ e^+e^- annihilation events is predicted to be

$$\frac{dN_{qg}^{(q\bar{q}g)}}{dN_{q\bar{q}}^{(q\bar{q}g)}} \simeq \frac{5N_c^2-1}{2N_c^2-4} = \frac{22}{7}. \tag{4.65}$$

Comparison of the denominator with the density of radiation in the $q\bar{q}$ valley in $q\bar{q}\gamma$ events with a gluon jet replaced by an energetic photon results in

$$\frac{dN_{q\bar{q}}^{(q\bar{q}\gamma)}}{dN_{q\bar{q}}^{(q\bar{q}g)}} \simeq \frac{2(N_c^2-1)}{N_c^2-2} = \frac{16}{7}. \tag{4.66}$$

Emitting an energetic gluon off the initial quark pair depletes accompanying radiation in the backward direction: color is *dragged* out of the $q\bar{q}$ valley. This destructive interference effect is so strong that the resulting multiplicity flow falls below that in the least favorable direction transverse to the three-jet event plane:

$$\frac{dN_{\perp}^{(q\bar{q}\gamma)}}{dN_{q\bar{q}}^{(q\bar{q}g)}} \simeq \frac{N_C+2C_F}{2(4C_F-N_c)} = \frac{17}{14}. \tag{4.67}$$

At the level of the PT accompanying gluon radiation (QCD radiophysics), such predictions are quite simple and straightforward to derive. The strange thing is, that these and many similar numbers are observed in experiments. The inter-jet particle flow we are discussing is dominated, at present energies, by pions with typical momenta in the 100–300 MeV range! The fact that even such soft junk follows the PT QCD rules is truly amazing.

Since, starting from the LEP-I epoch, the "predictions" of the humpbacked plateau and of the coherent string/drag effects stood up to scrutiny in e^+e^-, DIS and Tevatron experiments, we gained an important piece of knowledge. And this is not that theorists are capable of calculating things, even in the presence of quantum-mechanical effects (see below, in the Acknowledgements). More importantly, we have learned something interesting about the physics of hadronization, about confinement and thus about QCD.

4.5. SOFT CONFINEMENT

Honestly speaking, it makes little sense to treat few-hundred-MeV gluons as PT quanta. What hadron energy spectra and string/drag phenomena are trying to tell us is that the production of hadrons is driven by the strength of the underlying color fields generated by the system of energetic partons produced in a hard interaction. Pushing PT description down into the soft gluon domain is a mere tool for quantifying the strengths of the color field. We conclude:

- The *color field* that is developed by an ensemble of hard primary **partons** determines the structure of the final flow of **hadrons**.
- The Poynting vector of the color field translates into the hadron Poynting vector without visible reshuffling of particle momenta.

Mathematical similarity between the parton and hadron energy and angular distributions means that confinement is very soft and gentle. As far as the global characteristics of final states are concerned, there is no sign of strong forces at the hadronization stage.

When viewed globally, confinement is about *renaming* a flying-away quark into a flying-away pion rather than about forces *pulling* quarks together. To preserve this delicate correspondence is a challenge for the future quantitative theory of hadrons — the "whole QCD".

First steps have been made in this direction in recent years. But this is another story (see below, in the Complaints).

Acknowledgements and Complaints

Particle physics community consists of smart and clever people.

Theorists have to be clever to do the job. They should be clever enough to understand what their colleagues have measured, and doubly-clever to predict what they *will be measuring*. Triply-clever to suggest what experimenters *should measure*.

Experimenters not only need to be clever to do the job, they ought to be smart. They have to be smart to measure, doubly-smart not to listen to what they *will be measuring*, and triply-smart — to what they *should measure*.

Still, I dare to come up with a suggestion.

Organization of the ASI–2000 was perfect, thanks to Harrison Prosper, Misha Danilov and the ASI godfather Tom Ferbel. There is one serious complaint, however. It would be better if the Institute ran (at least) *annually*. The ASI would run out of popular lecturers faster, which would offer a chance to tell more interesting stories. Particularly on such a facinating subject as the physics of hadrons.

BARYON ASYMMETRY OF THE UNIVERSE

V. A. Rubakov
Institute for Nuclear Research of the Russian Academy of Sciences,
60-th October Anniversary Prospect 7a, Moscow 117312, Russia
rubakov@ms2.inr.ac.ru

Abstract We review basic ideas related to the problem of baryogenesis and discuss several specific mechanisms of the generation of the baryon asymmetry of the Universe.

1. INTRODUCTION

In his famous paper [1] Sakharov discussed for the first time the possibility of explaining the charge asymmetry of the Universe in terms of particle theory. The paper was submitted to JETP Letters in September 1966, two years after the discovery of CP-violation in K^0 decays [2] and one year after the microwave black-body radiation, predicted by the Big Bang theory [3], was found experimentally [4]. To explain baryon asymmetry, Sakharov proposed an approximate character for the baryon conservation law, i.e., baryon number non-conservation and proton decay. Three years later Kuzmin published a paper [5] where a different model leading to the baryon asymmetry was constructed. One of its consequences was another process with B non-conservation, namely, neutron-antineutron oscillations. Since that time the idea that baryon number may not be exactly conserved in Nature has been elaborated upon considerably, both in the context of the generation of the baryon asymmetry of the Universe [6, 7, 8, 9, 10, 11] (for reviews see refs. [12, 13, 14, 15, 16]) and because of theoretical developments that have lead to a unified picture of fundamental interactions. In the mid–70's, grand unified theories with inherent violation of baryon number were put forward [17, 18, 19, 20, 21]. Almost at the same time it was realized [22, 23] that non-perturbative effects related to instantons [24] and the complex structure of the gauge theory vacuum [22, 25, 26] lead to the non-conservation of baryon number even in the electroweak theory; it

H.B. Prosper and M. Danilov (eds.), Techniques and Concepts of High-Energy Physics, 97–141.

was later understood [27] that similar effects are relevant for the baryon asymmetry.

In his paper [1] Sakharov writes: "According to our hypothesis, the occurrence of C asymmetry is the consequence of violation of CP invariance in the nonstationary expansion of the hot Universe during the superdense stage, as manifest in the difference between the partial probabilities of the charge-conjugate reactions." Today, this short extract is usually dubbed as the three necessary Sakharov conditions for baryon asymmetry generation from the initial charge symmetric state in the hot Universe, namely:

(i) Baryon number non-conservation.

(ii) C and CP violation.

(iii) Deviations from thermal equilibrium.

All three conditions are easily understood. (i) If baryon number were conserved, and the initial baryonic charge of the Universe were zero, the Universe today would be symmetric, rather than asymmetric [1]. The statement of the necessity of the baryon number non-conservation was quite revolutionary at that time. Today it is very natural theoretically; still, lacking positive results from experiments searching for B non-conservation, the baryon asymmetry of the Universe is a unique observational evidence in favour of it.

(ii) If C or CP were conserved, then the rate of reactions with particles would be the same as the rate of reactions with antiparticles. If the initial state of the Universe was C- or CP- symmetric, then no charge asymmetry could develop from it. In more formal language, this follows from the fact that if the initial density matrix of the system ρ_0 commutes with C- or CP-operations, and the Hamiltonian of the system is C- or CP-invariant, then at any time the density matrix $\rho(t)$ is C- or CP-invariant, so that the average of any C- or CP-odd operator is zero.

(iii) Thermal equilibrium means that the system is stationary (no time dependence at all). Hence, if the initial baryon number is zero, then it is zero forever.

Clearly, the issue of the baryon asymmetry generation requires the development of many different areas of theoretical physics, such as model building, study of perturbative and non-perturbative effects leading to B-

[1] Of course, there is a loop-hole in this argument, which Sakharov knew. The Universe may be globally symmetric, but locally asymmetric, with the size of the baryonic cluster of matter large enough (necessarily larger than the present horizon size [28]). The inflationary models of the Universe expansion, together with specific models of particle interactions may provide a mechanism of the local asymmetry generation, keeping the conservation of the baryon number intact [13].

violation, finite temperature field theory and non-equilibrium statistical mechanics, and the theory of phase transitions.

The quantitative measure of the baryon asymmetry of the Universe is the dimensionless ratio of the baryon number density to entropy density,

$$\Delta_B = \frac{B}{s}. \tag{1.1}$$

This quantity is almost constant during the expansion of the Universe at the stages when baryon number is conserved, and its present value is

$$\Delta_B = (3 - 10) \times 10^{-11}. \tag{1.2}$$

This estimate of the value of Δ_B in the late Universe comes from the theory of Big Bang nucleosynthesis: the primordial abundance of light elements depends crucially on Δ_B.

It is worth noting that in the relatively early Universe, at temperatures of order 1 GeV or higher, the cosmic plasma was full of quark–anti-quark pairs. The baryon asymmetry Δ_B at those times characterized the excess of quarks over anti-quarks: the net baryon number density was equal to

$$B = \frac{1}{3}\left(n_q - n_{\bar{q}}\right),$$

where n_q and $n_{\bar{q}}$ are number densities of quarks and anti-quarks, respectively. On the other hand, the number density of quark–anti-quark pairs was equal to the entropy density, up to a factor of order 1,

$$n_q - n_{\bar{q}} \sim s.$$

Hence, in terms of quantities characterizing this relatively early epoch, the baryon asymmetry may be also expressed as follows,

$$\Delta_B \sim \frac{n_q - n_{\bar{q}}}{n_q + n_{\bar{q}}}.$$

Its value (1.2) tells us that there was one extra quark per about 10 billion quark–anti-quark pairs. It is this tiny excess that is responsible for the entire baryonic matter in the present Universe, including us. The theory of generation of the baryon asymmetry aims at explaining this excess.

There are two well-motivated mechanisms of baryon number non-conservation. The first one appears in grand unified theories and is due to the exchange of supermassive particles. It is very similar to the mechanism, say, of charm non-conservation in weak interactions which occurs via the exchange of heavy W-bosons. The scale of this new, baryon number non-conserving interactions is nothing but the grand unification scale, and is presumably of order 10^{16} GeV.

Another mechanism is non-perturbative and is related to the triangle anomaly in the baryonic current. It exists already in the Standard Model and, possibly with slight modifications, emerges in all its extensions. The two main features of this mechanism, as applied to the early Universe, are that it operates over a wide range of temperatures, roughly 100 GeV $< T < 10^{11}$ GeV and that it conserves $(B - L)$.

Even though realistic mechanisms of baryon number non-conservation are not numerous, several possibilities to generate the baryon asymmetry are discussed in the literature, which differ by the characteristic temperature at which the asymmetry is produced.

(i) *Temperature of grand unification*, $T \sim 10^{15} - 10^{16}$ GeV.

A viable possibility is that the observed baryon asymmetry is generated by baryon number violating interactions of grand unified theories. The most commonly discussed source of the baryon asymmetry in this context are the B- and CP-violating decays of ultra-heavy particles. At later times, the baryon number is still violated by electroweak processes, whose effect is basically that $(B + L)$, generated at grand unified temperatures, is washed out at some later time (recall that $(B - L)$ is conserved by anomalous electroweak processes). The asymmetry may survive from the grand unification epoch only if a large $(B - L)$ asymmetry is generated at $T \sim 10^{15} - 10^{16}$ GeV, and there are no strong lepton number violating interactions at intermediate temperatures, 100 GeV $< T < 10^{11}$ GeV (otherwise all fermionic quantum numbers are violated at these temperatures, and the baryon asymmetry is washed out). The first requirement points to non-standard $(B - L)$ violating modes of proton decay, though this indication is not strong.

(ii) *Intermediate temperatures*, 1 TeV $\ll T \ll 10^{15}$ GeV.

An interesting possibility is that there exist *lepton* number violating interactions at intermediate scales, and these interactions generate a lepton asymmetry of the Universe at intermediate temperatures. Then this lepton asymmetry is partially reprocessed into baryon asymmetry by anomalous electroweak interactions [29]. Possible manifestations of this scenario are Majorana neutrino masses and/or lepton number violating processes like $\mu \to e\gamma$, $\mu \to eee$ and μ-e conversion. Interestingly, the range of Majorana neutrino masses compatible with this mechanism includes the masses inferred from the atmospheric and solar neutrino experiments.

Another mechanism capable of generating the baryon asymmetry at intermediate temperatures [30] deals with coherent production of scalar fields carrying baryon number. At a later stage the "scalar" baryon number stored in scalar fields is transferred into an ordinary one. The consideration of this interesting possibility in the framework of the Supersymmetric Standard Model can be found, e.g., in ref. [31]; see also ref. [16]

(iii) *Electroweak temperatures*, $T \sim$ (a few) $\times$ 100 GeV.

The remaining possibility is that the observed baryon asymmetry is generated by anomalous electroweak interactions themselves. Since the Universe expands slowly during the electroweak epoch, a considerable departure from equilibrium (the third Sakharov condition) is possible only due to the first order phase transition. Indeed, this transition, which proceeds through the nucleation, expansion and collisions of the bubbles of the new phase, is quite a violent phenomenon.

A necessary condition for the electroweak baryogenesis is that the baryon asymmetry created during the electroweak phase transition should not be washed out after the phase transition completes. In other words, the rate of the electroweak B-violating transitions has to be negligible immediately after the phase transition. As we discuss in detail later on, the latter requirement *is not fulfilled* in the Minimal Standard Model, so the electroweak baryogenesis is only possible in extensions of the MSM. Extending the minimal model is useful in yet another respect: it generally provides extra sources of CP violation beyond the Kobayashi–Maskawa mechanism, so that the second Sakharov condition is satisfied more easily The phenomenological consequences of these extra sources of CP violation are electric dipole moments of the neutron and electron [32, 33], whose values are expected, on the basis of the considerations of baryogenesis [34], to be close to existing experimental limits.

Several specific mechanisms of electroweak baryogenesis will be outlined below. The outcome is that the observed baryon asymmetry may be naturally explained within extended versions of the Standard Model. This result is particularly fascinating as the physics involved is probed at LEP and the Tevatron and will be explored at the LHC relatively soon.

2. NON-CONSERVATION OF BARYON NUMBER

2.1. GRAND UNIFIED THEORIES

In grand unified theories, quarks and leptons are grouped together to form multiplets of the grand unified gauge group. The simplest example is the $SU(5)$ grand unified theory where quarks and leptons are accomodated into $\bar{\mathbf{5}}$- and $\mathbf{10}$-plets,

$$\bar{\mathbf{5}}: \ \bar{d}^i_L, \ e^-_L, \ \nu_{eL} \ ,$$

$$\mathbf{10}: \ u^i_L, \ \bar{u}^i_L, \ d^i_L, \ e^+_L \ ,$$

and similarly for other generations. Here $i = 1, 2, 3$ is the color index, L refers to a left-handed component of a fermion and the bar denotes the anti-particle. The emission of a gauge boson leads to the transformation of one particle into another within the same multiplet. A well known example is the emission of a gluon by $\bar{d}_L$ of the $\bar{\mathbf{5}}$-plet, $\bar{d}_L^1 \to G + \bar{d}_L^2$. Similarly, the emission of a W-boson transforms e_L^- into ν_{eL} of $\bar{\mathbf{5}}$-plet, $e_L^- \to W^- + \nu_{eL}$. As leptons and quarks are now in one and the same multiplet under the gauge group, grand unified theories predict new gauge bosons whose emission leads to the transformation of quarks into leptons and vice versa. In particular, the transformation within $\bar{\mathbf{5}}$-plet occurs as follows,

$$\begin{aligned} e_L^- &\to X + \bar{d}_L \ , \\ \nu_{eL} &\to Y + \bar{d}_L \ , \end{aligned} \tag{1.3}$$

where the new X and Y bosons are color triplets and carry electric charges $(-4/3)$ and $(-1/3)$, respectively.

Likewise, the processes involving X- and Y-bosons transform quarks and leptons within **10**-plet,

$$\begin{aligned} \bar{u}_L &\to X + u_L \ , \\ \bar{u}_L &\to Y + d_L \ , \\ u_L &\to Y + e_L^+ \ , \end{aligned} \tag{1.4}$$

etc. These elementary processes obviously violate baryon number. A similar picture holds in grand unified theories based on groups other than $SU(5)$.

The new gauge bosons are extremely heavy: their mass is determined by the scale of grand unification and in most theories is of order 10^{15} GeV or higher. At lower energies they occur in the intermediate state, but still mediate baryon number non-conservation. An example is the process due to Y-boson exchange, which is the combination of the above elementary processes,

$$u + d \to Y \to e^+ + \bar{u} \ .$$

This is one of the processes responsible for proton decay,

$$p(uud) \to e^+ + \pi^0(\bar{u}u) \ .$$

The extremely long proton life-time is naturally explained by the very large mass of the gauge bosons.

Besides the gauge bosons, grand unified theories predict other super-heavy particles that mediate baryon number non-conservation. In the first place, these are new Higgs bosons. In supersymmetric theories there exist also superpartners of the new gauge and Higgs bosons; in fact, the exchange of these superpartners is often the dominant mechanism of baryon number non-conservation at low energies (nucleon decay).

2.2. ANOMALOUS ELECTROWEAK NON-CONSERVATION OF FERMION QUANTUM NUMBERS

The mechanism of electroweak baryon number non-conservation may be less familiar. Let us discuss this mechanism in the context of a simplified model. The model has the gauge group $SU(2)$ and contains n_L massless left-handed fermionic doublets $\psi_L^{(i)}$, $i = 1, \ldots, n_L$. The absence of global anomaly [35] requires that n_L is even. We also add a Higgs doublet ϕ that breaks the $SU(2)$ symmetry completely. This theory is a simplified version of the electroweak sector of the Minimal Standard Model. All relevant features of the Standard Model are present in this simplified theory; later on we shall comment on minor complications due to $U(1)_Y$ gauge symmetry, right-handed fermions and Yukawa interactions leading to fermion masses. One may regard the simplified theory as the Standard Model in the approximation where $\sin\theta_W$ and all fermion masses are set to zero; for three families of quarks and leptons one has $n_L = 12$ and

$$\psi_L^{(i)} = \{q_L^{f,\alpha}, l_L^f\}, \tag{1.5}$$

where $f = 1,2,3$ is the family index and $\alpha = 1,2,3$ labels the color of quarks.

At the classical level, there exist n_L conserved global $U(1)$ currents,

$$J_\mu^{(i)} = \bar{\psi}_L^{(i)} \gamma^\mu \psi_L^{(i)},$$

which correspond to the conservation of the number of each fermionic species. At the quantum level these currents are no longer conserved due to the triangle anomaly [36, 37, 38]

$$\partial_\mu J_\mu^{(i)} = \frac{1}{32\pi^2} \mathrm{Tr}\left(F_{\mu\nu}\tilde{F}_{\mu\nu}\right). \tag{1.6}$$

where $F_{\mu\nu} = \frac{g}{2i}\tau^a F_{\mu\nu}^a$ is the (matrix-valued) gauge field strength, g is the gauge coupling constant and $\tilde{F}_{\mu\nu} = (1/2)\varepsilon_{\mu\nu\lambda\rho}F_{\lambda\rho}$. Therefore, one expects that fermion numbers $N_F^{(i)} = \int d^3x \; J_0^{(i)}$ are not conserved in any process where the gauge field evolves in such a way that

$$N[A] = \frac{1}{32\pi^2} \int d^4x \; \mathrm{Tr}F_{\mu\nu}\tilde{F}_{\mu\nu} \neq 0. \tag{1.7}$$

Namely,

$$\Delta N_F^{(i)} = N[A], \quad i = 1, \ldots, n_L. \tag{1.8}$$

It is clear from eq. (1.7) that in weakly coupled theories, one has to deal with strong fields: the field $F_{\mu\nu} = \frac{g}{2i}\tau^a F^a_{\mu\nu}$ should be of order 1, and $A^a_\mu = O(g^{-1})$. So, it is natural that the (semi)classical treatment of bosonic fields is often reliable.

Equation (1.8) may be viewed as the selection rule: the number of fermions changes by the same amount for every species. In terms of the assignment (1.5) it implies, in particular,

$$\Delta N_e = \Delta N_\mu = \Delta N_\tau = N[A],$$

$$\Delta B = \frac{1}{3} \cdot 3 \cdot 3 \cdot N[A], \tag{1.9}$$

where the factor 1/3 comes from the baryon number of a quark, while the factor $3 \cdot 3$ is due to the color and the number of generations. So, the amounts of non-conservation of baryon and lepton numbers are related:

$$\Delta N_e = \Delta N_\mu = \Delta N_\tau = \frac{1}{3}\Delta B;$$

$(B - L)$ is conserved while $(B + L)$ is violated.

The analysis of gauge field configurations with the non-zero topological number (1.7) is conveniently performed in the gauge

$$A_0 = 0.$$

In this gauge, there exists a discrete set of classical vacua, i.e. pure gauge configurations

$$A_i = \omega \partial_i \omega^{-1},$$

$$\phi = \omega \phi_0,$$

where $\phi_0 = (0, v/\sqrt{2})$ is the Higgs field in the trivial vacuum. The gauge functions ω depend only on spatial coordinates $\omega = \omega(\mathbf{x})$ and are characterized by an integer:

$$n[\omega] = -\frac{1}{24\pi^2} \int d^3x \; \epsilon^{ijk} \mathrm{Tr}(\omega \partial_i \omega^{-1} \cdot \omega \partial_j \omega^{-1} \cdot \omega \partial_k \omega^{-1}).$$

The vacua with different $n[\omega]$ cannot be continuously deformed into each other without generating non-vacuum gauge fields, so these vacua are separated by a potential barrier. Therefore, the gauge–Higgs system is similar to a particle in periodic potential, as shown in fig. 1. An explicit construction of the minimum energy path connecting the neighboring vacua was carried out in ref. [39], and the fermion sea contribution to this path was evaluated in ref. [40]. The topological number density entering eq. (1.7) is a total derivative,

$$\frac{1}{32\pi^2}\mathrm{Tr}(F_{\mu\nu}\tilde{F}_{\mu\nu}) = \partial_\mu K_\mu,$$

where

$$K_\mu = \epsilon^{\mu\nu\lambda\rho}\mathrm{Tr}\left(F_{\nu\lambda}A_\rho - \frac{2}{3}A_\nu A_\lambda A_\rho\right).$$

If one is interested in vacuum–vacuum transitions, then

$$N[A] = \int d^3x dt\ \partial_\mu K_\mu = \left[\int d^3x\ K_0\right]_{t=-\infty}^{t=+\infty}$$

$$= n[\omega_{t=+\infty}] - n[\omega_{t=-\infty}].$$

So, the topological number of the gauge field is non-zero for transitions between the distinct vacua.

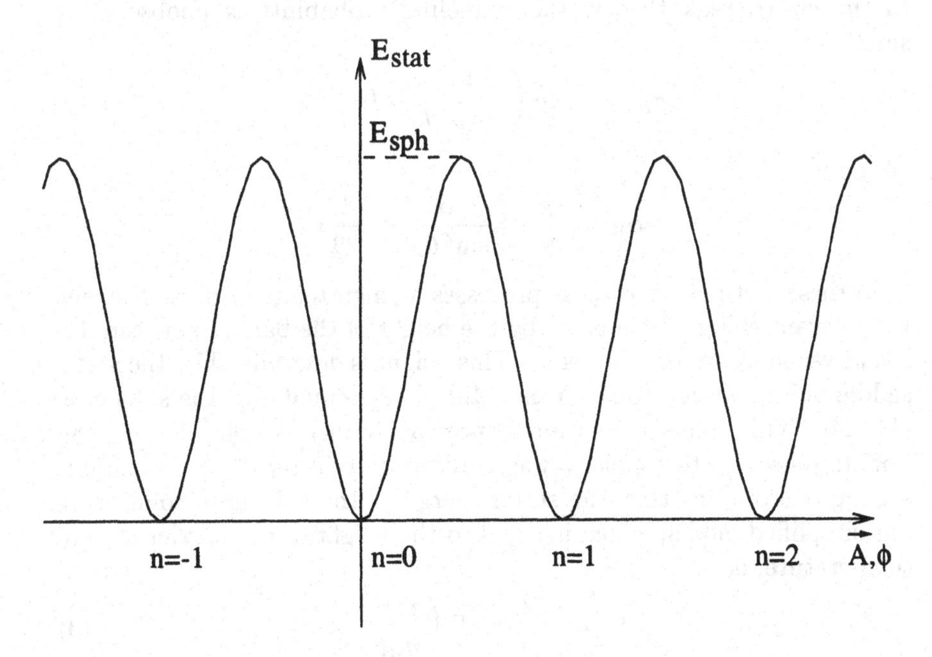

Figure 1 Schematic plot of static energy of gauge and Higgs fields. The minima correspond to the classical vacua.

At zero energies and temperatures, the transition between vacua with different $n[\omega]$ is a tunneling event which is described by instantons [24] (constrained instantons in theories with the Higgs mechanism [41]). In pure Yang–Mills theory an instanton is the solution to the Euclidean field equations, which is an absolute minimum of the Euclidean action in the sector $N[A] = 1$. Properties of instantons are reviewed in ref. [42]. The instanton field, up to gauge transformations, is

$$A_\mu^a = \frac{1}{g}\eta_{\mu\nu a}\frac{2x^\nu}{x^2+\rho^2}, \tag{1.10}$$

where $\eta_{\mu\nu a}$ are the 't Hooft symbols, and ρ is an arbitrary scale to be integrated over. The instanton action is

$$S_{inst} = \frac{8\pi^2}{g^2}$$

and the tunneling amplitude is proportional to

$$A_{inst} \propto \mathrm{e}^{-S_{inst}}. \tag{1.11}$$

In the electroweak theory, the tunneling probability is unobservably small,

$$\sigma_{inst} \propto \exp\left(-\frac{4\pi}{\alpha_W}\right) \sim 10^{-170}, \tag{1.12}$$

where

$$\alpha_W = \frac{g^2}{4\pi} = \frac{\alpha}{\sin^2\theta_W} \approx \frac{1}{29}.$$

In these lectures we discuss processes at high temperatures. Naively, the relevant energy scale is set by the height of the barrier between different vacua as sketched in fig. 1. This height is determined by the static saddle point solution to the Yang–Mills–Higgs equations, the sphaleron [43, 44]. This solution was found previously in refs. [45, 46, 47, 48], but its relevance to topology was realized only in ref. [44]. By simple scaling one obtains that the static energy of the sphaleron solution in our simplified model, which is equal to the height of the barrier at zero temperature, is

$$E_{sph} = \frac{2m_W}{\alpha_W}B\left(\frac{m_H}{m_W}\right), \tag{1.13}$$

where m_H is the mass of the Higgs boson. The function $B(m_H/m_W)$ has been evaluated numerically [44]; it varies from 1.56 to 2.72 as m_H/m_W

varies from zero to infinity [2]. So, the height of the barrier in the electroweak theory is of order 10 TeV.

There are at least two possible ways to see that fermion quantum numbers are indeed violated in instanton-like processes. One of them [22, 23] makes use of zero fermion modes in Euclidean background fields with $N[A] \neq 0$. This approach is reviewed in ref. [42] and its Minkowskian counterpart is considered in ref. [51]. A more intuitive way [52, 51] is related to the phenomenon of level crossing, which is as follows. Consider left-handed fermions in the background field $\mathbf{A}(\mathbf{x}, t)$ which changes in time from one vacuum at $t = -\infty$ to another vacuum at $t = +\infty$ (we again use the gauge $A_0 = 0$). At each time t one can evaluate the fermionic spectrum, i.e. the set of eigenvalues of the Dirac Hamiltonian in the static background $\mathbf{A}(\mathbf{x}, t)$ where t is viewed as a parameter. The spectrum varies with t; some levels cross zero from below and some cross zero from above, as shown in fig. 2. The relevant quantity is the net change of the number of positive energy levels, which is the difference between the total number of levels that cross zero from above and from below in the course of the entire evolution from $t = -\infty$ to $t = +\infty$. A general mathematical theorem [53] says that this difference is related to the topological number of the gauge field,

$$N_+ - N_- = n[\omega_{t=+\infty}] - n[\omega_{t=-\infty}] = N[A]. \tag{1.14}$$

Recall now that at vacuum values of $\mathbf{A}$, the ground state of the fermionic system has all negative energy levels filled and all positive energy levels empty. A real fermion corresponds to filled positive energy level and antifermion is an unoccupied negative energy level. As the energy levels cross zero, the number of real fermions changes, and the net change in the fermion number of each left-handed doublet is

$$\Delta N_F^{(i)} = N_+ - N_-.$$

Combining this relation with eq.(1.14) we see that the fermion number is indeed not conserved, and the amount of non-conservation is in perfect agreement with the anomaly relation, eq. (1.8).

We note in passing that the level crossing picture for right-handed fermions is reversed, as shown in fig. 2. So, in vector-like theories the total fermion number $(N_L + N_R)$ is conserved, whereas chirality $(N_L - N_R)$ is violated. This explains why there is no strong baryon number non-conservation: QCD interactions conserve baryon number but strongly and explicitly break axial $U(1)$.

[2] At very large m_H/m_W the situation is more complicated [49, 50], but the estimate (1.13) remains valid.

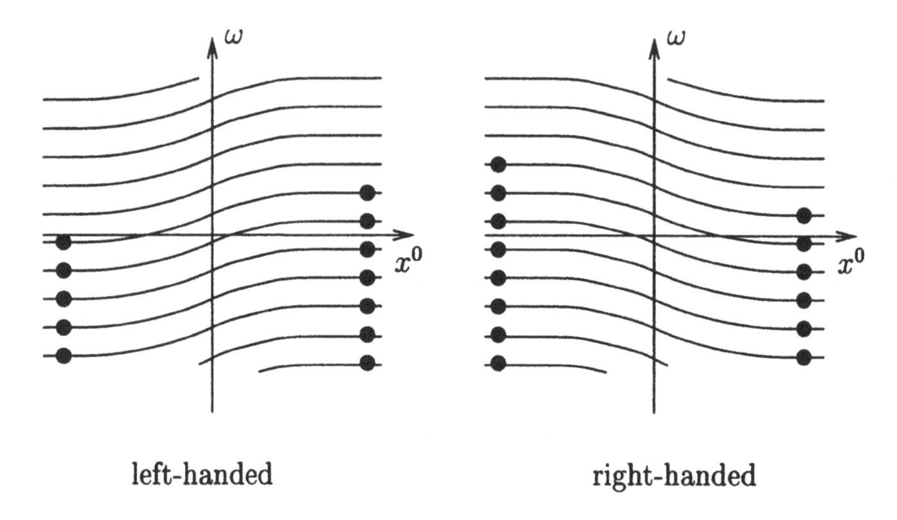

Figure 2

Although the above discussion was for massless fermions, the results remain valid for the standard electroweak theory where fermions acquire masses via the Yukawa coupling to the Higgs field [54, 55, 56]. Indeed, the triangle anomaly for baryon and lepton currents remains valid in the Standard Model, so that relation (1.9) must hold. The counting of fermion zero modes in the instanton background confirms this expectation [54, 56, 59]. Also, the level crossing phenomenon has been explicitly found in theories of this type [57, 58, 59]. So, the complications due to right-handed fermions and fermion masses do not change the picture of baryon and lepton number non-conservation.

Finally, the presence of $U(1)_Y$ gauge symmetry in the Standard Model does not modify the analysis to any considerable extent either. There are no instantons of the $U(1)_Y$ gauge field, while the effect of the $U(1)_Y$ interactions on the $SU(2)$ instantons is tiny. Also, the energy of the $SU(2)$ sphaleron is still given by eq.(1.13) where the factor B depends also on $\sin^2\theta_W$. For the actual value $\sin^2\theta_W = 0.23$, the deviation of B from its $SU(2)$ values is numerically small [44, 60, 61].

In hot Big Bang cosmology, there is an epoch of particular interest from the point of view of the electroweak physics. This is the epoch of the electroweak phase transition, the relevant temperatures being of the order of a few hundred GeV [62, 63, 64, 65]. Before the phase transition (high temperatures), the Higgs expectation value is zero, while after the phase transition the Higgs field develops a non-vanishing expectation value. The critical temperature T_c depends on the parameters of the electroweak theory; in the Minimal Standard Model (MSM) the only grossly unknown parameter is the mass of the Higgs boson, m_H. In extensions of the MSM, there are more parameters that determine T_c.

At sufficiently small m_H in MSM, the phase transition is of the first order, while at large m_H smooth cross over has been observed [66, 67]. It is important that the masses of W- and Z-bosons immediately after the phase transition, $m_W(T_c)$ and $m_Z(T_c)$, be smaller than their zero temperature values; the precise behaviour of $m_W(T)$ and $m_Z(T)$ again depends on the parameters of the model (on m_H in the MSM). Generally speaking, the stronger the first order phase transition, the larger $m_W(T_c)$ and $m_Z(T_c)$. The electroweak phase transition is considered in more detail in section 6.

Let us now turn to the rate of the electroweak baryon number non-conservation at high temperatures. While at zero temperatures the B non-conservation comes from tunneling and is unobservably small because of the tunneling exponent, it may proceed at high temperatures via thermal jumps over the barrier shown in fig. 1 [27]. At temperatures below the critical one, $T < T_c$, the probability to find the system at the saddle point separating the topologically distinct vacua is still suppressed, but now by the Boltzmann factor,

$$\Gamma_{sph} \propto \exp\left(-\frac{E_{sph}(T)}{T}\right), \tag{1.15}$$

where

$$E_{sph}(T) = \frac{2m_W(T)}{\alpha_W} B\left(\frac{m_H}{m_W}\right)$$

is the free energy of the sphaleron. Once the system jumps up to the saddle point (i.e. once the sphaleron is thermally created), the system may roll down to the neighbouring vacuum, and the baryon and lepton numbers may be violated. Therefore, the factor (1.15) is also the suppression factor for the rate of the electroweak baryon number non-conservation at $T < T_c$.

At $T > T_c$, the exponential suppression of the baryon number non-conserving transitions is absent. A simple power-counting estimate of the rate per unit time per unit volume in the unbroken phase gives [68, 69] $\Gamma = \text{const} \cdot (\alpha_W T)^4$. More refined estimates [70, 71, 72] show that the rate is smaller by a factor α_W, i.e., up to logarithms,

$$\Gamma_{sph} = \text{const} \cdot \alpha_W^5 T^4, \tag{1.16}$$

where the constant is of order 1. This estimate of the rate has been confirmed by numerical simulations [73]. In the cosmological context, this rate is quite large.

3. HOT BIG BANG

The expansion of the homogeneous and isotropic Universe with negligible spatial curvature[3] is governed by the Friedmann equation, which is all that remains of the Einstein equations in this simple case,

$$H^2 = \frac{8\pi}{3} G\rho. \tag{1.17}$$

Here $H = \dot{a}/a$ is the expansion rate (the Hubble parameter), a is the scale factor and ρ is the total energy density. During the hot phase of the evolution of the Universe, the cosmic plasma consisted of ultra-relativistic particles, so that $\rho \propto T^4$. Hence, there exists a simple relation between the expansion rate and temperature,

$$H = \frac{T^2}{M_{Pl}^*}. \tag{1.18}$$

Here $M_{Pl}^* = M_{Pl}/(1.66\sqrt{g_*})$, where g_* is the effective number of degrees of freedom (roughly speaking, the number of states of all particles with masses smaller than T) and $M_{Pl} = 1/\sqrt{G} \approx 10^{19}$ GeV is the Planck mass. The quantity M_{Pl}^* changes in time slightly, but at temperatures we will be interested in, it is always of order

$$M_{Pl}^* \sim 10^{18}\,\text{GeV}.$$

As pointed out in the introduction, the generation of baryon asymmetry requires departure from thermal equilibrium. To see that this requirement is non-trivial, let us compare the rates of processes (1.3), (1.4) with the expansion rate of the Universe just before the grand unified phase transition. On dimensional grounds, the former rates are of order

$$\Gamma \sim \alpha_{GUT} T.$$

With $\alpha_{GUT} = (a\ few) \cdot 10^{-2}$, one has $\Gamma > H$ provided that $T < 10^{16}$ GeV. This means that in most grand unified theories, baryon number violating interactions are in thermal equilibrium before the grand unified phase transition (if the Universe was ever so hot).

Obviously, thermal equilibrium gets even better at lower temperatures: QCD and perturbative electroweak processes are in thermal equilibrium until QCD and electroweak phase transition temperatures, respectively, and, in fact, long afterwards.

Similarly, one may wonder whether non-perturbative electroweak baryon number violating processes were in thermal equilibrium in the early Uni-

[3]The effects of spatial curvature are indeed negligibly small at the early stages of the evolution of the Universe.

verse. One compares the rate (1.16), multiplied by the characteristic volume T^{-3}, with the expansion rate (1.18) and finds that the rate of electroweak baryon number violation exceeds the expansion rate provided that

$$T < \text{const} \cdot \alpha_W^5 M_{Pl}^* \sim 10^{11}\,\text{GeV}.$$

In a wide range of temperatures, from 10^{11} GeV down to the electroweak phase transition, $T_{EW} \sim 100$ GeV, the electroweak baryon number violating processes are in thermal equilibrium. What happens immediately after the electroweak phase transition is a separate and important issue which will be discussed later.

At the electroweak epoch, $T \sim 100$ GeV, the Universe expands quite slowly. Indeed, the expansion time, $t_U \sim H^{-1}$ (which coincides with the age of the Universe, up to a factor of order one) is at that epoch of order

$$t_U \sim \frac{M_{Pl}^*}{(100\ \text{GeV})^2} \sim 10^{-10}\,\text{s},$$

which is very large by high energy physics standards. This is why a sizeable departure from thermal equilibrium at the electroweak epoch requires dramatic phenomena such as a first order phase transition.

4. GRAND UNIFIED BARYOGENESIS

4.1. BARYOGENESIS IN DECAYS OF ULTRA-HEAVY PARTICLES

One of the viable scenarios of baryogenesis is the generation of the baryon asymmetry due to baryon number violating interactions of grand unified theories. This mechanism is well understood and described in numerous reviews [12, 13, 14]. The source of the baryon asymmetry is out-of-equilibrium decays of ultra-heavy particles which violate both baryon number and CP. As an example, there are decay channels of X-particles discussed in section 2, with different baryon numbers of the final states (cross processes to those listed in eqs. (1.3), (1.4)),

$$X \to e^- d\,,\ B = \frac{1}{3},$$

$$X \to \bar{u}\bar{u}\,,\ B = -\frac{2}{3}. \tag{1.19}$$

Of course, anti-particles have conjugate decay channels,

$$\bar{X} \to e^+ \bar{d}\,,$$

$$\bar{X} \to uu.$$

If CP is not conserved in these decays, then the partial widths of the conjugate channels are in general different,

$$\Gamma_{X \to e^- d} \neq \Gamma_{\bar{X} \to e^+ \bar{d}},$$

although the total widths of X and $\bar{X}$ are equal due to CPT.

Let us consider a simplified case when eq. (1.19) exhausts all decay channels, and define the microscopic asymmetry,

$$\delta = \frac{\Gamma_{X \to e^- d} - \Gamma_{\bar{X} \to e^+ \bar{d}}}{\Gamma_{X,tot}}.$$

Let us assume that the temperature of the Universe exceeded m_X at a certain epoch. Then at this initial stage, there is an equal number of X and $\bar{X}$ in the cosmic plasma; the baryon number violating processes, including decays (1.19) and inverse decays (production of X-particles in the inverse processes) are in thermal equilibrium. As the temperature drops somewhat below m_X, the production of X-particles gets suppressed, and X-particles experience out-of-equilibrium decays. The produced baryon asymmetry is of order

$$\Delta_B \sim \frac{\delta \cdot n_X^{out}}{s},$$

where n_X^{out} is the number density of X-particles decaying out of equilibrium, and s is again the entropy density. The value of n_X^{out} depends crucially on the ratio

$$\left(\frac{\Gamma_{X,tot}}{H} \right)_{T \sim m_X} . \tag{1.20}$$

The most favorable case for baryogenesis is when this ratio is *very small.* Then at a temperature somewhat below m_X, the X-particles are no longer produced, but existing X-particles would not have had time to decay. Their number density remains the same as the number denisty of a massless species, and all X-particles eventually decay in an out-of-equilibrium way. In this case $n_X^{out} = n_{massless}$, so that

$$\Delta_B \sim \frac{\delta}{g_*},$$

where we took into account that the entropy density is of order $n_{massless} \cdot g_*$ with g_* being the effective number of massless species. With $g_* \sim 10^2$, a realistic value of the baryon asymmetry is obtained at fairly small CP-asymmetry, $\delta \sim 10^{-8}$.

If the ratio (1.20) is *much greater than one*, the baryon number violating processes involving either real or virtual X-particles are in thermal equilibrium at temperatures well below m_X. This means that the production of the baryon asymmetry is strongly suppressed. At very large values of the ratio (1.20) the suppression is exponential, and the resulting baryon asymmetry is far too small.

To estimate the ratio (1.20), we write

$$\Gamma_{X,tot} = \alpha m_X, \tag{1.21}$$

and obtain that the best case for the generation of the asymmetry is

$$m_X > \alpha M^*_{Pl}. \tag{1.22}$$

For grand unified gauge bosons one finds $m_X > (a\ few) \cdot 10^{16}$ GeV, which is not very realistic. Grand unified Higgs bosons are better, as their coupling constant α entering eq.(1.21) may be much smaller than the gauge coupling constant.

It is worth noting that eq.(1.22) suggests that the mechanism just described naturally invokes ultra-heavy particles. This fits rather well into the picture of grand unification not that far below the Planck (or string) scale.

In spite of its naturalness, the grand unified mechanism of baryogenesis has its problems. First, it requires that the hot stage of the evolution of the Universe began at very high temperatures, most naturally $T \sim 10^{16}$ GeV or higher. This is in conflict (at least, potentially) with inflation, as many inflationary models predict the reheat temperature several orders of magnitude lower than 10^{16} GeV. Also, in supersymmetric theories, the very high reheat temperature leads to the gravitino problem [74, 75, 76, 14]. An interesting way to avoid the problem of the high reheat temperature is to invoke preheating [77], a stage after inflation in which the Universe is still out of thermal equilibrium. The ultra-heavy particles may be abundantly produced during or immediately after this stage [78, 79], and the baryon asymmetry may be generated in their decays [79, 80] (for a review of this approach see ref. [16]).

Another problem is that the baryon asymmetry tends to be washed out by electroweak baryon number violating processes well after the grand unification epoch. Since the electroweak processes conserve $(B - L)$ but violate $(B + L)$, this wash-out is avoided only if a sufficiently large $(B - L)$ is produced at the grand unification temperature (or due to a preheating stage). Furthermore, lepton number violation should not be sizeable at the intermediate epoch. Let us discuss the latter point in some detail, as it is relevant to other mechanisms of baryogenesis as well.

4.2. SURVIVAL OF PRIMORDIAL BARYON ASYMMETRY

The anomalous electroweak processes are rapid at sufficiently high temperatures. Their rate Γ_{sph} exceeds the rate of the Universe's expansion $\frac{T^2}{M_0}$ in the standard Big Bang scenario in the following interval of temperatures:

$$100 \text{ GeV} \sim T_{EW} < T < T^{**} \simeq \alpha_W^5 M_{Pl} \simeq 10^{11} \text{ GeV}, \qquad (1.23)$$

Clearly, the equilibrium character of B-violating reactions has an important impact on the survival of the primordial baryon asymmetry. Several different cases can be distinguished, depending on initial conditions and on the rate of B and L non-conservation due to processes other than those associated with electroweak sphalerons.

(i) Suppose that the Universe is asymmetric with respect to the anomaly free fermionic charges $\Delta_i = L_i - \frac{1}{n_f}B$ of the Standard Model at $T > T^{**}$, and assume that at $T < T^{**}$ there is no B or L violating interactions besides the electroweak anomalous processes. The origin of the primordial asymmetry is not essential here. Then anomalous reactions convert the initial asymmetry to the baryonic one at $T = T_{EW}$. For the Minimal Standard Model the relationship is given by [81, 69] (see also [82, 83]),

$$\Delta_0 = \frac{8n_f + 4}{22n_f + 13}\Delta_{B-L} - K\frac{4}{13\pi^2}\sum_{i=1}^{n_f}\frac{m_i^2(T^*)}{(T^*)^2}\Delta_i. \qquad (1.24)$$

where m_i^2 is the lepton mass of a given generation, and $K \approx 1$. The first term in the right hand side of this equation shows that $(B-L)$-asymmetry is reprocessed into the baryon asymmetry, while $(B+L)$ tends to be washed out; the second term is the correction coming from slightly different behavior of quarks with different masses in the plasma. If the initial value of $(B-L)$ is non-zero (coming, say, from the GUT physics) then the baryon asymmetry, up to a possible contribution from the EW phase transition (see below), has a primordial character. If, on the contrary, the initial $(B-L)$-asymmetry is absent, we can rely only on the second term in (1.24). For three lepton generations one gets a suppression $\Delta_0 \simeq 3 \times 10^{-6}\Delta_3$. So, to have a non-negligible effect, the initial asymmetry Δ_3 must be very large, or the standard theory should be extended by adding heavy leptons.

(ii) Suppose now that there are some reactions, which do not conserve all Δ_i, and which are in thermal equilibrium for some period between T_{EW} and T^{**}. At this intermediate epoch B and L are non-conserved separately, and according to the third Sakharov condition all baryonic and leptonic asymmetries are washed out. Hence, the existence of these reactions is fatal for the primordial baryon asymmetry. If the baryon asymmetry is not produced at a later time, the requirement of the absence of these reactions may appear to be a powerful tool for constraining the properties of new particle interactions [84, 82, 85, 86]. However, some time ago it was realized that most of these constraints are drastically weakened due to the smallness of some Yukawa coupling constants in the Standard Model or its supersymmetric extensions [87, 88, 89, 90, 91, 92].

Let us discuss the main idea of these estimates using the example of lepton number violating interactions, leading to the Majorana neutrino masses m_{ij} [91]. We take for simplicity the Minimal Standard Model and add to it lepton number violating interactions. The $SU(2) \times U(1)$ symmetric low energy Lagrangian with $\Delta L = 2$ has the form

$$\frac{1}{v^2} m_{ij} (\bar{L}_i \phi)(\tilde{\phi}^\dagger L_j^c), \tag{1.25}$$

where L_i and L_j^c are lepton doublet and its charge conjugate, respectively, ϕ is the scalar doublet, v is the vacuum expectation value of the Higgs field, $v = 246$ GeV. The rate of L non-conserving reactions $L\phi \to L^c\phi^*$ at high temperatures has been found in ref. [91]

$$\Gamma \simeq \frac{9}{\pi^5} \frac{T^3}{v^4} \bar{m}_\nu^2, \tag{1.26}$$

where $\bar{m}_\nu$ is an average Majorana neutrino mass,

$$\bar{m}_\nu^2 = \frac{5}{3}|m_{ee}|^2 + |m_{e\mu}|^2 + |m_{e\tau}|^2. \tag{1.27}$$

These reactions were initially required [82] to be out of thermal equilibrium at $T < T^{**}$; this leads to a very stringent constraint

$$\bar{m}_\nu < \frac{v^2}{\sqrt{M_0 T^{**}}} \simeq 10^{-2} \text{eV}. \tag{1.28}$$

An implicit assumption in the derivation above is that the set of conserved numbers Δ_i is a complete one below $T = T^{**}$. In fact, this is not true due to the smallness of the right-handed electron Yukawa coupling constant. In the limit when this constant is zero, the right-handed electron number is conserved, and the asymmetry in it propagates to the asymmetry in baryon number. The rate of reactions not conserving the right-handed electron number (say, $e_L H \to e_R W$) is of the order of

$$\Gamma_R \sim \alpha_W f_e^2 T, \tag{1.29}$$

where f_e is the electron Yukawa coupling constant. These reactions are out of equilibrium at $T > T_R \simeq 3$ TeV [91]. It is this temperature which should be used in eq.(1.28) instead of T^{**}. Thus, we arrive at much weaker constraint [90, 91, 93]

$$m_\nu < 8\text{KeV}, \tag{1.30}$$

which must be satisfied in any case because of the known laboratory limits and other cosmological considerations.

The same sort of considerations apply to other possible interactions breaking lepton and baryon numbers. The general conclusion is that the initial charge asymmetry can survive during the epoch at which anomalous reactions are in thermal equilibrium. Moreover, initial asymmetries in fermionic quantum numbers, different from the baryon number, are usually transferred to baryon asymmetry towards the end of the equilibrium sphaleron period.

We barely know the history of the Universe at very high temperatures (say, at $T \gg 1$ TeV). It may well be that the Universe was symmetric with respect to all fermion charges at $T > 10^{12}$ GeV. This assumption, being a bit arbitrary, may be in fact a natural consequence of inflation, which exponentially dilutes the densities of all global quantum numbers (e.g. baryonic or leptonic). If true, baryon asymmetry should be produced at relatively late stages of the Universe expansion. As pointed out in the introduction, this may happen either at intermediate temperatures (1 TeV $< T < 10^{12}$ GeV) or at the electroweak temperature ($T \sim$ (a few)$\times$100 GeV).

5. LEPTOGENESIS

Let us now discuss an alternative mechanism for the generation of the baryon asymmetry of the Universe. The idea [29] is to generate the *lepton* asymmetry at an intermediate stage of the evolution, 100 GeV $= T_{EW} < T < T_{GUT} \sim 10^{16}$ GeV, by processes violating lepton number. The anomalous electroweak processes then convert (a substantial part of) this asymmetry into a baryon asymmetry, as we discussed in the previous section. The simplest mechanism that generates the lepton asymmetry is similar to the GUT mechanism discussed above: the lepton asymmetry is generated in decays of heavy particles. Natural candidates for the latter are heavy Majorana neutrinos.

Recently, this idea has got substantial support from solar and atmospheric neutrino experiments that indicate that ordinary neutrinos have small but non-zero masses. A nice way to explain these masses is provided by the see-saw mechanism that invokes heavy Majorana neutrinos. As we will see in a moment, see-saw parameters required to explain solar and atmospheric neutrino physics are in the range required to produce a realistic value of the baryon asymmetry.

In explaining the mechanism of leptogenesis, we follow ref. [94]. The most general lagrangian for couplings and masses of charged leptons and neutrinos in a model with one Higgs field reads

$$L_Y = -\bar{l_L}\,\tilde{\phi}\,g_l\,e_R - \bar{l_L}\,\phi\,g_\nu\,\nu_R - \frac{1}{2}\bar{\nu_R^C}\,M\,\nu_R + \text{ h.c.}\,. \qquad (1.31)$$

The vacuum expectation value v of the Standard Model Higgs field ϕ generates Dirac masses m_l and m_D for charged leptons and neutrinos, $m_l = g_l v$ and $m_D = g_\nu v$, respectively, which are assumed to be much smaller than the Majorana masses M. This yields light and heavy neutrino mass eigenstates

$$\nu \simeq K^\dagger \nu_L + \nu_L^C K \quad , \qquad N \simeq \nu_R + \nu_R^C\,, \qquad (1.32)$$

with masses

$$m_\nu \simeq -K^\dagger m_D \frac{1}{M} m_D^T K^* \quad , \quad m_N \simeq M\,. \qquad (1.33)$$

Here K is a unitary matrix which relates weak and mass eigenstates.

The right-handed neutrinos, whose exchange may erase any lepton asymmetry, can also generate a lepton asymmetry by means of out-of-equilibrium decays. The decay width of the heavy neutrino N_i is,

$$\begin{aligned} \Gamma_{Di} &= \Gamma\left(N^i \to \phi^c + l\right) + \Gamma\left(N^i \to \phi + l^c\right) \\ &= \frac{1}{8\pi}\frac{(m_D^\dagger m_D)_{ii}}{v^2} M_i\,. \end{aligned} \qquad (1.34)$$

In the same spirit as in the theory of GUT baryogenesis, from the decay width one obtains an upper bound on the light neutrino masses via the out-of-equilibrium condition[95]. Requiring for the lightest Majorana neutrino that $\Gamma_1 < H|_{T=M_1}$ one finds the constraint

$$\tilde{m}_1 = \frac{(m_D^\dagger m_D)_{11}}{M_1} < 10^{-3}\,\text{eV}\,. \qquad (1.35)$$

This already suggests that the masses of ordinary neutrinos are in the right ballpark.

Interference between the tree-level amplitude and the one-loop self-energy and vertex corrections yields the CP asymmetry[96, 97, 98]

$$\begin{aligned} \delta_L &= \frac{\Gamma(N_1 \to l\phi^c) - \Gamma(N_1 \to l^c\phi)}{\Gamma(N_1 \to l\phi^c) + \Gamma(N_1 \to l^c\phi)} \\ &\simeq \frac{3}{16\pi v^2}\frac{1}{\left(m_D^\dagger m_D\right)_{11}} \sum_{i=2,3} \text{Im}\left[\left(m_D^\dagger m_D\right)_{1i}^2\right]\frac{M_1}{M_i}\,. \end{aligned} \qquad (1.36)$$

Here we have assumed $M_1 \ll M_2, M_3$.

The CP asymmetry (1.36) leads to the generated lepton asymmetry [12, 99],

$$\Delta_L = \frac{n_L - n_{\bar{L}}}{s} = \kappa \frac{\delta_L}{g_*} . \tag{1.37}$$

Here the factor $\kappa < 1$ represents the effect of washout processes. In order to determine κ one has to solve the full Boltzmann equations. Typically, one has $\kappa \simeq 0.1 \ldots 0.01$.

Note, that according to eqs. (1.34) and (1.36) the CP asymmetry is determined by the mixings and phases present in the product $m_D^\dagger m_D$. These mixings and phases are generally different from those entering the mixing matrix K in the leptonic charged current, which determines CP-violation and mixings of the light leptons. This implies that there exists no direct connection between CP violation and generation mixing which are relevant at high energies and at low energies, respectively. In other words, the study of oscillations of ordinary neutrinos does not directly confirm or disprove the viability of baryogenesis via leptogenesis.

In many models the quark and lepton mass hierarchies and mixings are parametrised in terms of a common mixing parameter $\lambda \sim 0.1$. Assuming a hierarchy for the right-handed neutrino masses similar to the one satisfied by up-type quarks,

$$\frac{M_1}{M_2} \sim \frac{M_2}{M_3} \sim \lambda^2 , \tag{1.38}$$

and a corresponding CP asymmetry

$$\delta_L \sim \frac{\lambda^4}{16\pi} \frac{m_3^2}{v^2} \sim 10^{-6} \frac{m_3^2}{v^2} , \tag{1.39}$$

one obtains indeed the correct order of magnitude for the baryon asymmetry[100] if one chooses $m_3 \simeq m_t \simeq 175$ GeV, as expected, for instance, in theories with Pati-Salam symmetry. Using as a constraint the value for the ν_μ-mass which is preferred by the MSW explanation [101] of the solar neutrino deficit[102], $m_{\nu_\mu} \simeq 3 \cdot 10^{-3}$ eV, the ansatz[100] implies for the other light and the heavy neutrino masses

$$m_{\nu_e} \simeq 8 \cdot 10^{-6} \text{ eV} , \quad m_{\nu_\tau} \simeq 0.15 \text{ eV} , \quad M_3 \simeq 2 \cdot 10^{14} \text{ GeV} . \tag{1.40}$$

Consequently, one has $M_1 \simeq 2 \cdot 10^{10}$ GeV and $M_2 \simeq 2 \cdot 10^{12}$ GeV. The solution of the Boltzmann equations then yields the baryon asymmetry

$$\Delta_B \simeq 1 \cdot 10^{-10} , \tag{1.41}$$

which is indeed the correct order of magnitude. The precise value depends on unknown phases.

The large mass M_3 of the heavy Majorana neutrino N_3 (cf. (1.40)), suggests that $(B-L)$ is already broken at the unification scale $M_{GUT} \sim 10^{16}$ GeV, without any intermediate scale of symmetry breaking. This large value of M_3 is a consequence of the choice $m_3 \simeq m_t$. This is indeed necessary in order to obtain sufficiently large CP asymmetry.

The recently reported atmospheric neutrino anomaly[103] may be due to ν_μ-ν_τ oscillations. The required mass difference is $\Delta m^2_{\nu_\mu \nu_\tau} \sim 10^{-3}$ eV^2, together with a large mixing angle $\sin^2 2\Theta \sim 1$. In the case of hierarchical neutrinos this corresponds to a τ-neutrino mass $m_{\nu_\tau} \sim (0.1-0.01)$ eV. Within the theoretical uncertainties this is consistent with the τ-neutrino mass (1.40) obtained from baryogenesis. The ν_τ-ν_μ mixing angle is not constrained by leptogenesis and is therefore a free parameter, in principle.

Thus, baryogenesis via leptogenesis is a viable possibility. Its interesting feature is the rather unexpected connection to neutrino physics. This mechanism and its relation to supersymmetry, the dark matter problem, etc., is a subject of intense investigations (see, e.g., ref. [104] and references therein).

6. ELECTROWEAK BARYOGENESIS

6.1. PRELIMINARIES

The most intriguing possibility is that the baryon asymmetry of the Universe is generated by the electroweak baryon number violating processes themselves. This scenario implies that the baryogenesis occured at the relatively late electroweak epoch of the evolution of the Universe, at temperatures of order $T_{EW} \sim 100$ GeV. As we already pointed out, at this epoch the Universe expands quite slowly, so the departure from thermal equilibrium necessary for successful baryogenesis may occur only in the process of the first order phase transition.

Hence, one arrives at the following general picture. At temperatures somewhat higher than T_{EW}, the Universe was in the symmetric phase with vanishing Higgs expectation value[4]. The anomalous electroweak processes that violate baryon number are in thermal equilibrium at these temperatures. As the Universe cools down, the first order phase transi-

[4]The notions of "symmetric" and "asymmetric" phases with "zero and non-zero Higgs expectation values", respectively, are, in fact, loose. There is no local gauge-invariant order parameter (see, however, ref.[105]) that distinguishes between the two phases. Still, in a certain region of the parameter space, the first order phase transition indeed takes place in the Standard Model and its extensions.

tion occurs, and the Higgs field of the Standard Model (or its extension) develops a non-vanishing expectation value. The phase transition proceeds through the spontaneous formation of small bubbles of the new phase, which then rapidly grow and eventually fill the entire Universe. This later process is highly out-of-equilibrium, and the baryon asymmetry may be generated. Soon after the phase transition completes, thermal equilibrium is established again. The baryon number violating processes must switch off at this moment, i.e., their rate must be smaller than the expansion rate immediately after the phase transition, otherwise they would wash out the generated baryon asymmetry. If they indeed switch off, the baryon asymmetry generated in the process of electroweak phase transition survives until the present epoch.

The microscopic physics involved in this scenario is the physics at the electroweak energy scale. It is being probed at LEP and the Tevatron, and will be probed at the LHC. As we will see, the experimental information already available rules out this scenario in the Minimal Standard Model, almost rules it out in the MSSM but still allows for electroweak baryogenesis in more complicated extensions of the Standard Model.

To see whether electroweak baryogenesis works or not, one has to understand a number of dynamical issues, such as

(i) Is the electroweak phase transition indeed first order?

(ii) Do electroweak baryon number violating processes switch off immediately after the phase transition?

(iii) What mechanism generates the baryon asymmetry in the process of the phase transition? What is the relevant source of CP-violation and how does it work?

Some of these issues are understood reasonably well. With complete knowledge of microscopic physics in 100 GeV – 1 TeV energy range, electroweak baryogenesis will become either an irrelevant exercise or an established fact.

6.2. ELECTROWEAK PHASE TRANSITION

To describe the high temperature phase transitions in any given theory it is very important to have a relevant calculational formalism. The traditional tool is the effective potential for the scalar field ϕ. It is defined as the value of the free energy of the system (pressure with a minus sign) in a uniform background field ϕ. The minima of this potential correspond to the (meta)stable states of the system. The system undergoes a first order phase transition if there are two degenerate minima of this potential, separated by an energy barrier. In general, the effective potential is a gauge dependent quantity; perturbative calculations often produce complex terms. However, the values of the potential at its minima are gauge invariant; this allows for the gauge-invariant definition of the critical temperature and latent heat.

Simple estimates are obtained by making use of the one loop approximation. An example is provided by the Minimal Standard Model. Here the one-loop effective potential in the high temperature approximation is (for simplicity, we take the case when the Higgs boson is sufficiently light, and neglect the effects of the $U(1)_Y$ interactions):

$$V(\phi, T) = \frac{1}{2}\gamma(T^2 - T_-^2)\phi^2 - \frac{1}{3}\alpha T\phi^3 + \frac{1}{4}\lambda\phi^4. \tag{1.42}$$

For the Standard Model with the top quark mass m_{top}

$$\alpha = 9g^3/(32\pi), \gamma = \frac{3}{16}g^2 + \frac{1}{2}\lambda + \frac{m_{\text{top}}^2}{2v^2}, \tag{1.43}$$

and the lower metastability temperature T_- is related to the Higgs mass $m_H^2 = 2\lambda v^2$ through

$$T_- = \frac{m_H}{\sqrt{2\gamma}}. \tag{1.44}$$

Clearly, at $T < T_-$ the symmetric phase $\phi = 0$ is absolutely unstable.

Owing to the presence of the cubic term, the potential predicts the first order transition with the critical temperature

$$T_c = \frac{T_-}{\sqrt{1 - \frac{2}{9}\frac{\alpha^2}{\lambda\gamma}}} > T_-. \tag{1.45}$$

At $T = T_c$ the effective potential has two degenerate minima, at $\phi = 0$ and $\phi = \phi(T_c) \neq 0$, which correspond to the symmetric and asymmetric phases, respectively. Below T_c the minimum at $\phi \neq 0$ has lower free energy, and the first order phase transition takes place. The jump of the order parameter is

$$\frac{\phi(T_c)}{T_c} = \frac{2}{3}\frac{\alpha}{\lambda}. \tag{1.46}$$

The phase transition gets weaker when the scalar self-coupling increases. This is seen from the behaviour of the order parameter, latent heat, and the surface tension of the bubble wall, all of which decrease with the increase of λ.

Qualitatively, the one-loop description gives correct results, but concrete numbers may be quite different from those obtained by a more refined treatment. This is discussed in detail in ref.[15] and references therein.

6.3. ELECTROWEAK SPHALERONS AFTER THE PHASE TRANSITION

A necessary condition for successful electroweak baryogenesis is that electroweak processes violating baryon number switch off immediately after the phase transition. As we discussed in subsection 2.2, the rate of these processes in the asymmetric phase is predominantly determined by the sphaleron free energy. The sphaleron rate, with fermionic and bosonic determinants included, has been calculated in refs.[68, 106, 107], but for qualitative estimates a simple formula

$$\Gamma_{sph} \sim T \exp\left(-\frac{E_{sph}(T)}{T}\right), \tag{1.47}$$

is sufficient. Requiring that this rate be lower than the expansion rate at $T = T_c$, one obtains a constraint on the parameters of a theory with successful electroweak baryogenesis,

$$\frac{E_{sph}(T_c)}{T_c} > 45. \tag{1.48}$$

As we will now see, this constraint is not easy to satisfy.

Consider, as an example, the Minimal Standard Model. The only unknown parameter here is the mass of the Higgs boson, m_H, so the constraint (1.48) implies a constraint on m_H. As an estimate, let us make use of the one loop approximation described in the previous subsection. We have

$$\frac{E_{sph}(T_c)}{T_c} \propto \frac{m_W(T_c)}{\alpha_W T_c} \propto \frac{\phi(T_c)}{g_W T_c}.$$

The latter ratio is estimated from eq.(1.46),

$$\frac{E_{sph}(T_c)}{T_c} \propto \frac{g_W^2}{\lambda} \propto \frac{m_W^2}{m_H^2}.$$

We see that the exponential for the sphaleron rate is inversely proportional to m_H^2 so eq.(1.48) gives an upper bound on the Higgs boson mass. Numerically,

$$m_H < 45\,\text{GeV},$$

which is experimentally excluded. Hence, our simple estimates show that electroweak baryogenesis does not work in the Minimal Standard Model: the electroweak baryon number violating processes are in thermal equilibrium after the phase transition. More refined estimates (see, e.g., ref.[15] and references therein) do not chnage this conclusion.

The two Higgs doublet model has more freedom, and the results of refs. [108, 109] show that the constraint (1.48) can be satisfied there. The extensions of the Standard Model (supersymmetric or not), including scalar singlets, can also help [110, 111, 112].

According to refs. [113, 114, 115] the phase transition in the MSSM, in most part of the parameter space, occurs in the same way as it does in the MSM. Here the MSSM also fails in preserving the baryon asymmetry after the phase transition. However, in a series of papers [116, 117, 118] a specific portion of the parameter space of the MSSM, where electroweak baryogenesis is possible, has been found. What is most interesting is that quite strong constraints on the masses of the Higgs boson and squarks were derived.

In order to explain the idea of ref. [116] in the simplest way let us add to the Minimal Standard Model an $SU(2)$ singlet but color triplet scalar field (scalar quark) χ with the potential

$$U(\chi, \Phi) = -\frac{1}{2}m_H^2\Phi^\dagger\Phi + \lambda(\Phi^\dagger\Phi)^2 + m^2\chi^*\chi + 2h\chi^*\chi\Phi^\dagger\Phi + \lambda_s(\chi^*\chi)^2. \tag{1.49}$$

Assume now that the expectation value of the field χ is zero at all temperatures (this is possible at some particular choice of parameters). Then the contribution of this field to the effective high temperature Higgs potential is

$$-\frac{2 \cdot 3}{12\pi}(m^2(T) + h\phi^2)^{\frac{3}{2}}. \tag{1.50}$$

Now, if the effective high temperature mass $m(T)$ is small near the electroweak phase transition, $m^2(T_c) \simeq 0$, then this term increases the magnitude of the cubic coupling α in the effective potential (1.42), $\alpha \to 9g^3/(32\pi) + 3h^{\frac{3}{2}}/(2\pi)$. This, in turn, makes the phase transition more strongly first order, and the value $\phi(T_c)/T_c$ (see eq. (1.46)), crucial for the electroweak baryogenesis, increases.

In the case of the MSSM the role of the SU(2) singlet is played by the right handed light stop [116, 117, 118]. Its high temperature effective mass $m^2(T)$ contains two essential contributions. The first one is the soft supersymmetry breaking mass, and the second is a positive temperature contribution $\sim g_s^2T^2$, where g_s is the strong gauge coupling constant. To make the idea work, the soft SUSY breaking mass must be negative and approximately equal to the high temperature contribution at the critical temperature. Previously, the negative values of that mass have not been considered because of the danger of color breaking; the authors of [116, 117, 118] have shown that it is possible to satisfy simultaneously the requirements of the absence of color symmetry breaking, a strong enough first order phase transition together with experimental bounds on SUSY particles. The region of parameters allowing for electroweak baryogenesis requires that the Higgs mass be quite small, and the lightest stop mass be smaller than or of order of the top mass. This mass pattern, though somewhat contrived, does appear in realistic models incorporating supersymmetry breaking [119]. It is worth noting that this range of masses is accessible for experimental search at LEP2 and the Tevatron.

6.4. SOURCES OF CP-VIOLATION IN THE EW THEORY AND ITS EXTENSIONS

To produce the baryon asymmetry, the particle interactions must break C and CP symmetry. C symmetry is broken due to the chiral character of electroweak interactions. In the Minimal Standard Model, the conventional source of CP violation is that associated with Kobayashi–Maskawa (KM) mixing of quarks. The Yukawa interaction of quarks with the Higgs boson in the MSM has the following form,

$$\mathcal{L}_Y = \frac{g_W}{\sqrt{2}M_W}\{\bar{Q}_L K M_d D_R \phi + \bar{Q}_L M_u U_R \tilde{\phi} + \text{h.c.}\}, \tag{1.51}$$

where M_u and M_d are diagonal mass matrices of up and down quarks, K is the KM mixing matrix, containing one CP violating phase δ_{CP}. The MSM contains yet another source of CP-violation, associated with the QCD vacuum angle θ. It is constrained experimentally, $\theta < 10^{-9}$.

A popular extension of the MSM is a model with two Higgs doublets, φ_1 and φ_2. In order to suppress flavour changing neutral currents, the interaction of Higgs bosons with fermions is chosen in such a way that φ_1 couples only to right-handed up quarks while φ_2 couples only to down quarks. The other possibility is that φ_2 decouples from fermions completely and φ_1 gives masses to all the fermions. In addition to the KM mixing, this model contains CP violation in the Higgs sector. The scalar potential has the form [120]:

$$\begin{aligned} V = {} & \lambda_1(\varphi_1^\dagger\varphi_1 - v_1^2)^2 + \lambda_2(\varphi_2^\dagger\varphi_2 - v_2^2)^2 + \\ & \lambda_3[(\varphi_1^\dagger\varphi_1 - v_1^2) + (\varphi_2^\dagger\varphi_2 - v_2^2)]^2 + \\ & \lambda_4[(\varphi_1^\dagger\varphi_1)(\varphi_2^\dagger\varphi_2) - (\varphi_1^\dagger\varphi_2)(\varphi_2^\dagger\varphi_1)] + \\ & \lambda_5[\mathrm{Re}(\varphi_1^\dagger\varphi_2) - v_1 v_2\cos\xi]^2 + \lambda_6[\mathrm{Im}(\varphi_1^\dagger\varphi_2) - v_1 v_2 \sin\xi]^2, \end{aligned} \tag{1.52}$$

ξ being a CP-violating phase.

In the supersymmetric extensions of the Standard Model the Higgs potential is CP invariant and CP is violated by the soft supersymmetry breaking terms. In the simplest version of the MSSM there are two extra CP phases and the relevant interaction has the form [121], (for a review see ref. [122])

$$[\mu \hat{H}\hat{H}']_F + m_g[A(\hat{U}\xi_U\hat{Q}\hat{H} + \hat{D}\xi_D\hat{Q}\hat{H}' + \hat{E}\xi_E\hat{L}\hat{H}') + \mu_B\hat{H}\hat{H}']_A + \text{h.c.} \tag{1.53}$$

where $\hat{U}$, $\hat{D}$, $\hat{Q}$, $\hat{L}$, $\hat{E}, \hat{H}$ and $\hat{H}'$ are the quark, lepton and Higgs superfields, respectively; parameters μ and A are complex and flavour matrices ξ are assumed to be real, and m_g is the gravitino mass. In this model extra CP-violating phases appear in the vertices containing superpartners of ordinary particles.

6.5. UNIFORM SCALAR FIELDS

A good theoretical laboratory, allowing an understanding of physical processes giving rise to the charge asymmetry, is the consideration of the uniform but time dependent scalar fields. Probably, this situation is never realized, but this case is much simpler than that of the bubble wall propagation.

Suppose that we have a kind of spinodial decomposition phase transition, in which case the scalar field is initially near $\phi = 0$ and the system is in the symmetric phase. Sphaleron processes are in thermal equilibrium. Then the scalar field uniformly rolls down to the true vacuum, where the $SU(2) \times U(1)$ symmetry is broken and sphaleron processes are suppressed. The first rough estimates of the baryon asymmetry in this case were given in ref. [123], and a lot of work on this subject has been done in refs. [124, 125, 126, 127, 128, 129] and many others, for reviews see refs. [130, 131, 15] and references therein.

We will consider the main idea using the example of the two Higgs doublet model. Our scalar fields φ_1 and φ_2 are uniform in space but change from $\varphi = 0$ to $\varphi = \varphi_c$ during time Δt of the spinodial decomposition phase transition. Suppose that this time is small enough, $\Delta t/\tau_{top} \ll 1$ where τ_{top} is the typical time of top quark chirality flip (the top quark is most important since it has the largest Yukawa coupling constant). Then the top quark distribution has no time to adjust itself to the changing scalar field. So, it may be integrated out with the use of the equilibrium Matsubara technique. This was carried out in ref. [127] with the result that the effective action has the following form:

$$S_P = \mu N_{CS}, \tag{1.54}$$

where

$$\mu = -i\frac{7}{4}\zeta(3)\left(\frac{m_t}{\pi T}\right)^2 \frac{2}{v_1^2}\mathcal{O}(\varphi_1), \quad \mathcal{O}(\varphi_1) = (\varphi_1^\dagger \mathcal{D}_0\varphi_1 - (\mathcal{D}_0\varphi_1)^\dagger\varphi_1), \tag{1.55}$$

and m_t is the mass of the top quark, ζ is the Rieman ζ-function.

The effective bosonic action now breaks P and CP simultaneously, with CP violation in the scalar field potential, and P violation in the term (1.54). This allows us to generate the non-zero value of the topological charge Q, which is P- and CP-odd [5]. Note also that in more complicated models the operator μ may appear to be P- and CP-even (e.g., $\mu \sim \partial_0(\varphi^\dagger\varphi)$). In the latter case the effective action (1.54) itself breaks P and CP simultaneously. Estimates of the baryon asymmetry produced in this case were made in refs. [123, 128].

The zero-temperature bosonic effective action of this model also contains the odd parity term $\theta(x)q(x)$, where $\theta(x)$ is the relative phase of the scalar fields. This important fact was discovered by Turok and Zadrozny and applied to baryogenesis in ref. [125].

If Δt is not too small, $\Delta t \cdot m(T) \gg 1$, where $m(T)$ is the typical mass scale at high temperatures, then the term (1.54) may be considered as the chemical potential for the Chern–Simons number and the number density of fermions created during the transition is

$$n_B = n_f \int_0^\infty dt \Gamma_{sph}(t)\mu(t), \tag{1.56}$$

where Γ_{sph} is the time-dependent rate of the sphaleron transitions, and $n_f = 3$ is the number of fermion generations. The sphaleron rate Γ_{sph} rapidly decreases when the mass of the vector boson increases. A natural way to estimate it in the entire range of W-boson masses is to use eq. (1.47) for some $M_W > M_{crit}$ and (1.16) for the opposite case where $M_{crit} \simeq 7\alpha_W T$ is found[6] from the relation $\Gamma_{br} = \Gamma_{sym}$. In this approximation,

$$n_B \simeq n_f(\alpha_W T)^4 \mu(t^*), \tag{1.57}$$

where $\mu(t^*)$ is the chemical potential at the "freezing" time $m_W(t^*) = M_{crit}$. The asymmetry was estimated in ref. [127] and lately corrected in the detailed analysis of ref. [34] for the spinodial decomposition phase transition. The asymmetry reads

$$\Delta \sim \frac{45}{2\pi^2 N_{eff}} \kappa n_f \alpha_W^6 \sin^3 2\alpha \lambda_{CP} \frac{m_t^2 T_c^2}{v_1^3 v_2}, \tag{1.58}$$

[5]The presence of fermions is essential here. The purely bosonic tree action conserves P, and a net topological charge cannot be produced.

[6]A factor 2-3 instead of 7 was found in ref. [132] from other considerations. Clearly, these estimates are qualitative rather than quantitative.

with N_{eff} being the number of effectively massless degrees of freedom, $\lambda_{CP} = (\lambda_5 - \lambda_6)\sin 2\xi_0$, and

$$\tan\alpha = \frac{m_1^2(T_c)}{m_2^2(T_c)}, \tag{1.59}$$

with $m_i(T_c)$ being the temperature-dependent scalar masses at the moment of the phase transition (see, e.g. ref. [108]). An analogous dependence on the coupling constants was found in ref. [132]. In spite of the rather high power of the coupling constants, this estimate can give an asymmetry consistent with observations [7].

This consideration can be easily generalized to more complicated models. First, one calculates an effective bosonic action, which breaks P and CP and defines an effective potential for the CS number. Then, an estimate of the net production of fermions is given by (1.56).

An essential assumption in the estimates presented above is that the time of the phase transition is shorter than that of kinetic reactions. An "exact" solution to the problem in the opposite case for quite a specific situation has been suggested in ref. [129]. Since this example is instructive, we reproduce here the main idea of this paper, using correct coefficients from ref. [134].

Let us take again the two Higgs doublet model. For simplicity, we set all Yukawa couplings, except for that of the t-quark, to zero, i.e. the Yukawa interaction is assumed to be

$$L_Y = f_t \bar{Q}_3 U_3 \varphi_1. \tag{1.60}$$

Here Q_i are the left-handed fermion doublets, U_i, D_i are right-handed quark fields, and i is the generation index. We also neglect the reactions with quark chirality flip associated with strong sphalerons.

We put $\lambda_5 = \lambda_6 = 0$ in eq. (1.52); with these couplings the potential has an extra global U(1) symmetry, which is spontaneously broken. Let us consider this model in an external Higgs background of a special form, namely $\theta = \dot{\theta} t$ at $t > 0$ and $\theta = 0$ at $t < 0$, where θ is the Goldstone mode,

$$\tan\theta = \frac{\mathrm{Im}(\varphi_1^\dagger \varphi_2)}{\mathrm{Re}(\varphi_1^\dagger \varphi_2)}. \tag{1.61}$$

Suppose that at $t < 0$ the system was in thermal equilibrium and was charge symmetric. We want to determine the baryon number of the system at $t \to \infty$. The density matrix $\rho(t)$ of the system obeys the

[7]One can obtain a similar estimate for the asymmetry from different consideration [123, 125, 126] dealing with non-perturbative fluctuations of the Chern-Simons number in the symmetric phase [133].

Liouville equation

$$i\frac{\partial \rho(t)}{dt} = [H(t), \rho(t)], \tag{1.62}$$

where $H(t)$ is the time-dependent Hamiltonian of the system in the background field. Now, one can make an anomaly free hypercharge rotation of the fermion fields in such a way that the time dependence disappears from the Yukawa coupling (1.60). Because of the global U(1) symmetry this converts the time dependent Hamiltonian to a time independent one, $H(t) \to H_{eff} = H - \dot{\theta} Y_F$ where Y_F is the fermionic hypercharge operator,

$$Y_F = \sum_{i=1}^{3} \left[\frac{1}{3}\bar{Q}_i\gamma_0 Q_i + \frac{4}{3}\bar{U}_i\gamma_0 U_i - \frac{2}{3}\bar{D}_i\gamma_0 D_i - \bar{L}_i\gamma_0 L_i - 2\bar{E}_i\gamma_0 E_i \right]. \tag{1.63}$$

At $t \to \infty$ the system must be in thermal equilibrium, $\frac{\partial \rho(t)}{dt} = 0$. Since in the new representation the Hamiltonian is time independent, the density matrix is

$$\rho(\infty) = \frac{1}{Z} \exp\left[-\frac{1}{T}(H_{eff} - \mu_i X_i) \right], \tag{1.64}$$

where X_i is a *complete* set of conserved charges (operators commuting with the Hamiltonian). Their average must be equal to zero. This requirement fixes the chemical potentials μ_i and allows the unambiguous determination of the baryon number of the system. A complete list of the conserved charges can be found in ref. [134], and we quote here the final result for the baryon number only,

$$\langle B \rangle = \frac{n_s}{6 + 11 n_s} T^2 \dot{\theta} (1 + O(m_t^2/t^2)), \tag{1.65}$$

where $n_s = 2$ is the number of the scalar doublets.

The result (1.65) is quite amazing. It does not contain the Yukawa coupling constant, the scalar vacuum expectation value, or the rate of sphaleron transitions. One might even think that it is wrong, since if, say, Γ_{sph} is zero, then, obviously, one must have $\langle B \rangle = 0$. Nevertheless, it is correct. The key point is that the time at which the asymptotic value of the baryon number is reached tends to infinity when Γ_{sph} tends to zero. Many conclusions based on the straightforward analysis of the perturbation theory break down at large times, when the application of the kinetic theory is essential, and this is one of the examples. For typical values of the parameters, the top quark chirality equilibration time is $\tau_t \sim 30/T$, and the B non-conservation time is $\tau_{sph} \sim 10^5/T$; the result (1.65) is valid only for $t > \tau_{sph}$. The discussion of intermediate time $\tau_t \ll t \ll \tau_{sph}$ is contained in ref. [129, 134].

We conclude this discussion by remarking that the use of a U(1) global symmetry was essential in the derivation of eq. (1.65). Without it the hypercharge rotation would not remove the time dependence from the Hamiltonian, and the solution of the Liouville equation could not be found so straightforwardly. This is discussed in more detail in refs. [132, 144].

6.6. ASYMMETRY FROM FERMION–DOMAIN WALL INTERACTIONS

In reality, though, the phase transition goes through the bubble nucleation rather than as spinodial decomposition. This is an additional challenge, since the baryon number (or, in general, asymmetries in particle number densities) can now be distributed in a non-uniform way and depend on the distance from the domain wall. Correspondingly, the analysis of the kinetic equations is much more complicated.

Two different cases are usually considered, depending on the relation between the mean free paths of particles and domain wall thickness. The physics of thick wall baryogenesis was originally considered in refs. [127, 128] and has much in common with the quasi-adiabatic case of uniform fields discussed in the previous subsection. P or CP non-invariant interaction of fermions with the moving domain wall together with CP-breaking scalar dynamics induces P- and CP-odd terms in the bosonic effective action, which bias the sphaleron transitions inside the domain wall. The excess of quarks generated in this way is absorbed then by the expanding bubble. (Another, equivalent way to say this [129] is that the particle densities of fermions gradually adapt to the scalar background, changing in space and time in such a way that an excess of right quarks and left antiquarks is created. Left antiquarks are destroyed by the sphaleron reactions while the right fermion number is intact and is converted into baryon number at the end.)

A nice physical picture of thin domain wall baryogenesis was suggested by Cohen, Kaplan and Nelson [135, 136, 137]. Since the masses of fermions are different in the symmetric and broken phases, they scatter on domain walls (are reflected or transmitted). CP violation manifests itself in different reflection coefficients for particles and antiparticles. So, the moving domain wall acts like a separator for different types of fermion numbers, filling the bubble with fermions and outer space with antifermions (or vice versa, depending on the sign of CP violation). Of course, the interactions of fermions with domain wall conserve the fermion number, i.e. the number of fermions flying into the broken phase is equal to the number of anti-fermions moving into the symmetric phase. Antifermions, injected into the symmetric phase, participate in the anomalous reactions that change fermion number, while fermions injected into the broken phase do not. As a result, non-zero baryon and lepton asymmetries are established in the broken phase.

Clearly, reliable calculations of the effect in realistic theories are quite complicated because of the large number of different particle species participating in interactions. Moreover, a number of effects, discussed above, should be taken into account. A number of papers is devoted to the study of the origin of the CP-violating fermion currents and their propagation in front of domain walls [138, 139, 140, 141]; the most recent (and probably most elaborate) treatment can be found in [142, 143, 144, 145, 146].

Below we discuss the qualitative features of the domain wall baryogenesis. Our consideration is by no means complete, and the reader may consult the original papers for details.

At sufficiently small velocities v of the domain walls we can divide the problem in two parts. The first one is the microscopic calculation of various fermion currents *at the* domain wall. The second one is the consideration of the diffusion of the particle number densities in front of the wall and their dissipation in different processes.

We begin with the first part [135, 136, 137]. Let us ignore for a moment any high temperature effects. The simplest case is that of the thin domain wall moving with some constant velocity v. Let us choose the rest frame of the wall and consider scattering of fermions on it. For example, left-handed fermions incident in the symmetric phase may be reflected back to the symmetric phase as right-handed fermions (because of the spin conservation) or can be transmitted through. The transmission and reflection coefficients r_{ij} (i is the label of an incident fermion and j is that of the final state) can be found from the Dirac equation:

$$\begin{pmatrix} \omega + i\frac{\partial}{\partial x} & M \\ M^\dagger & \omega - i\frac{\partial}{\partial x} \end{pmatrix} \cdot \begin{pmatrix} L \\ R \end{pmatrix} = 0 \tag{1.66}$$

with appropriate boundary conditions. Here L and R correspond to up and down components of two-dimensional Weyl spinors. The x dependent matrix M is complex, giving rise to CP-violation. In general, r_{ij} for particles are different from $\bar{r}_{ij}$ for anti-particles. This leads to non-zero fermionic currents:

$$\langle J_i \rangle = \int \frac{d\omega k_{||} dk_{||}}{(2\pi)^2} (n_F(\omega_+) - n_F(\omega_-)) \left[r^\dagger r - \bar{r}^\dagger \bar{r} \right]_i , \tag{1.67}$$

where n_F^i is the Fermi distribution for the incident particles, $\omega_\pm = \omega \pm vp_t$, p_t and $p_{||}$ are the momenta of fermions tangential and parallel, respectively, to the domain wall. This expression vanishes if the domain wall is at rest ($v = 0$) or if there is no CP violation. The total baryonic current $J_{CP} = J_L + J_R$ which results from the solution of the Dirac equation (1.66) vanishes, but the currents of left-handed (J_L) and right-handed (J_R) fermions are non-zero.

The construction of the Dirac equation for quasiparticles, accounting for leading high temperature effects, was done in ref. [147, 148], where generalized expressions for the particle currents can be found. The major qualitative effect of high temperature corrections is that the currents of left-handed and right-handed fermions do not compensate each other and the total baryon current is produced. Physically, this happens because left-handed particles participate in the weak interactions but right-handed particles do not, and $J_{CP} \sim \alpha_W J_L$.

The thin wall description of the fermion scattering is applicable only if the mean free path of fermions at high temperatures is much larger than the domain wall thickness. This allows us to use distribution functions of fermions undisturbed by the domain wall and impose ordinary boundary conditions for the scattering problem at spatial infinity. The thick wall case (the mean free path is small compared with the thickness) is much more complicated. Clearly, the scattering description is not adequate in that case. The physical phenomenon to be taken into account is the modification of the particle distributions across the domain wall. A number of interesting effects arising in the latter situation are discussed in ref. [143, 144, 142].

The second problem is the particle transport. Several approaches were applied to the consideration of it. The first one is that of Monte Carlo simulations of the injected flux of particles [137], the second one uses diffusion equations [148, 144, 145, 142, 146]. Limitations of the diffusion approximation were considered in ref. [149]. In the discussion below we closely follow ref. [148] where the analytical approximation to the problem was constructed for a simple case.

Given the flavour and chirality structure of the fermionic currents, the diffusion equations should be written for all particle species. Fermions participate in many processes on both sides of the wall with different time scales. In order to understand what the relevant time scales are, let us consider the fate of a particle after it has been reflected from the domain wall towards the unbroken phase. Roughly, its typical distance from the bubble wall is given by $\sqrt{Dt} - vt$, D is the diffusion coefficient. The first term describes the random walk of the particle in the rest frame of the plasma and the second term describes the motion of the bubble wall. This particle will be trapped by the bubble after the time interval $t_D \sim D/v^2$, so that all processes with characteristic time $\tau < t_D$ must be taken into account. The examples of the relevant processes include B-violation, the elastic scattering of quarks and gluons, chirality flipping transitions of heavy quarks, strong sphalerons.

In order to get a better feeling of the physics involved, consider the simplest case when the total baryonic current originated from CP non-invariant interactions is not zero and neglect all processes besides the elastic scattering of fermions and anomalous B and L non-conservation. Let us take a planar domain wall which moves through the plasma with sufficiently small velocity v (we shall see below how small it should be in order that this consideration works). We take a reference frame associated with the domain wall; let the broken phase be to the right and the symmetric one be to the left, and x be the distance from the domain wall. Assume that the thickness of the domain wall is small enough (again, we shall see below what this means). We denote by $n_B(x,t)$ and $n_L(x,t)$ the densities of baryon and lepton numbers in the rest frame of the wall. The diffusion equations for the broken phase, where sphalerons do not operate, are

$$\frac{\partial}{\partial t}\begin{pmatrix} n_B \\ n_L \end{pmatrix} = \begin{pmatrix} D_B\frac{\partial^2}{\partial x^2} - v\frac{\partial}{\partial x} & 0 \\ 0 & D_L\frac{\partial^2}{\partial x^2} - v\frac{\partial}{\partial x} \end{pmatrix}\begin{pmatrix} n_B \\ n_L \end{pmatrix}. \tag{1.68}$$

For $x < 0$ we have

$$\frac{\partial}{\partial t}\begin{pmatrix} n_B \\ n_L \end{pmatrix} = \begin{pmatrix} D_B\frac{\partial^2}{\partial x^2} - v\frac{\partial}{\partial x} - \frac{3}{2}\Gamma & -\Gamma \\ -\frac{3}{2}\Gamma & D_L\frac{\partial^2}{\partial x^2} - v\frac{\partial}{\partial x} - \Gamma \end{pmatrix}\begin{pmatrix} n_B \\ n_L \end{pmatrix}, \tag{1.69}$$

where $\Gamma = 9\Gamma_{sph}/T^3$, D_B and D_L are diffusion constants for quarks and leptons respectively. An estimate of these gives [144, 145] $D_B \sim \frac{6}{T}$, $D_L \sim \frac{100}{T}$.

We are looking for a steady state (time independent) solution to these equations. In the broken phase the only solution consistent with the boundary conditions is constant density,

$$n_B = n_L = \text{const} = B_+. \tag{1.70}$$

In the symmetric phase, the solution is a combination of dying exponentials. We present it in two limiting cases. The first one deals with "large" velocities,

$$\rho = 3D_B\Gamma/v^2 \ll 1.$$

Physically, this corresponds to the situation when an extra antibaryon, injected into the symmetric phase, is exposed to the B-violating reactions for short times $t \sim D_B/v^2$, before it is trapped by the moving domain wall. One finds [148]

$$n_B = C_1 \exp\left(\frac{vx}{D_B}\right), \quad n_L = C_2 \exp\left(\frac{vx}{D_L}\right), \tag{1.71}$$

corresponding to the diffusion of quarks and leptons to the distances D_B/v and D_L/v, respectively. In the opposite case of low velocity, the transitions from quarks to leptons due to sphaleron processes are essential, and the solution reads

$$n_B = C_3 \exp\left(\frac{3vx}{5D_L}\right) + C_4 \exp\left(\sqrt{\frac{5\Gamma}{2D_B}}x\right), \tag{1.72}$$

$$n_L = -\frac{3}{2}C_3 \exp\left(\frac{vx}{D_L}\right) + C_4 \frac{D_B}{D_L} \exp\left(\sqrt{\frac{5\Gamma}{2D_B}}x\right).$$

One of the requirements for the validity of the diffusion approximation is that the diffusion tail (the shortest one is for the quarks) is much longer than the domain wall thickness l, namely $l \ll D_B/v$.

The constants C_1–C_4 can be determined from the matching conditions at the domain wall. If we denote by J_{CP} the total CP-odd baryonic current originating from interactions of quarks with the domain wall and assume that CP asymmetry in the leptonic current is zero, then

$$B_+ = \frac{12}{5} J_{CP} f_{sph}(\rho), \tag{1.73}$$

where $f_{sph}(\rho) = 1$ for $\rho \gg 1$ and $f_{sph}(\rho) = \frac{5}{6}\rho$ for $\rho \ll 1$.

The asymmetry inside the bubble has an interesting velocity dependence. If $J_{CP} \sim v$, as in the quantum-mechanical consideration of the thin wall case, then the maximum asymmetry is produced at $\rho \sim 1$, i.e. $v \sim \sqrt{3\Gamma D_B} \simeq 0.01$. It is worth noting that these small velocities are quite possible at the final stage of the phase transition if the Universe is reheated up to the critical temperature. The analysis of the consequences of this scenario can be found in ref. [150]. For a more realistic thick wall case one effectively has $J_{CP} \sim v^2$ [143, 144, 145], and the asymmetry is velocity independent. The same conclusion has been reached also in ref. [142].

In the realistic case of many particle species this consideration must be generalized. Instead of the CP-violating flavour independent baryonic current considered above, many CP-odd currents appear, resembling the flavour dependent interaction of fermions with the domain wall. The left- and right-handed currents must be distinguished, since particles of different chiralities have different interactions with the heat bath and sphalerons. Quantitatively, the results are model dependent. For example, in some schemes the lepton interactions with domain walls produce more asymmetry than quark interactions [151]. Serious investigations of realistic models have been carried out in several papers [142, 145, 144, 146], and we refer to them for more detail.

In supersymmetric extensions of the Standard Model, such as the MSSM, fermions other than ordinary quarks and leptons may play a crucial role in electroweak baryogenesis. Importantly, interactions of charginos and neutralinos with the moving bubble walls may produce a net excess in the higgsino number which then gets reprocessed into the baryon asymmetry. This mechanism has been studied in detail in refs. [152, 117, 153, 154, 155, 156] (for a review see ref. [16]) with the result that the production of an acceptable baryon asymmetry is possible in a fairly large region of the parameter space of the MSSM. The relevant source of CP-violation in this case are the new CP-phases appearing in the MSSM according to eq. (1.53); the constraints on these phases coming from the bounds on the electric dipole moments of neutron and electron do not rule out this mechanism.

7. CONCLUSIONS

In these lectures we discussed only a few of the proposed mechanisms of baryogenesis. There exist a number of other proposals that attract considerable attention. Among them is the Affleck–Dine mechanism [30] which is most naturally implemented in supersymmetric theories with flat directions of the scalar potential; the mechanisms that assume the existence of relatively light right-handed neutrinos [157, 158]; the scenarios invoking topological defects [159, 160, 161, 162, 163], electroweak baryogenesis due to preheating after inflation [164, 165, 166, 167], etc.

Shall we ever be able to tell what mechanism of baryogenesis actually worked in our Universe? Different scenarios have different degrees of testability, depending on the energy scale of the physics involved. There is little doubt that electroweak baryogenesis will be tested at colliders of the next generation. The firm proof of neutrino oscillations, the experimental study of neutrino masses and mixings will, to a certain extent, support or disfavor leptogenesis. The GUT mechanism is most difficult to test, and it is not likely that even indirect evidence for or against the GUT scenario will be obtained in the foreseeable future. With luck, the time to uncover the origin of baryons in our Universe is a matter of a decade, if not less; with no luck, the baryon asymmetry will remain an outstanding mystery, forever.

References

[1] A. D. Sakharov. *JETP Lett.*, 6:24, 1967.

[2] J.H. Christenson, J.W. Cronin, V.L. Fitch, and R. Turlay. *Phys. Rev. Lett.*, 13:138, 1964.

[3] G. Gamov. *Phys. Rev.*, 70:572, 1946.

[4] A.A. Penzias and R.W. Wilson. *Astrophys. J.*, 142:419, 1965.

[5] V.A. Kuzmin. *Pisma ZhETF*, 13:335, 1970.

[6] A.Yu. Ignatiev, N.V. Krasnikov, V.A. Kuzmin, and A.N. Tavkhelidze. *Phys. Lett.*, 76B:436, 1978.

[7] M. Yoshimura. *Phys. Rev. Lett.*, 41:281, 1978;42:476(E)1979.

[8] S. Dimopoulos and L. Susskind. *Phys. Rev.*, D18:4500, 1978.

[9] S. Weinberg. *Phys. Rev. Lett.*, 42:850, 1979.

[10] J. Ellis, M. Gaillard, and D. Nanopoulos. *Phys. Lett.*, B80:360, 1979.

[11] A.Yu. Ignatiev, V.A. Kuzmin, and M.E. Shaposhnikov. *Phys. Lett.*, 87B:114, 1979.

[12] A.D. Dolgov and Ya.B. Zeldovich. *Rev. Mod. Phys.*, 53:1, 1981.

[13] A.D. Dolgov. *Physics Reports*, 222:309, 1992.

[14] E.W. Kolb and M.S. Turner. *The Early Universe.* Addison-Wesley, Reading, MA, 1990.

[15] V. A. Rubakov and M. E. Shaposhnikov. *Uspekhi Fiz. Nauk*, 166:493, 1996.

[16] A. Riotto and M. Trodden, *Rev. Nucl. Part. Sci.*, 49:35, 1999.

[17] J.C. Pati and A. Salam. *Phys. Rev.*, D8:1240, 1973.

[18] J.C. Pati and A. Salam. *Phys. Rev. Lett.*, 31:661, 1973.

[19] H. Georgi and S.L. Glashow. *Phys. Rev. Lett.*, 32:438, 1974.

[20] H. Georgi, H.R. Quinn, and S. Weinberg. *Phys. Rev. Lett.*, 33:451, 1974.

[21] H. Fritzsch and P. Minkowski. *Ann. Phys. (NY)*, 93:193, 1975.

[22] G. 't Hooft. *Phys. Rev. Lett.*, 37:8, 1976.

[23] G. 't Hooft. *Phys. Rev.*, D14:3432, 1976.

[24] A.A Belavin, A.M. Polyakov, A.S. Schwarz, and Yu.S. Tyupkin. *Phys. Lett.*, 59B:85, 1975.

[25] C. G. Callan, R.F. Dashen, and D.J. Gross. *Phys. Lett.*, 63B:334, 1976.

[26] R. Jackiw and C. Rebbi. *Phys. Rev. Lett.*, 37:172, 1976.

[27] V. A. Kuzmin, V. A. Rubakov, and M. E. Shaposhnikov. *Phys. Lett.*, 155B:36, 1985.

[28] A. G. Cohen, A. De Rujula and S. L. Glashow. *Astrophys. J.*, 495:539, 1998.

[29] M. Fukugita and T. Yanagida. *Phys. Lett.*, 174B:45, 1986.

[30] I. Affleck and M. Dine. *Nucl. Phys.*, B249:361, 1985.

[31] M. Dine, L. Randall, and S. Thomas. *hep-ph*, (9507453), 1995.

[32] S.M. Barr and William J. Marciano. *BNL-41939*, 1988.

[33] S.M. Barr. *Phys. Rev.*, D47:2025, 1993.

[34] A.M. Kazarian, S.V. Kuzmin, and M.E. Shaposhnikov. *Phys. Lett.*, B276:131, 1992.

[35] E. Witten. *Phys. Lett.*, B117:324, 1982.

[36] S. Adler. *Phys. Rev.*, 177:2426, 1969.

[37] J.S. Bell and R.Jackiw. *Nuovo Cim.*, 51:47, 1969.

[38] W.A. Bardeen. *Phys. Rev.*, 184:1841, 1969.

[39] T. Akiba, H. Kikuchi and T. Yanagida. *Phys. Rev.*, D38:1937, 1988.

[40] D. Diakonov, M. Polyakov, P. Sieber, J. Schaldach and K. Goeke. *Phys. Rev.* D49:6864, 1994.

[41] I. Affleck. *Nucl. Phys.*, B191:455, 1981.

[42] A.I.Vainshtein, V.I. Zakharov, V.A. Novikov, and M.A. Shifman. *Usp. Fiz. Nauk*, 136:553, 1982.

[43] N.S. Manton. *Phys. Rev.*, D28:2019, 1983.

[44] F.R. Klinkhamer and N.S. Manton. *Phys. Rev.*, D30:2212, 1984.

[45] R. Dashen, B. Hasslacher, and A. Neveu. *Phys. Rev.*, D10:4138, 1974.

[46] V. Soni. *Phys. Lett.*, 93B:101, 1980.

[47] J. Boguta. *Phys. Rev. Lett.*, 50:148, 1983.

[48] P. Forgacs and Z. Horvath. *Phys. Lett.*, 138B:397, 1984.

[49] L. Yaffe. *Phys. Rev.*, D40:3463, 1989.

[50] Y. Brihaye and J. Kunz. *Mod. Phys. Lett.*, A4:2723, 1989.

[51] N. Christ. *Phys. Rev.*, D21:1591, 1980.

[52] C.G. Callan, R.F. Dashen, and D.J. Gross. *Phys. Rev.*, D17:2717, 1978.

[53] M.F. Atiyah, V. Patodi, and I.M. Singer. *Bull. London Math. Soc.*, 5:229, 1973.

[54] N.V. Krasnikov, V.A. Rubakov, and V.F. Tokarev. *J. Phys.*, A12:L343, 1979.

[55] A. Ringwald. *Phys. Lett.*, B213:61, 1988.

[56] A.A. Anselm and A.A. Johansen. *Nucl. Phys.*, B412:553, 1994.

[57] V.A. Rubakov. *Nucl. Phys.*, B256:509, 1985.

[58] J. Kunz and Y. Brihaye. *Phys. Rev.*, D50:1051, 1994.

[59] M. Axenides, A. Johansen and H.B. Nielsen. *Nucl. Phys.*, B414:53, 1994.

[60] F.R. Klinkhamer and R. Laterveer. *Z. Phys.*, C53:247, 1992.

[61] Y. Brihaye and J. Kunz. *Phys. Rev.*, D47:4789, 1993.

[62] D.A. Kirzhnitz. *JETP Lett.*, 15:529, 1972.

[63] D.A. Kirzhnitz and A.D. Linde. *Phys. Lett.*, 72B:471, 1972.

[64] L. Dolan and R.Jackiw. *Phys. Rev.*, D9:3320, 1974.

[65] S. Weinberg. *Phys. Rev.*, D9:3357, 1974.

[66] K. Kajantie, M. Laine, K. Rummukainen and M. Shaposhnikov. *Phys. Rev. Lett.*, 77:2887, 1996.

[67] F. Karsch, T. Neuhaus, A. Patkos and J. Rank. *Nucl. Phys. Proc. Suppl.*, 53:623, 1997.

[68] P. Arnold and L. McLerran. *Phys. Rev.*, D36:581, 1987.

[69] S.Yu. Khlebnikov and M.E. Shaposhnikov. *Nucl. Phys.*, B308:885, 1988.

[70] P. Arnold, D. Son and L. G. Yaffe. *Phys. Rev.*, D55:6264, 1997.

[71] D. Bodeker. *Phys. Lett.*, B426:351, 1998.

[72] P. Arnold and L. G. Yaffe. "Non-perturbative dynamics of hot non-Abelian gauge fields: Beyond leading log approximation," hep-ph/9912305.

[73] D. Bodeker, G. D. Moore and K. Rummukainen. *Nucl. Phys. Proc. Suppl.*, 83:583, 2000.

[74] M. Yu. Khlopov and A. D. Linde. *Phys. Lett.*, B138:265, 1984.

[75] J. Ellis, J. E. Kim and D. V. Nanopoulos. *Phys. Lett.*, B145:181, 1984.

[76] J. Ellis, D. V. Nanopoulos and S. Sarkar. *Nucl. Phys.* , B259:175, 1985.

[77] L. Kofman, A. Linde and A. A. Starobinsky. *Phys. Rev. Lett.*, 73:3195, 1994.

[78] I. I. Tkachev. *Phys. Lett.*, B376:35, 1996.

[79] E. W. Kolb, A. Linde and A. Riotto. *Phys. Rev. Lett.*, 77:4290, 1996.

[80] E. W. Kolb, A. Riotto and I. I. Tkachev. *Phys. Lett.*, B423:348, 1998.

[81] V.A. Kuzmin, V.A. Rubakov, and M.E. Shaposhnikov. *Phys. Lett.*, 191B:171, 1987.

[82] J.A. Harvey and M.S. Turner. *Phys. Rev.*, D42:3344, 1990.

[83] H. Dreiner and G.G. Ross. *Nucl. Phys.*, B410:188, 1993.

[84] M. Fukugita and T. Yanagida. *Phys. Rev.*, D42:1285, 1990.

[85] A.E. Nelson and S.M. Barr. *Phys. Lett.*, B246:141, 1990.

[86] W. Fischler, G.F. Giudice, R.G. Leigh, and S. Paban. *Phys. Lett.*, B258:45, 1991.

[87] L. E. Ibanez and F. Quevedo. *Phys. Lett.*, B283:261, 1992.

[88] B.A. Campbell, S. Davidson, J. Ellis, and K. A. Olive. *Phys. Lett.*, B297:118, 1992.

[89] B.A. Campbell, S. Davidson, J. Ellis, and K. A. Olive. *Astropart. Phys.*, 1:77, 1992.

[90] J. M. Cline, K. Kainulainen, and K. A. Olive. *Phys. Rev. Lett.*, 71:2372, 1993.

[91] J. M. Cline, K. Kainulainen, and K. A. Olive. *Phys. Rev.*, D49:6394, 1994.

[92] J. M. Cline, K. Kainulainen, and K. A. Olive. *Astropart. Phys.*, 1:387, 1993.

[93] K. Kainulainen. *Nucl. Phys. B (Proc. Suppl.)*, 43:291, 1995.

[94] W. Buchmuller. 'Baryogenesis above the Fermi scale," *hep-ph/9812447*, 1998.

[95] W. Fischler, G. F. Giudice, R. G. Leigh, S. Paban *Phys. Lett.*, B258:45, 1991.

[96] L. Covi, E. Roulet, F. Vissani. *Phys. Lett.*, B384:169, 1996.

[97] M. Flanz, E. A. Paschos, U. Sarkar. *Phys. Lett.*, B345:248, 1995.

[98] W. Buchmüller, M. Plümacher. *Phys. Lett.*, B431:354, 1998.

[99] E. W. Kolb, S. Wolfram. *Nucl. Phys.*, 172B:224, 1980.

[100] W. Buchmüller, M. Plümacher. *Phys. Lett.*, B389:73, 1996.

[101] S. P. Mikheev and A. Y. Smirnov. *Sov. J. Nucl. Phys.*, 42:913, 1985;
L. Wolfenstein. *Phys. Rev.*, D17:2369, 1978.

[102] N. Hata, P. Langacker. *Phys. Rev.*, D56:6107, 1997.

[103] Super-Kamiokande Collaboration, Y. Fukuda et al. *Phys. Rev. Lett.*, 81:1562, 1998;
Super-Kamiokande Collaboration, S. Fukuda et al.. "Tau neutrinos favored over sterile neutrinos in atmospheric muon neutrino oscillations," *hep-ex/0009001*, 2000.

[104] W. Buchmuller and M. Plumacher. "Neutrino masses and the baryon asymmetry," *hep-ph/0007176*, 2000.

[105] M. Laine and M. Shaposhnikov. *Phys. Lett.*, B463:280, 1999.

[106] D. Diakonov, M. Polyakov, P. Pobylitsa, P. Sieber, J. Schaldach, and K. Goeke. *Phys. Lett.*, B336:457–463, 1994.

[107] D. Diakonov, M. Polyakov, P. Sieber, J. Schaldach, and K. Goeke. *Phys. Rev.*, D53:3366, 1996.

[108] A.I. Bochkarev, S.V. Kuzmin, and M.E. Shaposhnikov. *Phys. Rev.*, D43:369, 1991.

[109] N. Turok and J. Zadrozny. *Nucl. Phys.*, B369:729, 1992.

[110] G.W. Anderson and L.J. Hall. *Phys. Rev.*, D45:2685, 1992.

[111] J. Choi and R.R. Volkas. *Phys. Lett.*, B317:385, 1993.

[112] J.R. Espinosa and M. Quiros. *Phys. Lett.*, B305:98, 1993.

[113] G. F. Giudice. *Phys. Rev.*, D45:3177, 1992.

[114] J.R. Espinosa, M. Quiros, and F. Zwirner. *Phys. Lett.*, B307:106, 1993.

[115] A. Brignole, J.R. Espinosa, M. Quiros, and F. Zwirner. *Phys. Lett.*, B324:181, 1994.

[116] M. Carena, M. Quiros and C. E. Wagner *Phys. Lett.*, B380:81, 1996.

[117] M. Carena, M. Quiros, A. Riotto, I. Vilja and C. E. Wagner. *Nucl. Phys.*, B503:387, 1997.

[118] M. Carena, M. Quiros and C. E. Wagner *Nucl. Phys.*, B524:3, 1998.

[119] D. S. Gorbunov. *Mod. Phys. Lett.*, A15:207, 2000.

[120] J.F. Gunion and Howard E. Haber. *Nucl. Phys.*, B272:1, 1986.

[121] J. Ellis, S. Ferrara, and D.V. Nanopoulos. *Phys. Lett.*, 114B:231, 1982.

[122] H.P. Nilles. *Phys. Rept.*, 110:1, 1984.

[123] M.E. Shaposhnikov. *Nucl. Phys.*, B299:797, 1988.

[124] L. McLerran. *Phys. Rev. Lett.*, 62:1075, 1989.

[125] N. Turok and J. Zadrozny. *Phys. Rev. Lett.*, 65:2331, 1990.

[126] N. Turok and J. Zadrozny. *Nucl. Phys.*, B358:471, 1991.

[127] L. McLerran, M. Shaposhnikov, N. Turok, and M. Voloshin. *Phys. Lett.*, B256:451, 1991.

[128] M. Dine, P. Huet, R. Singleton, and L. Susskind. *Phys. Lett.*, 257B:351, 1991.

[129] A.G. Cohen, D.B. Kaplan, and A.E. Nelson. *Phys. Lett.*, B263:86, 1991.

[130] N. Turok. *In: Perspectives in Higgs Physics.* World Scientific, Singapore, 1992.

[131] A.G. Cohen, D.B. Kaplan, and A.E. Nelson. *Annu. Rev. Nucl. Part. Sci.*, 43:27, 1993.

[132] M. Dine and S. Thomas. *Phys. Lett.*, B328:73, 1994.

[133] J. Ambjørn, M.L. Laursen, and M.E. Shaposhnikov. *Nucl. Phys.*, B316:483, 1989.

[134] G.F. Giudice and M. Shaposhnikov. *Phys. Lett.*, B326:118, 1994.

[135] A.G. Cohen, D.B. Kaplan, and A.E. Nelson. *Phys. Lett.*, B245:561, 1990.

[136] A.G. Cohen, D.B. Kaplan, and A.E. Nelson. *Nucl. Phys.*, B349:727, 1991.

[137] A.E. Nelson, D.B. Kaplan, and A.G. Cohen. *Nucl. Phys.*, B373:453, 1992.

[138] A. Ayala, J. Jalilian-Marian, L. McLerran, and A. P. Vischer. *Phys. Rev.*, D49:5559, 1994.

[139] G. R. Farrar and J.W. McIntosh. *Phys. Rev.*, D51:5889, 1995.

[140] K. Funakubo, A. Kakuto, S. Otsuki, K. Takenaga, and F. Toyoda. *Phys. Rev.*, D50:1105, 1994.

[141] S.Yu. Khlebnikov. *Phys. Rev.*, D52:702, 1995.

[142] A.G. Cohen, D.B. Kaplan, and A.E. Nelson. *Phys. Lett.*, B336:41, 1994.

[143] M. Joyce, T. Prokopec, and N. Turok. *Phys. Rev. Lett.*, 75:1695, 1995.

[144] M. Joyce, T. Prokopec, and N. Turok. *Phys. Rev.*, D53:2958, 1996.

[145] M. Joyce, T. Prokopec, and N. Turok. *Phys. Rev.*, D53:2930, 1996.

[146] J. M. Cline, K. Kainulainen, and A. P. Vischer. *Phys. Rev.*, D54:2451, 1996.

[147] G.R. Farrar and M.E. Shaposhnikov. *Phys. Rev. Lett.*, 70:2833, 1993.

[148] G.R. Farrar and M.E. Shaposhnikov. *Phys. Rev.*, D50:774, 1994.

[149] J. M. Cline. *Phys. Lett.*, B338:263, 1994.

[150] A. F. Heckler. *Phys. Rev.*, D51:405, 1995.

[151] M. Joyce, T. Prokopec, and N. Turok. *Phys. Lett.*, B338:269, 1994.

[152] P. Huet and A. E. Nelson, *Phys. Rev.*, D53:4578, 1996.

[153] A. Riotto, *Nucl. Phys.*, B518:339, 1998.

[154] A. Riotto, *Phys. Rev.*, D58:095009, 1998.

[155] J. M. Cline, M. Joyce and K. Kainulainen, *Phys. Lett.*, B417:79, 1998.

[156] J. M. Cline, M. Joyce and K. Kainulainen, *JHEP*, 0007:018, 2000.

[157] E. K. Akhmedov, V. A. Rubakov and A. Y. Smirnov, *Phys. Rev. Lett.*, 81:1359, 1998.

[158] K. Dick, M. Lindner, M. Ratz and D. Wright. *Phys. Rev. Lett.*, 84:4039, 2000.

[159] R. H. Brandenberger, A.-C. Davis, and A. M. Matheson. *Phys. Lett.*, B218:304, 1989.

[160] R. H. Brandenberger, A.-C. Davis, and M. Hindmarsh. *Phys. Lett.*, B263:239, 1991.

[161] R. H. Brandenberger and A.-C. Davis. *Phys. Lett.*, B308:79, 1993.

[162] R. Brandenberger, A.-C. Davis, and M. Trodden. *Phys. Lett.*, B335:123, 1994.

[163] M. Trodden, A.-C. Davis, and R. Brandenberger. *Phys. Lett.*, B349:131, 1995.

[164] J. Garcia-Bellido, D. Y. Grigoriev, A. Kusenko and M. Shaposhnikov. *Phys. Rev.*, D60:123504, 1999.

[165] M. Trodden. "Making baryons below the electroweak scale," *hep-ph/0001026*, 2000.

[166] J. M. Cornwall and A. Kusenko. *Phys. Rev.*, D61:103510, 2000.

[167] S. Davidson, M. Losada and A. Riotto. *Phys. Rev. Lett.*, 84:4284, 2000.

INTRODUCTION TO SUPERSTRING THEORY

John H. Schwarz
California Institute of Technology
Pasadena, CA 91125, USA *

Abstract These four lectures, addressed to an audience of graduate students in experimental high energy physics, survey some of the basic concepts in string theory. The purpose is to convey a general sense of what string theory is and what it has achieved. Since the characteristic scale of string theory is expected to be close to the Planck scale, the structure of strings probably cannot be probed directly in accelerator experiments. The most accessible experimental implication of superstring theory is supersymmetry at or below the TeV scale.

1. INTRODUCTION

Tom Ferbel has presented me with a large challenge: explain string theory to an audience of graduate students in experimental high energy physics. The allotted time is four 75-minute lectures. This should be possible, if the goals are realistic. One goal is to give a general sense of what the subject is about, and why so many theoretical physicists are enthusiastic about it. Perhaps you should regard these lectures as a cultural experience providing a window into the world of abstract theoretical physics. Don't worry if you miss some of the technical details in the second and third lectures. There is only one message in these lectures that is important for experimental research: low-energy supersymmetry is very well motivated theoretically, and it warrants the intense effort that is being made to devise ways of observing it. There are other facts that are nice to know, however. For example, consistency of quantum theory and gravity is a severe restriction, with farreaching consequences.

As will be explained, string theory requires supersymmetry, and therefore string theorists were among the first to discover it. Supersymmetric string theories are called superstring theories. At one time there seemed

*Work supported in part by the U.S. Dept. of Energy under Grant No. DE-FG03-92-ER40701.

H.B. Prosper and M. Danilov (eds.), Techniques and Concepts of High-Energy Physics, 143–187.

to be five distinct superstring theories, but it was eventually realized that each of them is actually a special limiting case of a completely unique underlying theory. This theory is not yet fully formulated, and when it is, we might decide that a new name is appropriate. Be that as it may, it is clear that we are exploring an extraordinarily rich structure with many deep connections to various branches of fundamental mathematics and theoretical physics. Whatever the ultimate status of this theory may be, it is clear that these studies have already been a richly rewarding experience.

To fully appreciate the mathematical edifice underlying superstring theory requires an investment of time and effort. Many theorists who make this investment really become hooked by it, and then there is no turning back. Well, hooking you in this way is not my goal, since you are engaged in other important activities; but hopefully these lectures will convey an idea of why many theorists find the subject so enticing. For those who wish to study the subject in more detail, there are two standard textbook presentations [1, 2].

The plan of these lectures is as follows: The first lecture will consist of a general non-technical overview of the subject. It is essentially the current version of my physics colloquium lecture. It will describe some of the basic concepts and issues without technical details. If successful, it will get you sufficiently interested in the subject that you are willing to sit through some of the basic nitty-gritty analysis that explains what we mean by a relativistic string, and how its normal modes are analyzed. Lecture 2 will present the analysis for the bosonic string theory. This is an unrealistic theory, with bosons only, but its study is a pedagogically useful first step. It involves many, but not all, of the issues that arise for superstrings. In lecture 3 the extension to incorporate fermions and supersymmetry is described. There are two basic formalisms for doing this (called RNS and GS). Due to time limitations, only the first of these will be presented here. The final lecture will survey some of the more recent developments in the field. These include various nonperturbative dualities, the existence of an 11-dimensional limit (called M-theory) and the existence of extended objects of various dimensionalities, called p-branes. As will be explained, a particular class of p-branes, called D-branes, plays an especially important role in modern research.

2. LECTURE 1: OVERVIEW AND MOTIVATION

Many of the major developments in fundamental physics of the past century arose from identifying and overcoming contradictions between existing ideas. For example, the incompatibility of Maxwell's equations and Galilean invariance led Einstein to propose the special theory of relativity. Similarly, the inconsistency of special relativity with Newtonian gravity led him to develop the general theory of relativity. More recently, the reconciliation of special relativity with quantum mechanics led to the development of quantum field theory. We are now facing another crisis of the same character. Namely, general relativity appears to be incompatible with quantum field theory. Any straightforward attempt to "quantize" general relativity leads to a nonrenormalizable theory. In my opinion, this means that the theory is inconsistent and needs to be modified at short distances or high energies. The way that string theory does this is to give up one of the basic assumptions of quantum field theory, the assumption that elementary particles are mathematical points, and instead to develop a quantum field theory of one-dimensional extended objects, called strings, There are very few consistent theories of this type, but superstring theory shows great promise as a unified quantum theory of all fundamental forces including gravity. There is no realistic string theory of elementary particles that could serve as a new standard model, since there is much that is not yet understood. But that, together with a deeper understanding of cosmology, is the goal. This is still a work in progress.

Even though string theory is not yet fully formulated, and we cannot yet give a detailed description of how the standard model of elementary particles should emerge at low energies, there are some general features of the theory that can be identified. These are features that seem to be quite generic irrespective of how various details are resolved. The first, and perhaps most important, is that general relativity is necessarily incorporated in the theory. It gets modified at very short distances/high energies but at ordinary distance and energies it is present in exactly the form proposed by Einstein. This is significant, because it is arising within the framework of a consistent quantum theory. Ordinary quantum field theory does not allow gravity to exist; string theory requires it! The second general fact is that Yang–Mills gauge theories of the sort that comprise the standard model naturally arise in string theory. We do not understand why the specific $SU(3) \times SU(2) \times U(1)$ gauge theory of the standard model should be preferred, but (anomaly-free) theories of this general type do arise naturally at ordinary energies. The third general feature of string theory solutions is supersymmetry. The mathematical consistency of string theory depends crucially on supersymmetry, and it is very hard to find consistent solutions (quantum vacua) that do not preserve at least a portion of this supersymmetry. This prediction of string theory differs from the other two (general relativity and gauge theories) in that it really is a prediction. It is a generic feature of string theory that has not yet been discovered experimentally.

2.1. SUPERSYMMETRY

Even though supersymmetry is a very important part of the story, the discussion here will be very brief, since it will be discussed in detail by other lecturers. There will only be a few general remarks. First, as we have just said, supersymmetry is the major prediction of string theory that could appear at accessible energies, that has not yet been discovered. A variety of arguments, not specific to string theory, suggest that the characteristic energy scale associated to supersymmetry breaking should be related to the electroweak scale, in other words in the range 100 GeV – 1 TeV. The symmetry implies that all known elementary particles should have partner particles, whose masses are in this general range. This means that some of these superpartners should be observable at the CERN Large Hadron Collider (LHC), which will begin operating in the middle part of this decade. There is even a chance that Fermilab Tevatron experiments could find superparticles earlier than that.
In most versions of phenomenological supersymmetry there is a multiplicatively conserved quantum number called R-parity. All known particles have even R-parity, whereas their superpartners have odd R-parity. This implies that the superparticles must be pair-produced in particle collisions. It also implies that the lightest supersymmetry particle (or LSP) should be absolutely stable. It is not known with certainty which particle is the LSP, but one popular guess is that it is a "neutralino." This is an electrically neutral fermion that is a quantum-mechanical mixture of the partners of the photon, Z^0, and neutral Higgs particles. Such an LSP would interact very weakly, more or less like a neutrino. It is of considerable interest, since it is an excellent dark matter candidate. Searches for dark matter particles called WIMPS (weakly interacting massive particles) could discover the LSP some day. Current experiments might not have sufficient detector volume to compensate for the exceedingly small cross sections.

There are three unrelated arguments that point to the same mass range for superparticles. The one we have just been discussing, a neutralino LSP as an important component of dark matter, requires a mass of order 100 GeV. The precise number depends on the mixture that comprises the LSP, what their density is, and a number of other details. A second argument is based on the famous hierarchy problem. This is the fact that standard model radiative corrections tend to renormalize the Higgs mass to a very high scale. The way to prevent this is to extend the standard model to a supersymmetric standard model and to have the supersymmetry be broken at a scale comparable to the Higgs mass, and hence to the electroweak scale. The third argument that gives an estimate of the susy-breaking scale is grand unification. If one accepts the notion that the standard model gauge group is embedded in a larger gauge group such as $SU(5)$ or $SO(10)$, which is broken at a high mass scale, then the three standard model coupling constants should unify at that mass scale. Given the spectrum of particles, one can compute the evolution of the couplings as a function of energy using renormalization group equations. One finds that if one only includes the standard model particles this unification fails quite badly. However, if one also includes all the supersymmetry particles required by the minimal supersymmetric extension of the standard model, then the couplings do unify at an energy of about 2×10^{16} GeV. For this agreement to take place, it is necessary that the masses of the superparticles are less than a few TeV.

There is other support for this picture, such as the ease with which supersymmetric grand unification explains the masses of the top and bottom quarks and electroweak symmetry breaking. Despite all these indications, we cannot be certain that this picture is correct until it is demonstrated experimentally. One could suppose that all this is a giant coincidence, and the correct description of TeV scale physics is based on something entirely different. The only way we can decide for sure is by doing the experiments. As I once told a newspaper reporter, in order to be sure to be quoted: discovery of supersymmetry would be more profound than life on Mars.

2.2. BASIC IDEAS OF STRING THEORY

In conventional quantum field theory the elementary particles are mathematical points, whereas in perturbative string theory the fundamental objects are one-dimensional loops (of zero thickness). Strings have a characteristic length scale, which can be estimated by dimensional analysis. Since string theory is a relativistic quantum theory that includes gravity it must involve the fundamental constants c (the speed of light), $\hbar$ (Planck's constant divided by 2π), and G (Newton's gravitational constant). From these one can form a length, known as the

Planck length

$$\ell_p = \left(\frac{\hbar G}{c^3}\right)^{3/2} = 1.6 \times 10^{-33}\,\text{cm}. \tag{1.1}$$

Similarly, the Planck mass is

$$m_p = \left(\frac{\hbar c}{G}\right)^{1/2} = 1.2 \times 10^{19}\,\text{GeV}/c^2. \tag{1.2}$$

Experiments at energies far below the Planck energy cannot resolve distances as short as the Planck length. Thus, at such energies, strings can be accurately approximated by point particles. From the viewpoint of string theory, this explains why quantum field theory has been so successful.

As a string evolves in time it sweeps out a two-dimensional surface in spacetime, which is called the world sheet of the string. This is the string counterpart of the world line for a point particle. In quantum field theory, analyzed in perturbation theory, contributions to amplitudes are associated to Feynman diagrams, which depict possible configurations of world lines. In particular, interactions correspond to junctions of world lines. Similarly, string theory perturbation theory involves string world sheets of various topologies. A particularly significant fact is that these world sheets are generically smooth. The existence of interaction is a consequence of world-sheet topology rather than a local singularity on the world sheet. This difference from point-particle theories has two important implications. First, in string theory the structure of interactions is uniquely determined by the free theory. There are no arbitrary interactions to be chosen. Second, the ultraviolet divergences of point-particle theories can be traced to the fact that interactions are associated to world-line junctions at specific spacetime points. Because the string world sheet is smooth, string theory amplitudes have no ultraviolet divergences.

2.3. A BRIEF HISTORY OF STRING THEORY

String theory arose in the late 1960's out of an attempt to describe the strong nuclear force. The inclusion of fermions led to the discovery of supersymmetric strings — or superstrings — in 1971. The subject fell out of favor around 1973 with the development of QCD, which was quickly recognized to be the correct theory of strong interactions. Also, string theories had various unrealistic features such as extra dimensions and massless particles, neither of which are appropriate for a hadron theory.

Among the massless string states there is one that has spin two. In 1974, it was shown by Scherk and me [3], and independently by Yoneya [4], that this particle interacts like a graviton, so the theory actually includes general relativity. This led us to propose that string theory should be used for unification rather than for hadrons. This implied, in particular, that the string length scale should be comparable to the Planck length, rather than the size of hadrons (10^{-13} cm) as we had previously assumed.

In the "first superstring revolution," which took place in 1984–85, there were a number of important developments (described later) that convinced a large segment of the theoretical physics community that this is a worthy area of research. By the time the dust settled in 1985 we had learned that there are five distinct consistent string theories, and that each of them requires spacetime supersymmetry in the ten dimensions (nine spatial dimensions plus time). The theories, which will be described later, are called type I, type IIA, type IIB, $SO(32)$ heterotic, and $E_8 \times E_8$ heterotic.

2.4. COMPACTIFICATION

In the context of the original goal of string theory – to explain hadron physics – extra dimensions are unacceptable. However, in a theory that incorporates general relativity, the geometry of spacetime is determined dynamically. Thus one could imagine that the theory admits consistent quantum solutions in which the six extra spatial dimensions form a compact space, too small to have been observed. The natural first guess is that the size of this space should be comparable to the string scale and the Planck length. Since the equations must be satisfied, the geometry of this six-dimensional space is not arbitrary. A particularly appealing possibility, which is consistent with the equations, is that it forms a type of space called a Calabi–Yau space [5].

Calabi–Yau compactification, in the context of the $E_8 \times E_8$ heterotic string theory, can give a low-energy effective theory that closely resembles a supersymmetric extension of the standard model. There is actually a lot of freedom, because there are very many different Calabi–Yau spaces, and there are other arbitrary choices that can be made. Still, it is interesting that one can come quite close to realistic physics. It is also interesting that the number of quark and lepton families that one obtains is determined by the topology of the Calabi–Yau space. Thus, for suitable choices, one can arrange to end up with exactly three families. People were very excited by the picture in 1985. Nowadays, we tend to make a more sober appraisal that emphasizes all the arbitrariness that is involved, and the things that don't work exactly right. Still, it would not be surprising if some aspects of this picture survive as part of the story when we understand the right way to describe the real world.

2.5. PERTURBATION THEORY

Until 1995 it was only understood how to formulate string theories in terms of perturbation expansions. Perturbation theory is useful in a quantum theory that has a small dimensionless coupling constant, such as quantum electrodynamics, since it allows one to compute physical quantities as power series expansions in the small parameter. In QED the small parameter is the fine-structure constant $\alpha \sim 1/137$. Since this is quite small, perturbation theory works very well for QED. For a physical quantity $T(\alpha)$, one computes (using Feynman diagrams)

$$T(\alpha) = T_0 + \alpha T_1 + \alpha^2 T_2 + \dots . \tag{1.3}$$

It is the case generically in quantum field theory that expansions of this type are divergent. More specifically, they are asymptotic expansions with zero radius convergence. Nonetheless, they can be numerically useful if the expansion parameter is small. The problem is that there are various non-perturbative contributions (such as instantons) that have the structure

$$T_{NP} \sim e^{-(const./\alpha)}. \tag{1.4}$$

In a theory such as QCD, there are regimes where perturbation theory is useful (due to asymptotic freedom) and other regimes where it is not. For problems of the latter type, such as computing the hadron spectrum, nonperturbative methods of computation, such as lattice gauge theory, are required.

In the case of string theory the dimensionless string coupling constant, denoted g_s, is determined dynamically by the expectation value of a scalar field called the dilaton. There is no particular reason that this number should be small. So it is unlikely that a realistic vacuum could be analyzed accurately using perturbation theory. More importantly, these theories have many qualitative properties that are inherently nonperturbative. So one needs nonperturbative methods to understand them.

2.6. THE SECOND SUPERSTRING REVOLUTION

Around 1995 some amazing and unexpected "dualities" were discovered that provided the first glimpses into nonperturbative features of string theory. These dualities were quickly recognized to have three major implications.

The dualities enabled us to relate all five of the superstring theories to one another. This meant that, in a fundamental sense, they are all equivalent to one another. Another way of saying this is that there is a unique underlying theory, and what we had been calling five theories are better viewed as perturbation expansions of this underlying theory about five different points (in the space of consistent quantum vacua). This was a profoundly satisfying realization, since we really didn't want five theories of nature. That there is a completely unique theory, without any dimensionless parameters, is the best outcome one could have hoped for. To avoid confusion, it should be emphasized that even though the theory is unique, it is entirely possible that there are many consistent quantum vacua. Classically, the corresponding statement is that a unique equation can admit many solutions. It is a particular solution (or quantum vacuum) that ultimately must describe nature. At least, this is how a particle physicist would say it. If we hope to understand the origin and evolution of the universe, in addition to properties of elementary particles, it would be nice if we could also understand cosmological solutions.

A second crucial discovery was that the theory admits a variety of nonperturbative excitations, called p-branes, in addition to the fundamental strings. The letter p labels the number of spatial dimensions of the excitation. Thus, in this language, a point particle is a 0-brane, a string is a 1-brane, and so forth. The reason that p-branes were not discovered in perturbation theory is that they have tension (or energy density) that diverges as $g_s \to 0$. Thus they are absent from the perturbative theory.

The third major discovery was that the underlying theory also has an eleven-dimensional solution, which is called M-theory. Later, we will explain how the eleventh dimension arises.

One type of duality is called S duality. (The choice of the letter S is a historical accident of no great significance.) Two string theories (let's call them A and B) are related by S duality if one of them evaluated at strong coupling is equivalent to the other one evaluated at weak coupling. Specifically, for any physical quantity f, one has

$$f_A(g_s) = f_B(1/g_s). \tag{1.5}$$

Two of the superstring theories — type I and $SO(32)$ heterotic — are related by S duality in this way. The type IIB theory is self-dual. Thus S duality is a symmetry of the IIB theory, and this symmetry is unbroken if $g_s = 1$. Thanks to S duality, the strong-coupling behavior of each of these three theories is determined by a weak-coupling analysis. The remaining two theories, type IIA and $E_8 \times E_8$ heterotic, behave very differently at strong coupling. They grow an eleventh dimension!

Another astonishing duality, which goes by the name of T duality, was discovered several years earlier. It can be understood in perturbation theory, which is why it was found first. But, fortunately, it often continues to be valid even at strong coupling. T duality can relate different compactifications of different theories. For example, suppose theory A has a compact dimension that is a circle of radius R_A and theory B has a compact dimension that is a circle of radius R_B. If these two theories are related by T duality this means that they are equivalent provided that

$$R_A R_B = (\ell_s)^2, \tag{1.6}$$

where ℓ_s is the fundamental string length scale. This has the amazing implication that when one of the circles becomes small the other one becomes large. In a later lecture, we will explain how this is possible. T duality relates the two type II theories and the two heterotic theories. There are more complicated examples of the same phenomenon involving compact spaces that are more complicated than a circle, such as tori, K3, Calabi–Yau spaces, etc.

2.7. THE ORIGINS OF GAUGE SYMMETRY

There are a variety of mechanisms than can give rise to Yang–Mills type gauge symmetries in string theory. Here, we will focus on two basic possibilities: Kaluza–Klein symmetries and brane symmetries.

The basic Kaluza–Klein idea goes back to the 1920's, though it has been much generalized since then. The idea is to suppose that the 10- or 11-dimensional geometry has a product structure $M \times K$, where M is Minkowski spacetime and K is a compact manifold. Then, if K has symmetries, these appear as gauge symmetries of the effective theory defined on M. The Yang–Mills gauge fields arise as components of the gravitational metric field with one direction along K and the other along M. For example, if the space K is an n-dimensional sphere, the symmetry group is $SO(n+1)$, if it is CP^n — which has $2n$ dimensions — it is $SU(n+1)$, and so forth. Elegant as this may be, it seems unlikely that a realistic K has any such symmetries. Calabi–Yau spaces, for example, do not have any.

A rather more promising way of achieving realistic gauge symmetries is via the brane approach. Here the idea is that a certain class of p-branes (called D-branes) have gauge fields that are restricted to their world volume. This means that the gauge fields are not defined throughout the 10- or 11-dimensional spacetime but only on the $(p+1)$-dimensional hypersurface defined by the D-branes. This picture suggests that the world we observe might be a D-brane embedded in a higher-dimensional space. In such a scenario, there can be two kinds of extra dimensions: compact dimensions along the brane and compact dimensions perpendicular to the brane.

The traditional viewpoint, which in my opinion is still the best bet, is that all extra dimensions (of both types) have sizes of order 10^{-30} to 10^{-32} cm corresponding to an energy scale of $10^{16} - 10^{18}$ GeV. This makes them inaccessible to direct observation, though their existence would have definite low-energy consequences. However, one can and should ask "what are the experimental limits?" For compact dimensions along the brane, which support gauge fields, the nonobservation of extra dimensions in tests of the standard model implies a bound of about 1 TeV. The LHC should extend this to about 10 TeV. For compact dimensions "perpendicular to the brane," which only support excitations with gravitational strength forces, the best bounds come from Cavendish-type experiments, which test the $1/R^2$ structure of the Newton force law at short distances. No deviations have been observed to a distance of about 1 mm, so far. Experiments planned in the near future should extend the limit to about 100 microns. Obviously, observation of any deviation from $1/R^2$ would be a major discovery.

2.8. CONCLUSION

This introductory lecture has sketched some of the remarkable successes that string theory has achieved over the past 30 years. There are many others that did not fit in this brief survey. Despite all this progress, there are some very important and fundamental questions whose answers are unknown. It seems that whenever a breakthrough occurs, a host of new questions arise, and the ultimate goal still seems a long way off. To convince you that there is a long way to go, let us list some of the most important questions:

- What is the theory? Even though a great deal is known about string theory and M theory, it seems that the optimal formulation of the underlying theory has not yet been found. It might be based on principles that have not yet been formulated.

- We are convinced that supersymmetry is present at high energies and probably at the electroweak scale, too. But we do not know how or why it is broken.

- A very crucial problem concerns the energy density of the vacuum, which is a physical quantity in a gravitational theory. This is characterized by the cosmological constant, which observationally appears to have a small positive value — so that the vacuum energy of the universe is comparable to the energy in matter. In Planck units this is a tiny number ($\Lambda \sim 10^{-120}$). If supersymmetry were unbroken, we could argue that $\Lambda = 0$, but if it is broken at the 1 TeV scale, that would seem to suggest $\Lambda \sim 10^{-60}$, which is very far from the truth. Despite an enormous amount of effort and ingenuity, it is not yet clear how superstring theory will conspire to break supersymmetry at the TeV scale and still give a value for Λ that is much smaller than 10^{-60}. The fact that the desired result is about the square of this might be a useful hint.

- Even though the underlying theory is unique, there seem to be many consistent quantum vacua. We would very much like to formulate a theoretical principle (not based on observation) for choosing among these vacua. It is not known whether the right approach to the answer is cosmological, probabilistic, anthropic, or something else.

3. LECTURE 2: STRING THEORY BASICS

In this lecture we will describe the world-sheet dynamics of the original bosonic string theory. As we will see this theory has various unrealistic and unsatisfactory properties. Nonetheless it is a useful preliminary before describing supersymmetric strings, because it allows us to introduce many of the key concepts without simultaneously addressing the added complications associated with fermions and supersymmetry.

We will describe string dynamics from a first-quantized point of view. This means that we focus on understanding it from a world-sheet sum-over-histories point of view. This approach is closely tied to perturbation theory analysis. It should be contrasted with "second quantized" string field theory which is based on field operators that create or destroy entire strings. Since the first-quantized point of view may be less familiar to you than second-quantized field theory, let us begin by reviewing how it can be used to describe a massive point particle.

3.1. WORLD-LINE DESCRIPTION OF A POINT PARTICLE

A point particle sweeps out a trajectory (or world line) in spacetime. This can be described by functions $x^\mu(\tau)$ that describe how the world line, parameterized by τ, is embedded in the spacetime, whose coordinates are denoted x^μ. For simplicity, let us assume that the spacetime is flat Minkowski space with a Lorentz metric

$$\eta_{\mu\nu} = \begin{pmatrix} -1 & 0 & 0 & 0 \\ 0 & 1 & 0 & 0 \\ 0 & 0 & 1 & 0 \\ 0 & 0 & 0 & 1 \end{pmatrix}. \tag{1.7}$$

Then, the Lorentz invariant line element is given by

$$ds^2 = -\eta_{\mu\nu} dx^\mu dx^\nu. \tag{1.8}$$

In units $\hbar = c = 1$, the action for a particle of mass m is given by

$$S = -m \int ds. \tag{1.9}$$

This could be generalized to a curved spacetime by replacing $\eta_{\mu\nu}$ by a metric $g_{\mu\nu}(x)$, but we will not do so here. In terms of the embedding functions, $x^\mu(t)$, the action can be rewritten in the form

$$S = -m \int d\tau \sqrt{-\eta_{\mu\nu}\dot{x}^\mu \dot{x}^\nu}, \tag{1.10}$$

where dots represent τ derivatives. An important property of this action is invariance under local reparametrizations. This is a kind of gauge invariance, whose meaning is that the form of S is unchanged under an arbitrary reparametrization of the world line $\tau \to \tau(\tilde{\tau})$. Actually, one should require that the function $\tau(\tilde{\tau})$ is smooth and monotonic $\left(\frac{d\tau}{d\tilde{\tau}} > 0\right)$. The reparametrization invariance is a one-dimensional analog of the four-dimensional general coordinate invariance of general relativity. Mathematicians refer to this kind of symmetry as diffeomorphism invariance.

The reparametrization invariance of S allows us to choose a gauge. A nice choice is the "static gauge"

$$x^0 = \tau. \tag{1.11}$$

In this gauge (renaming the parameter t) the action becomes

$$S = -m \int \sqrt{1 - v^2} dt, \tag{1.12}$$

where

$$\vec{v} = \frac{d\vec{x}}{dt}. \tag{1.13}$$

Requiring this action to be stationary under an arbitrary variation of $\vec{x}(t)$ gives the Euler–Lagrange equations

$$\frac{d\vec{p}}{dt} = 0, \tag{1.14}$$

where

$$\vec{p} = \frac{\delta S}{\delta \vec{v}} = \frac{m\vec{v}}{\sqrt{1-v^2}}, \tag{1.15}$$

which is the usual result. So we see that usual relativistic kinematics follows from the action $S = -m \int ds$.

3.2. WORLD-VOLUME ACTIONS

We can now generalize the analysis of the massive point particle to a p-brane of tension T_p. The action in this case involves the invariant $(p+1)$-dimensional volume and is given by

$$S_p = -T_p \int d\mu_{p+1}, \tag{1.16}$$

where the invariant volume element is

$$d\mu_{p+1} = \sqrt{-\det(-\eta_{\mu\nu}\partial_\alpha x^\mu \partial_\beta x^\nu)}d^{p+1}\sigma. \tag{1.17}$$

Here the embedding of the p-brane into d-dimensional spacetime is given by functions $x^\mu(\sigma^\alpha)$. The index $\alpha = 0, \ldots, p$ labels the $p+1$ coordinates σ^α of the p-brane world-volume and the index $\mu = 0, \ldots, d-1$ labels the d coordinates x^μ of the d-dimensional spacetime. We have defined

$$\partial_\alpha x^\mu = \frac{\partial x^\mu}{\partial \sigma^\alpha}. \tag{1.18}$$

The determinant operation acts on the $(p+1) \times (p+1)$ matrix whose rows and columns are labeled by α and β. The tension T_p is interpreted as the mass per unit volume of the p-brane. For a 0-brane, it is just the mass.

Exercise: Show that S_p is reparametrization invariant. In other words, substituting $\sigma^\alpha = \sigma^\alpha(\tilde{\sigma}^\beta)$, it takes the same form when expressed in terms of the coordinates $\tilde{\sigma}^\alpha$.

Let us now specialize to the string, $p = 1$. Evaluating the determinant gives

$$S[x] = -T \int d\sigma d\tau \sqrt{\dot{x}^2 x'^2 - (\dot{x} \cdot x')^2}, \tag{1.19}$$

where we have defined $\sigma^0 = \tau$, $\sigma^1 = \sigma$, and

$$\dot{x}^\mu = \frac{\partial x^\mu}{\partial \tau}, \quad x'^\mu = \frac{\partial x^\mu}{\partial \sigma}. \tag{1.20}$$

This action, called the Nambu–Goto action, was first proposed in 1970 [6, 7]. The Nambu–Goto action is equivalent to the action

$$S[x, h] = -\frac{T}{2} \int d^2\sigma \sqrt{-h} h^{\alpha\beta} \eta_{\mu\nu} \partial_\alpha x^\mu \partial_\beta x^\nu, \tag{1.21}$$

where $h_{\alpha\beta}(\sigma, \tau)$ is the world-sheet metric, $h = \det h_{\alpha\beta}$, and $h^{\alpha\beta}$ is the inverse of $h_{\alpha\beta}$. The Euler–Lagrange equation obtained by varying $h^{\alpha\beta}$ are

$$T_{\alpha\beta} = \partial_\alpha x \cdot \partial_\beta x - \frac{1}{2} h_{\alpha\beta} h^{\gamma\delta} \partial_\gamma x \cdot \partial_\delta x = 0. \tag{1.22}$$

Exercise: Show that $T_{\alpha\beta} = 0$ can be used to eliminate the world-sheet metric from the action, and that when this is done one recovers the Nambu–Goto action. (Hint: take the determinant of both sides of the equation $\partial_\alpha x \cdot \partial_\beta x = \frac{1}{2} h_{\alpha\beta} h^{\gamma\delta} \partial_\gamma x \cdot \partial_\delta x$.)

In addition to reparametrization invariance, the action $S[x, h]$ has another local symmetry, called conformal invariance (or Weyl invariance). Specifically, it is invariant under the replacement

$$\begin{aligned} h_{\alpha\beta} &\rightarrow \Lambda(\sigma, \tau) h_{\alpha\beta} \\ x^\mu &\rightarrow x^\mu. \end{aligned} \tag{1.23}$$

This local symmetry is special to the $p = 1$ case (strings).

The two reparametrization invariance symmetries of $S[x, h]$ allow us to choose a gauge in which the three functions $h_{\alpha\beta}$ (this is a symmetric 2×2 matrix) are expressed in terms of just one function. A convenient choice is the "conformally flat gauge"

$$h_{\alpha\beta} = \eta_{\alpha\beta} e^{\phi(\sigma, \tau)}. \tag{1.24}$$

Here, $\eta_{\alpha\beta}$ denoted the two-dimensional Minkowski metric of a flat world sheet. However, because of the factor e^ϕ, $h_{\alpha\beta}$ is only "conformally flat."

Classically, substitution of this gauge choice into $S[x,h]$ leaves the gauge-fixed action

$$S = \frac{T}{2} \int d^2\sigma \eta^{\alpha\beta} \partial_\alpha x \cdot \partial_\beta x. \tag{1.25}$$

Quantum mechanically, the story is more subtle. Instead of eliminating h via its classical field equations, one should perform a Feynman path integral, using standard machinery to deal with the local symmetries and gauge fixing. When this is done correctly, one finds that in general ϕ does not decouple from the answer. Only for the special case $d = 26$ does the quantum analysis reproduce the formula we have given based on classical reasoning [8]. Otherwise, there are correction terms whose presence can be traced to a conformal anomaly (i.e., a quantum-mechanical breakdown of the conformal invariance).

The gauge-fixed action is quadratic in the x's. Mathematically, it is the same as a theory of d free scalar fields in two dimensions. The equations of motion obtained by varying x^μ are simply free two-dimensional wave equations:

$$\ddot{x}^\mu - x''^\mu = 0. \tag{1.26}$$

This is not the whole story, however, because we must also take account of the constraints $T_{\alpha\beta} = 0$. Evaluated in the conformally flat gauge, these constraints are

$$\begin{aligned} T_{01} &= T_{10} = \dot{x} \cdot x' = 0 \\ T_{00} &= T_{11} = \frac{1}{2}(\dot{x}^2 + x'^2) = 0. \end{aligned} \tag{1.27}$$

Adding and subtracting gives

$$(\dot{x} \pm x')^2 = 0. \tag{1.28}$$

3.3. BOUNDARY CONDITIONS

To go further, one needs to choose boundary conditions. There are three important types. For a closed string one should impose periodicity in the spatial parameter σ. Choosing its range to be π (as is conventional)

$$x^\mu(\sigma, \tau) = x^\mu(\sigma + \pi, \tau). \tag{1.29}$$

For an open string (which has two ends), each end can be required to satisfy either Neumann or Dirichlet boundary conditions (for each value of μ).

$$\begin{aligned} \text{Neumann}: \quad \frac{\partial x^\mu}{\partial \sigma} &= 0 \quad \text{at } \sigma = 0 \text{ or } \pi & (1.30) \\ \text{Dirichlet}: \quad \frac{\partial x^\mu}{\partial \tau} &= 0 \quad \text{at } \sigma = 0 \text{ or } \pi. & (1.31) \end{aligned}$$

The Dirichlet condition can be integrated, and then it specifies a spacetime location on which the string ends. The only way this makes sense is if the open string ends on a physical object – it ends on a D-brane. (D stands for Dirichlet.) If all the open-string boundary conditions are Neumann, then the ends of the string can be anywhere in the spacetime. The modern interpretation is that this means that there are spacetime-filling D-branes present.

Let us now consider the closed-string case in more detail. The general solution of the 2d wave equation is given by a sum of "right-movers" and "left-movers":

$$x^\mu(\sigma, \tau) = x^\mu_R(\tau - \sigma) + x^\mu_L(\tau + \sigma). \tag{1.32}$$

These should be subject to the following additional conditions:

- $x^\mu(\sigma, \tau)$ is real
- $x^\mu(\sigma + \pi, \tau) = x^\mu(\sigma, \tau)$
- $(x'_L)^2 = (x'_R)^2 = 0$ (These are the $T_{\alpha\beta} = 0$ constraints in eq. (1.28).)

The first two of these conditions can be solved explicitly in terms of Fourier series:

$$\begin{aligned} x^\mu_R &= \frac{1}{2}x^\mu + \ell_s^2 p^\mu(\tau - \sigma) + \frac{i}{\sqrt{2}}\ell_s \sum_{n\neq 0} \frac{1}{n}\alpha^\mu_n e^{-2in(\tau-\sigma)} & (1.33) \\ x^\mu_L &= \frac{1}{2}x^\mu + \ell_s^2 p^\mu(\tau + \sigma) + \frac{i}{\sqrt{2}}\ell_s \sum_{n\neq 0} \frac{1}{n}\tilde{\alpha}^\mu_n e^{-2in(\tau+\sigma)}, \end{aligned}$$

where the expansion parameters α^μ_n, $\tilde{\alpha}^\mu_n$ satisfy

$$\alpha^\mu_{-n} = (\alpha^\mu_n)^\dagger, \quad \tilde{\alpha}^\mu_{-n} = (\tilde{\alpha}^\mu_n)^\dagger. \tag{1.34}$$

The center-of-mass coordinate x^μ and momentum p^μ are also real. The fundamental string length scale ℓ_s is related to the tension T by

$$T = \frac{1}{2\pi\alpha'}, \quad \alpha' = \ell_s^2. \tag{1.35}$$

The parameter α' is called the universal Regge slope, since the string modes lie on linear parallel Regge trajectories with this slope.

3.4. QUANTIZATION

The analysis of closed-string left-moving modes, closed-string right-moving modes, and open-string modes are all very similar. Therefore, to avoid repetition, we will focus on the closed-string right-movers. Starting with the gauge-fixed action in eq.(1.25), the canonical momentum of the string is

$$p^{\mu}(\sigma,\tau) = \frac{\delta S}{\delta \dot{x}^{\mu}} = T\dot{x}^{\mu}. \tag{1.36}$$

Canonical quantization (this is just free 2d field theory for scalar fields) gives

$$[p^{\mu}(\sigma,\tau), x^{\nu}(\sigma',\tau)] = -i\hbar\eta^{\mu\nu}\delta(\sigma-\sigma'). \tag{1.37}$$

In terms of the Fourier modes (setting $\hbar = 1$) these become

$$[p^{\mu}, x^{\nu}] = -i\eta^{\mu\nu} \tag{1.38}$$

$$\begin{aligned} [\alpha^{\mu}_m, \alpha^{\nu}_n] &= m\delta_{m+n,0}\eta^{\mu\nu}, \\ [\tilde{\alpha}^{\mu}_m, \tilde{\alpha}^{\nu}_n] &= m\delta_{m+n,0}\eta^{\mu\nu}, \end{aligned} \tag{1.39}$$

and all other commutators vanish.

Recall that a quantum-mechanical harmonic oscillator can be described in terms of raising and lowering operators, usually called $a^{\dagger}$ and a, which satisfy

$$[a, a^{\dagger}] = 1. \tag{1.40}$$

We see that, aside from a normalization factor, the expansion coefficients α^{μ}_{-m} and α^{μ}_m are raising and lowering operators. There is just one problem. Because $\eta^{00} = -1$, the time components are proportional to oscillators with the wrong sign ($[a, a^{\dagger}] = -1$). This is potentially very bad, because such oscillators create states of negative norm, which could lead to an inconsistent quantum theory (with negative probabilities, etc.). Fortunately, as we will explain, the $T_{\alpha\beta} = 0$ constraints eliminate the negative-norm states from the physical spectrum.

The classical constraint for the right-moving closed-string modes, $(x'_R)^2 = 0$, has Fourier components

$$L_m = \frac{T}{2}\int_0^{\pi} e^{-2im\sigma}(x'_R)^2 d\sigma = \frac{1}{2}\sum_{n=-\infty}^{\infty} \alpha_{m-n}\cdot\alpha_n, \tag{1.41}$$

which are called Virasoro operators. Since α_m^μ does not commute with α_{-m}^μ, L_0 needs to be normal-ordered:

$$L_0 = \frac{1}{2}\alpha_0^2 + \sum_{n=1}^{\infty} \alpha_{-n} \cdot \alpha_n. \tag{1.42}$$

Here $\alpha_0^\mu = \ell_s p^\mu/\sqrt{2}$, where p^μ is the momentum.

3.5. THE FREE STRING SPECTRUM

Recall that the Hilbert space of a harmonic oscillator is spanned by states $|n\rangle, n = 0, 1, 2, \ldots$, where the ground state, $|0\rangle$, is annihilated by the lowering operator ($a|0\rangle = 0$) and

$$|n\rangle = \frac{(a^\dagger)^n}{\sqrt{n!}}|0\rangle. \tag{1.43}$$

Then, for a normalized ground-state ($\langle 0|0\rangle = 1$), one can use $[a, a^\dagger] = 1$ repeatedly to prove that

$$\langle m|n\rangle = \delta_{m,n} \tag{1.44}$$

and

$$a^\dagger a|n\rangle = n|n\rangle. \tag{1.45}$$

The string spectrum (of right-movers) is given by the product of an infinite number of harmonic-oscillator Fock spaces, one for each α_n^μ, subject to the Virasoro constraints [9]

$$\begin{aligned} (L_0 - q)|\phi\rangle &= 0 \\ L_n|\phi\rangle &= 0, \quad n > 0. \end{aligned} \tag{1.46}$$

Here $|\phi\rangle$ denotes a physical state, and q is a constant to be determined. It accounts for the arbitrariness in the normal-ordering prescription used to define L_0. As we will see, the L_0 equation is a generalization of the Klein–Gordon equation. It contains $p^2 = -\partial \cdot \partial$ plus oscillator terms whose eigenvalue will determine the mass of the state.

It is interesting to work out the algebra of the Virasoro operators L_m, which follows from the oscillator algebra. The result, called the Virasoro algebra, is

$$[L_m, L_n] = (m - n)L_{m+n} + \frac{c}{12}(m^3 - m)\delta_{m+n,0}. \tag{1.47}$$

The second term on the right-hand side is called the "conformal anomaly term" and the constant c is called the "central charge."

Exercise: Verify the first term on the right-hand side. For extra credit, verify the second term, showing that each component of x^μ contributes $c = 1$, so that altogether $c = d$.

There are more sophisticated ways to describe the string spectrum (in terms of BRST cohomology), but they are equivalent to the more elementary approach presented here. In the BRST approach, gauge-fixing to the conformal gauge in the quantum theory requires the addition of world-sheet Faddeev-Popov ghosts, which turn out to contribute $c = -26$. Thus the total anomaly of the x^μ and the ghosts cancels for the particular choice $d = 26$, as we asserted earlier. Moreover, it is also necessary to set the parameter $q = 1$, so that mass-shell condition becomes

$$(L_0 - 1)|\phi\rangle = 0. \tag{1.48}$$

Since the mathematics of the open-string spectrum is the same as that of closed-string right movers, let us now use the equations we have obtained to study the open string spectrum. (Here we are assuming that the open-string boundary conditions are all Neumann, corresponding to spacetime-filling D-branes.) The mass-shell condition is

$$M^2 = -p^2 = -\frac{1}{2}\alpha_0^2 = N - 1, \tag{1.49}$$

where

$$N = \sum_{n=1}^{\infty} \alpha_{-n} \cdot \alpha_n = \sum_{n=1}^{\infty} n a_n^\dagger \cdot a_n. \tag{1.50}$$

The $a^\dagger$'s and a's are properly normalized raising and lowering operators. Since each $a^\dagger a$ has eigenvalues $0, 1, 2, \ldots$, the possible values of N are also $0, 1, 2, \ldots$. The unique way to realize $N = 0$ is for all the oscillators to be in the ground state, which we denote simply by $|0; p^\mu\rangle$, where p^μ is the momentum of the state. This state has $M^2 = -1$, which is a tachyon (p^μ is spacelike). Such a faster-than-light particle is certainly not possible in a consistent quantum theory, because the vacuum would be unstable. However, in perturbation theory (which is the framework we are implicitly considering) this instability is not visible. Since this string theory is only supposed to be a warm-up exercise before considering tachyon-free superstring theories, let us continue without worrying about it.

The first excited state, with $N = 1$, corresponds to $M^2 = 0$. The only way to achieve $N = 1$ is to excite the first oscillator once:

$$|\phi\rangle = \zeta_\mu \alpha_{-1}^\mu |0; p\rangle. \tag{1.51}$$

Here ζ_μ denotes the polarization vector of a massless spin-one particle. The Virasoro constraint condition $L_1|\phi\rangle = 0$ implies that ζ_μ must satisfy

$$p^\mu \zeta_\mu = 0. \tag{1.52}$$

This ensures that the spin is transversely polarized, so there are $d - 2$ independent polarization states. This agrees with what one finds for a massless Maxwell or Yang–Mills field.

At the next mass level, where $N = 2$ and $M^2 = 1$, the most general possibility has the form

$$|\phi\rangle = (\zeta_\mu \alpha_{-2}^\mu + \lambda_{\mu\nu} \alpha_{-1}^\mu \alpha_{-1}^\nu)|0; p\rangle. \tag{1.53}$$

However, the constraints $L_1|\phi\rangle = L_2|\phi\rangle = 0$ restrict ζ_μ and $\lambda_{\mu\nu}$. The analysis is interesting, but only the results will be described. If $d > 26$, the physical spectrum contains a negative-norm state, which is not allowed. However, when $d = 26$, this state becomes zero norm and decouples from the theory. This leaves a pure massive "spin two" (symmetric traceless tensor) particle as the only physical state at this mass level.

Let us now turn to the closed-string spectrum. A closed-string state is described as a tensor product of a left-moving state and a right-moving state, subject to the condition that the N value of the left-moving and the right-moving state is the same. The reason for this "level-matching" condition is that we have $(L_0 - 1)|\phi\rangle = (\tilde{L}_0 - 1)|\phi\rangle = 0$. The sum $(L_0 + \tilde{L}_0 - 2)|\phi\rangle$ is interpreted as the mass-shell condition, while the difference $(L_0 - \tilde{L}_0)|\phi\rangle = (N - \tilde{N})|\phi\rangle = 0$ is the level-matching condition.

Using this rule, the closed-string ground state is just

$$|0\rangle \otimes |0\rangle, \tag{1.54}$$

which represents a spin 0 tachyon with $M^2 = -2$. (The notation no longer displays the momentum p of the state.) Again, this signals an unstable vacuum, but we will not worry about it here. Much more important, and more significant, is the first excited state

$$|\phi\rangle = \zeta_{\mu\nu}(\alpha_{-1}^\mu|0\rangle \otimes \tilde{\alpha}_{-1}^\nu|0\rangle), \tag{1.55}$$

which has $M^2 = 0$. The Virasoro constraints $L_1|\phi\rangle = \tilde{L}_1|\phi\rangle = 0$ imply that $p^\mu \zeta_{\mu\nu} = 0$. Such a polarization tensor encodes three distinct spin states, each of which plays a fundamental role in string theory. The symmetric part of $\zeta_{\mu\nu}$ encodes a spacetime metric field $g_{\mu\nu}$ (massless spin two) and a scalar dilaton field ϕ (massless spin zero). The $g_{\mu\nu}$ field is the graviton field, and its presence (with the correct gauge invariances) accounts for the fact that the theory contains general relativity, which is a good approximation for $E \ll 1/\ell_s$. Its vacuum value determines the spacetime geometry. Similarly, the value of ϕ determines the string coupling constant ($g_s = < e^{\phi} >$).

$\zeta_{\mu\nu}$ also has an antisymmetric part, which corresponds to a massless antisymmetric tensor gauge field $B_{\mu\nu} = -B_{\nu\mu}$. This field has a gauge transformation of the form

$$\delta B_{\mu\nu} = \partial_\mu \Lambda_\nu - \partial_\nu \Lambda_\mu, \tag{1.56}$$

(which can be regarded as a generalization of the gauge transformation rule for the Maxwell field: $\delta A_\mu = \partial_\mu \Lambda$). The gauge-invariant field strength (analogous to $F_{\mu\nu} = \partial_\mu A_\nu - \partial_\nu A_\mu$) is

$$H_{\mu\nu\rho} = \partial_\mu B_{\nu\rho} + \partial_\nu B_{\rho\mu} + \partial_\rho B_{\mu\nu}. \tag{1.57}$$

The importance of the $B_{\mu\nu}$ field resides in the fact that the fundamental string is a source for $B_{\mu\nu}$, just as a charged particle is a source for the vector potential A_μ. Mathematically, this is expressed by the coupling

$$q \int B_{\mu\nu} dx^\mu \wedge dx^\nu, \tag{1.58}$$

which generalizes the coupling of a charged particle to a Maxwell field

$$q \int A_\mu dx^\mu \tag{1.59}$$

in a convenient notation.

3.6. THE NUMBER OF PHYSICAL STATES

The number of physical states grows rapidly as a function of mass. This can be analyzed quantitatively. For the open string, let us denote the number of physical states with $\alpha' M^2 = n - 1$ by d_n. These numbers are encoded in the generating function

$$G(w) = \sum_{n=0}^{\infty} d_n w^n = \prod_{m=1}^{\infty} (1 - w^m)^{-24}. \tag{1.60}$$

The exponent 24 reflects the fact that in 26 dimensions, once the Virasoro conditions are taken into account, the spectrum is exactly what one would get from 24 transversely polarized oscillators. It is easy to deduce from this generating function the asymptotic number of states for large n, as a function of n

$$d_n \sim n^{-27/4} e^{4\pi\sqrt{n}}. \tag{1.61}$$

Exercise: Verify this formula.
This asymptotic degeneracy implies that the finite-temperature partition function

$$\mathrm{tr}\,(e^{-\beta H}) = \sum_{n=0}^{\infty} d_n e^{-\beta M_n} \tag{1.62}$$

diverges for $\beta^{-1} = T > T_H$, where T_H is the Hagedorn temperature

$$T_H = \frac{1}{4\pi\sqrt{\alpha'}} = \frac{1}{4\pi\ell_s}. \tag{1.63}$$

T_H might be the maximum possible temperature or else a critical temperature at which there is a phase transition.

3.7. THE STRUCTURE OF STRING PERTURBATION THEORY

As we discussed in the first lecture, perturbation theory calculations are carried out by computing Feynman diagrams. Whereas in ordinary quantum field theory Feynman diagrams are webs of world lines, in the case of string theory they are two-dimensional surfaces representing string world sheets. For these purposes, it is convenient to require that the world-sheet geometry is Euclidean (i.e., the world-sheet metric $h_{\alpha\beta}$ is positive definite). The diagrams are classified by their topology, which is very well understood in the case of two-dimensional surfaces. The world-sheet topology is characterized by the number of handles (h), the number of boundaries (b), and whether or not they are orientable. The order of the expansion (i.e., the power of the string coupling constant) is determined by the Euler number of the world sheet M. It is given by $\chi(M) = 2 - 2h - b$. For example, a sphere has $h = b = 0$, and hence $\chi = 2$. A torus has $h = 1$, $b = 0$, and $\chi = 0$, a cylinder has $h = 0$, $b = 2$, and $\chi = 0$, and so forth. Surfaces with $\chi = 0$ admit a flat metric.

A scattering amplitude is given by a path integral of the schematic structure

$$\int Dh_{\alpha\beta}(\sigma) Dx^{\mu}(\sigma) e^{-S[h,x]} \prod_{i=1}^{n_c} \int_M V_{\alpha_i}(\sigma_i) d^2\sigma_i \prod_{j=1}^{n_o} \int_{\partial M} V_{\beta_j}(\sigma_j) d\sigma_j. \tag{1.64}$$

The action $S[h, x]$ is given in eq. (1.21). V_{α_i} is a vertex operator that describes emission or absorption of a closed-string state of type α_i from the interior of the string world sheet, and V_{β_j} is a vertex operator that describes emission of absorption of an open-string state of type β_j from the boundary of the string world sheet. There are lots of technical details that are not explained here. In the end, one finds that the conformally inequivalent world sheets of a given topology are described by a finite number of parameters, and thus these amplitudes can be recast as finite-dimensional integrals over these "moduli." (The momentum integrals are already done.) The dimension of the resulting integral turns out to be

$$N = 3(2h + b - 2) + 2n_c + n_o. \tag{1.65}$$

As an example consider the amplitude describing elastic scattering of two-open string ground states. In this case $h = 0$, $b = 1$, $n_c = 0$, $n_o = 4$, and therefore $N = 1$. In terms of the usual Mandelstam invariants $s = -(p_1 + p_2)^2$ and $t = -(p_1 - p_4)^2$, the result is

$$A(s,t) = g_s^2 \int_0^1 dx \;\; x^{-\alpha(s)-1}(1-x)^{-\alpha(t)-1}, \tag{1.66}$$

where the Regge trajectory $\alpha(s)$ is

$$\alpha(s) = 1 + \alpha' s. \tag{1.67}$$

This integral is just the Euler beta function

$$A(s,t) = g_s^2 B(-\alpha(s), -\alpha(t)) = g_s^2 \frac{\Gamma(-\alpha(s))\Gamma(-\alpha(t))}{\Gamma(-\alpha(s) - \alpha(t))}. \tag{1.68}$$

This is the famous Veneziano amplitude [10], which got the whole business started.

3.8. RECAPITULATION

This lecture described some of the basic facts of the 26-dimensional bosonic string theory. One significant point that has not yet been made clear is that there are actually a number of distinct theories depending on what kinds of strings one includes

- oriented closed strings only
- oriented closed strings and oriented open strings. In this case one can incorporate $U(n)$ gauge symmetry.

- unoriented closed strings only
- unoriented closed strings and unoriented open strings. In this case one can incorporate $SO(n)$ or $Sp(n)$ gauge symmetry.

As we have mentioned already, all the bosonic string theories are sick as they stand, because (in each case) the closed-string spectrum contains a tachyon. A tachyon means that one is doing perturbation theory about an unstable vacuum. This is analogous to the unbroken symmetry extremum of the Higgs potential in the standard model. In that case, we know that there is a stable minimum, where the Higgs fields acquires a vacuum value. It is conceivable that the closed-string tachyon condenses in an analogous manner, or else there might not be a stable vacuum. Recently, there has been success in demonstrating that open-string tachyons condense at a stable minimum, but the fate of closed-string tachyons is still an open problem.

4. LECTURE 3: SUPERSTRINGS

Among the deficiencies of the bosonic string theory is the fact that there are no fermions. As we will see, the addition of fermions leads quite naturally to supersymmetry and hence superstrings. There are two alternative formalisms that are used to study superstrings. The original one, which grew out of the 1971 papers by Ramond [11] and by Neveu and me [12], is called the RNS formalism. In this approach, the supersymmetry of the two-dimensional world-sheet theory plays a central role. The second approach, developed by Michael Green and me in the early 1980's [13], emphasizes supersymmetry in the ten-dimensional spacetime. Due to lack of time, only the RNS approach will be presented.

In the RNS formalism, the world-sheet theory is based on the d functions $x^\mu(\sigma,\tau)$ that describe the embedding of the world sheet in the spacetime, just as before. However, in order to supersymmetrize the world-sheet theory, we also introduce d fermionic partner fields $\psi^\mu(\sigma,\tau)$. Note that x^μ transforms as a vector from the spacetime viewpoint, but as d scalar fields from the two-dimensional world-sheet viewpoint. The ψ^μ also transform as a spacetime vector, but as world-sheet spinors. Altogether, x^μ and ψ^μ described d supersymmetry multiplets, one for each value of μ.

The reparametrization invariant world-sheet action described in the preceding lecture can be generalized to have local supersymmetry on the world sheet, as well. (The details of how that works are a bit too involved to describe here.) When one chooses a suitable conformal gauge ($h_{\alpha\beta} = e^\phi \eta_{\alpha\beta}$), together with an appropriate fermionic gauge condition, one ends up with a world-sheet theory that has global supersymmetry supplemented by constraints. The constraints form a super-Virasoro algebra. This means that in addition to the Virasoro constraints of the bosonic string theory, there are fermionic constraints, as well.

4.1. THE GAUGE-FIXED THEORY

The globally supersymmetric world-sheet action that arises in the conformal gauge takes the form

$$S = -\frac{T}{2}\int d^2\sigma(\partial_\alpha x^\mu \partial^\alpha x_\mu - i\bar{\psi}^\mu \rho^\alpha \partial_\alpha \psi_\mu). \tag{1.69}$$

The first term is exactly the same as in eq. (1.25) of the bosonic string theory. Recall that it has the structure of d free scalar fields. The second term that has now been added is just d free massless spinor fields, with Dirac-type actions. The notation is that ρ^α are two 2×2 Dirac matrices and $\psi = \binom{\psi_-}{\psi_+}$ is a two-component Majorana spinor. The Majorana condition simply means that ψ_+ and ψ_- are real in a suitable representation of Dirac algebra. In fact, a convenient choice is one for which

$$\bar{\psi}\rho^\alpha \partial_\alpha \psi = \psi_- \partial_+ \psi_- + \psi_+ \partial_- \psi_+, \tag{1.70}$$

where $\partial_\pm$ represent derivatives with respect to $\sigma^\pm = \tau \pm \sigma$. In this basis, the equations of motion are simply

$$\partial_+ \psi_-^\mu = \partial_- \psi_+^\mu = 0. \tag{1.71}$$

Thus ψ_-^μ describes right-movers and ψ_+^μ describes left-movers.

Concentrating on the right-movers ψ_-^μ, the global supersymmetry transformations, which are a symmetry of the gauge-fixed action, are

$$\begin{aligned} \delta x^\mu &= i\epsilon\psi_-^\mu \\ \delta\psi_-^\mu &= -2\partial_- x^\mu \epsilon. \end{aligned} \tag{1.72}$$

Exercise: Show that this is a symmetry of the action (1.69).
There is an analogous symmetry for the left-movers. (Accordingly, the world-sheet theory is said to have $(1,1)$ supersymmetry.) Continuing to focus on the right-movers, the Virasoro constraint is

$$(\partial_- x)^2 + \frac{i}{2}\psi_-^\mu \partial_- \psi_{\mu-} = 0. \tag{1.73}$$

The first term is what we found in the bosonic string theory, and the second term is an additional fermionic contribution. There is also an associated fermionic constraint

$$\psi_-^\mu \partial_- x_\mu = 0. \tag{1.74}$$

The Fourier modes of these constraints satisfy the super-Virasoro algebra. There is a second identical super-Virasoro algebra for the left-movers.

As in the bosonic string theory, the Virasoro algebra has conformal anomaly terms proportional to a central charge c. As in that theory, each component of x^μ contributes $+1$ to the central charge, for a total of d, while (in the BRST quantization approach) the reparametrization symmetry ghosts contribute -26. But now there are additional contributions. Each component of ψ^μ gives $+1/2$, for a total of $d/2$, and the local supersymmetry ghosts contribute $+11$. Adding all of this up, gives a grand total of $c = \frac{3d}{2} - 15$. Thus, we see that the conformal anomaly cancels for the specific choice $d = 10$. This is the preferred critical dimension for superstrings, just as $d = 26$ is the critical dimension for bosonic strings. For other values the theory has a variety of inconsistencies.

4.2. THE R AND NS SECTORS

Let us now consider boundary conditions for $\psi^\mu(\sigma, \tau)$. (The story for x^μ is exactly as before.) First, let us consider open-string boundary conditions. For the action to be well-defined, it turns out that one must set $\psi_+ = \pm\psi_-$ at the two ends $\sigma = 0, \pi$. An overall sign is a matter of convention, so we can set

$$\psi_+^\mu(0,\tau) = \psi_-^\mu(0,\tau), \tag{1.75}$$

without loss of generality. But this still leaves two possibilities for the other end, which are called R and NS:

$$\begin{aligned} \mathrm{R} &: \ \psi_+^\mu(\pi,\tau) = \psi_-^\mu(\pi,\tau) \\ \mathrm{NS} &: \ \psi_+^\mu(\pi,\tau) = -\psi_-^\mu(\pi,\tau). \end{aligned} \tag{1.76}$$

Combining these with the equations of motion $\partial_- \psi_+ = \partial_+ \psi_- = 0$, allows us to express the general solutions as Fourier series

$$\begin{aligned} \mathrm{R}: \quad \psi^\mu_- &= \frac{1}{\sqrt{2}} \sum_{n \in \mathbf{Z}} d^\mu_n e^{-in(\tau-\sigma)} \\ \psi^\mu_+ &= \frac{1}{\sqrt{2}} \sum_{n \in \mathbf{Z}} d^\mu_n e^{-in(\tau+\sigma)} \\ \mathrm{NS}: \quad \psi^\mu_- &= \frac{1}{\sqrt{2}} \sum_{r \in \mathbf{Z}+1/2} b^\mu_r e^{-ir(\tau-\sigma)} \\ \psi^\mu_+ &= \frac{1}{\sqrt{2}} \sum_{r \in \mathbf{Z}+1/2} b^\mu_r e^{-ir(\tau+\sigma)}. \end{aligned} \tag{1.77}$$

The Majorana condition implies that $d^\mu_{-n} = d^{\mu\dagger}_n$ and $b^\mu_{-r} = b^{\mu\dagger}_r$. Note that the index n takes integer values, whereas the index r takes half-integer values $(\pm\frac{1}{2}, \pm\frac{3}{2}, \ldots)$. In particular, only the R boundary condition gives a zero mode.

Canonical quantization of the free fermi fields $\psi^\mu(\sigma, \tau)$ is very standard and straightforward. The result can be expressed as anticommutation relations for the coefficients d^μ_m and b^μ_r:

$$\begin{aligned} \mathrm{R} &: \quad \{d^\mu_m, d^\nu_n\} = \eta^{\mu\nu} \delta_{m+n,0} \qquad m, n \in \mathbf{Z} \\ \mathrm{NS} &: \quad \{d^\mu_r, d^\nu_s\} = \eta^{\mu\nu} \delta_{r+s,0} \qquad r, s \in \mathbf{Z} + \frac{1}{2}. \end{aligned} \tag{1.78}$$

Thus, in addition to the harmonic oscillator operators α^μ_m that appear as coefficients in mode expansions of x^μ, there are fermionic oscillator operators d^μ_m or b^μ_r that appear as coefficients in mode expansions of ψ^μ. The basic structure $\{b, b^\dagger\} = 1$ is very simple. It describes a two-state system with $b|0\rangle = 0$, and $b^\dagger|0\rangle = |1\rangle$. The b's or d's with negative indices can be regarded as raising operators and those with positive indices as lowering operators, just as we did for the α^μ_n.

In the NS sector, the ground state $|0; p\rangle$ satisfies

$$\alpha^\mu_m |0; p\rangle = b^\mu_r |0; p\rangle = 0, \quad m, r > 0 \tag{1.79}$$

which is a straightforward generalization of how we defined the ground state in the bosonic string theory. All the excited states obtained by acting with the α and b raising operators are spacetime bosons. We will see later that the ground state, defined as we have done here, is again a tachyon. However, in this theory, as we will also see, there is a way by which this tachyon can (and must) be removed from the physical spectrum.

In the R sector there are zero modes that satisfy the algebra

$$\{d_0^\mu, d_0^\nu\} = \eta^{\mu\nu}. \tag{1.80}$$

This is the d-dimensional spacetime Dirac algebra. Thus the d_0's should be regarded as Dirac matrices and all states in the R sector should be spinors in order to furnish representation spaces on which these operators can act. The conclusion, therefore, is that whereas all string states in the NS sector are spacetime bosons, all string states in the R sector are spacetime fermions.

In the closed-string case, the physical states are obtained by tensoring right-movers and left-movers, each of which are mathematically very similar to the open-string spectrum. This means that there are four distinct sectors of closed-string states: NS⊗NS and R⊗R describe spacetime bosons, whereas NS⊗R and R⊗NS describe spacetime fermions. We will return to explore what this gives later, but first we need to explore the right-movers by themselves in more detail.

The zero mode of the fermionic constraint $\psi^\mu \partial_- x_\mu = 0$ gives a wave equation for (fermionic) strings in the Ramond sector, $F_0|\psi\rangle = 0$, which is called the Dirac–Ramond equation. In terms of the oscillators

$$F_0 = \alpha_0 \cdot d_0 + \sum_{n\neq 0} \alpha_{-n} \cdot d_n. \tag{1.81}$$

The zero-mode piece of F_0, $\alpha_0 \cdot d_0$, has been isolated, because it is just the usual Dirac operator, $\gamma^\mu \partial_\mu$, up to normalization. (Recall that $\alpha_{0\mu}$ is proportional to $p_\mu = -i\partial_\mu$, and d_0^μ is proportional to the Dirac matrices γ^μ.) The fermionic ground state $|\psi_0\rangle$, which satisfies

$$\alpha_n^\mu |\psi_0\rangle = d_n^\mu |\psi_0\rangle = 0, \quad n > 0, \tag{1.82}$$

satisfies the wave equation

$$\alpha_0 \cdot d_0 |\psi_0\rangle = 0, \tag{1.83}$$

which is precisely the massless Dirac equation. Hence the fermionic ground state is a massless spinor.

4.3. THE GSO PROJECTION

In the NS (bosonic) sector the mass formula is

$$M^2 = N - \frac{1}{2}, \tag{1.84}$$

which is to be compared with the formula $M^2 = N - 1$ of the bosonic string theory. This time the number operator N has contributions from the b oscillators as well as the α oscillators. (The reason that the normal-ordering constant is $-1/2$ instead of -1 works as follows. Each transverse α oscillator contributes $-1/24$ and each transverse b oscillator contributes $-1/48$. The result follows since the bosonic theory has 24 transverse directions and the superstring theory has 8 transverse directions.) Thus the ground state, which has $N = 0$, is now a tachyon with $M^2 = -1/2$.

This is where things stood until the 1976 work of Gliozzi, Scherk, and Olive [14]. They noted that the spectrum admits a consistent truncation (called the GSO projection) which is necessary for the consistency of the interacting theory. In the NS sector, the GSO projection keeps states with an odd number of b-oscillator excitations, and removes states with an even number of b-oscillator excitation. Once this rule is implemented the only possible values of N are half integers, and the spectrum of allowed masses are integral

$$M^2 = 0, 1, 2, \ldots . \tag{1.85}$$

In particular, the bosonic ground state is now massless. The spectrum no longer contains a tachyon. The GSO projection also acts on the R sector, where there is an analogous restriction on the d oscillators. This amounts to imposing a chirality projection on the spinors.

Let us look at the massless spectrum of the GSO-projected theory. The ground state boson is now a massless vector, represented by the state $\zeta_\mu b^\mu_{-1/2}|0;p\rangle$, which (as before) has $d - 2 = 8$ physical polarizations. The ground state fermion is a massless Majorana–Weyl fermion which has $\frac{1}{4} \cdot 2^{d/2} = 8$ physical polarizations. Thus there are an equal number of bosons and fermions, as is required for a theory with spacetime supersymmetry. In fact, this is the pair of fields that enter into ten-dimensional super Yang–Mills theory. The claim is that the complete theory now has spacetime supersymmetry.

If there is spacetime supersymmetry, then there should be an equal number of bosons and fermions at every mass level. Let us denote the number of bosonic states with $M^2 = n$ by $d_{NS}(n)$ and the number of fermionic states with $M^2 = n$ by $d_R(n)$. Then we can encode these numbers in generating functions, just as we did for the bosonic string theory

$$f_{NS}(w) = \sum_{n=0}^{\infty} d_{NS}(n) w^n = \frac{1}{2\sqrt{w}} \left(\prod_{m=1}^{\infty} \left(\frac{1 + w^{m-1/2}}{1 - w^m} \right)^8 - \prod_{m=1}^{\infty} \left(\frac{1 - w^{m-1/2}}{1 - w^m} \right)^8 \right) \tag{1.86}$$

$$f_R(w) = \sum_{n=0}^{\infty} d_R(n) w^n = 8 \prod_{m=1}^{\infty} \left(\frac{1 + w^m}{1 - w^m} \right)^8 . \tag{1.87}$$

The 8's in the exponents refer to the number of transverse directions in ten dimensions. The effect of the GSO projection is the subtraction of the second term in f_{NS} and reduction of coefficient in f_R from 16 to 8. In 1829, Jacobi discovered the formula

$$f_R(w) = f_{NS}(w). \tag{1.88}$$

(He used a different notation, of course.) For him this relation was an obscure curiosity, but we now see that it provides strong evidence for supersymmetry of this string theory in ten dimensions. A complete proof of supersymmetry for the interacting theory was constructed by Green and me five years after the GSO paper [13].

4.4. TYPE II SUPERSTRINGS

We have described the spectrum of bosonic (NS) and fermionic (R) string states. This also gives the spectrum of left-moving and right-moving closed-string modes, so we can form the closed-string spectrum by forming tensor products as before. In particular, the massless right-moving spectrum consists of a vector and a Majorana–Weyl spinor. Thus the massless closed-string spectrum is given by

$$(\text{vector } + \text{MW spinor}) \otimes (\text{vector } + \text{MW spinor}). \tag{1.89}$$

There are actually two distinct possibilities because two MW spinor can have either opposite chirality or the same chirality.

When the two MW spinors have opposite chirality, the theory is called type IIA superstring theory, and its massless spectrum forms the type IIA supergravity multiplet. This theory is left-right symmetric. In other words, the spectrum is invariant under mirror reflection. This implies that the IIA theory is parity conserving. When the two MW spinors have the same chirality, the resulting type IIB superstring theory is chiral, and hence parity violating. In each case there are two gravitinos, arising from vector $\otimes$ spinor and spinor $\otimes$ vector, which are gauge fields for local supersymmetry. (In four dimensions we would say that the gravitinos have spin 3/2, but that is not an accurate description in ten dimensions.) Thus, since both type II superstring theories have two gravitinos, they have local $\mathcal{N} = 2$ supersymmetry in the ten-dimensional sense. The supersymmetry charges are Majorana–Weyl spinors, which have 16 components, so the type II theories have 32 conserved supercharges. This is the same amount of supersymmetry as what is usually called $\mathcal{N} = 8$ in four dimensions.

The type II superstring theories contain only oriented closed strings (in the absence of D-branes). However, there is another superstring theory, called type I, which can be obtained by a projection of the type IIB theory, that only keeps the diagonal sum of the two gravitinos. Thus, this theory only has $\mathcal{N} = 1$ supersymmetry (16 supercharges). It is a theory of unoriented closed strings. However, it can be supplemented by unoriented open strings. This introduces a Yang–Mills gauge group, which classically can be $SO(n)$ or $Sp(n)$ for any value of n. Quantum consistency singles out $SO(32)$ as the unique possibility. This restriction can be understood in a number of ways. The way that it was first discovered was by considering anomalies.

4.5. ANOMALIES

Chiral (parity-violating) gauge theories can be inconsistent due to anomalies. This happens when there is a quantum mechanical breakdown of the gauge symmetry, which is induced by certain one-loop Feynman diagrams. (Sometimes one also considers breaking of global symmetries by anomalies, which does not imply an inconsistency. That is not what we are interested in here.) In the case of four dimensions, the relevant diagrams are triangles, with the chiral fields going around the loop and three gauge fields attached as external lines. In the case of the standard model, the quarks and leptons are chiral and contribute to a variety of possible anomalies. Fortunately, the standard model has just the right content so that all of the gauge anomalies cancel. If one discarded the quark or lepton contributions, it would not work.

In the case of ten-dimensional chiral gauge theories, the potentially anomalous Feynman diagrams are hexagons, with six external gauge fields. The anomalies can be attributed to the massless fields, and therefore they can be analyzed in the low-energy effective field theory. There are several possible cases in ten dimensions:

- $\mathcal{N} = 1$ supersymmetric Yang–Mills theory. This theory has anomalies for every choice of gauge group.
- Type I supergravity. This theory has gravitational anomalies.
- Type IIA supergravity. This theory is non-chiral, and therefore it is trivially anomaly-free.
- Type IIB supergravity. This theory has three chiral fields each of which contributes to several kinds of gravitational anomalies. However, when their contributions are combined, the anomalies all cancel. (This result was obtained by Alvarez–Gaumé and Witten in 1983 [15].)

- Type I supergravity coupled to super Yang–Mills. This theory has both gauge and gravitational anomalies for every choice of Yang-Mills gauge group except $SO(32)$ and $E_8 \times E_8$. For these two choices, all the anomalies cancel. (This result was obtained by Green and me in 1984 [16].)

As we mentioned earlier, at the classical level one can define type I superstring theory for any orthogonal or symplectic gauge group. Now we see that at the quantum level, the only choice that is consistent is $SO(32)$. For any other choice there are fatal anomalies. The term $SO(32)$ is used here somewhat imprecisely. There are several different Lie groups that have the same Lie algebra. It turns out that the precise Lie group that is appropriate is Spin $(32)/\mathbf{Z}_2$.

4.6. HETEROTIC STRINGS

The two Lie groups that are singled out — $E_8 \times E_8$ and Spin $(32)/\mathbf{Z}_2$ — have several properties in common. Each of them has dimension $= 496$ and rank $= 16$. Moreover, their weight lattices correspond to the only two even self-dual lattices in 16 dimensions. This last fact was the crucial clue that led Gross, Harvey, Martinec, and Rohm [17] to the discovery of the heterotic string soon after the anomaly cancellation result. One hint is the relation $10 + 16 = 26$. The construction of the heterotic string uses the $d = 26$ bosonic string for the left-movers and the $d = 10$ superstring the right movers. The sixteen extra left-moving dimensions are associated to an even self-dual 16-dimensional lattice. In this way one builds in the $SO(32)$ or $E_8 \times E_8$ gauge symmetry.

Thus, to recapitulate, by 1985 we had five consistent superstring theories, type I (with gauge group $SO(32)$), the two type II theories, and the two heterotic theories. Each is a supersymmetric ten-dimensional theory. The perturbation theory was studied in considerable detail, and while some details may not have been completed, it was clear that each of the five theories has a well-defined, ultraviolet-finite perturbation expansion, satisfying all the usual consistency requirements (unitarity, analyticity, causality, etc.) This was pleasing, though it was somewhat mysterious why there should be five consistent quantum gravity theories. It took another ten years until we understood that these are actually five special quantum vacua of a unique underlying theory.

4.7. T DUALITY

T duality, an amazing result obtained in the late 1980's, relates one string theory with a circular compact dimension of radius R to another string theory with a circular dimension of radius $1/R$ (in units $\ell_s = 1$).

This is very profound, because it indicates a limitation of our usual motions of classical geometry. Strings see geometry differently from point particles. Let us examine how this is possible.

The key to understanding T duality is to consider the kinds of excitations that a string can have in the presence of a circular dimension. One class of excitations, called Kaluza–Klein excitations, is a very general feature of any quantum theory, whether or not based on strings. The idea is that in order for the wave function e^{ipx} to be single valued, the momentum along the circle must be a multiple of $1/R$, $p = n/R$, where n is an integer. From the lower-dimension viewpoint this is interpreted as a contribution $(n/R)^2$ to the square of the mass.

There is a second type of excitation that is special to closed strings. Namely, a closed string can wind m times around the circular dimension, getting caught up on the topology of the space, contributing an energy given by the string tension times the length of the string

$$E_m = 2\pi R \cdot m \cdot T. \tag{1.90}$$

Putting $T = \frac{1}{2\pi}$ (for $\ell_s = 1$), this is just $E_m = mR$.

The combined energy-squared of the Kaluza–Klein and winding-mode excitations is

$$E^2 = \left(\frac{n}{R}\right)^2 + (mR)^2 + \dots, \tag{1.91}$$

where the dots represent string oscillator contributions. Under T duality

$$m \leftrightarrow n, \quad R \leftrightarrow 1/R. \tag{1.92}$$

Together, these interchanges leave the energy invariant. This means that what is interpreted as a Kaluza–Klein excitation in one string theory is interpreted as a winding-mode excitation in the T-dual theory, and the two theories have radii R and $1/R$, respectively. The two principle examples of T-dual pairs are the two type II theories and the two heterotic theories. In the latter case there are additional technicalities that explain how the two gauge groups are related. Basically, when the compactification on a circle to nine dimension is carried out in each case, it is necessary to include effects that we haven't explained (called Wilson lines) to break the gauge groups to $SO(16) \times SO(16)$, which is a common subgroup of $SO(32)$ and $E_8 \times E_8$.

5. LECTURE 4: FROM SUPERSTRINGS TO M THEORY

Superstring theory is currently undergoing a period of rapid development in which important advances in understanding are being achieved.

The focus in this lecture will be on explaining why there can be an eleven-dimensional vacuum, even though there are only ten dimensions in perturbative superstring theory. The nonperturbative extension of superstring theory that allows for an eleventh dimension has been named *M theory.* The letter M is intended to be flexible in its interpretation. It could stand for *magic, mystery,* or *meta* to reflect our current state of incomplete understanding. Those who think that two-dimensional supermembranes (the M2-brane) are fundamental may regard M as standing for *membrane.* An approach called *Matrix theory* is another possibility. And, of course, some view M theory as the *mother* of all theories.

In the first superstring revolution we identified five distinct superstring theories, each in ten dimensions. Three of them, the type I theory and the two heterotic theories, have $\mathcal{N} = 1$ supersymmetry in the ten-dimensional sense. Since the minimal 10d spinor is simultaneously Majorana and Weyl, this corresponds to 16 conserved supercharges. The other two theories, called type IIA and type IIB, have $\mathcal{N} = 2$ supersymmetry (32 supercharges). In the IIA case the two spinors have opposite handedness so that the spectrum is left-right symmetric (nonchiral). In the IIB case the two spinors have the same handedness and the spectrum is chiral.

In each of these five superstring theories it became clear, and was largely proved, that there are consistent perturbation expansions of on-shell scattering amplitudes. In four of the five cases (heterotic and type II) the fundamental strings are oriented and unbreakable. As a result, these theories have particularly simple perturbation expansions. Specifically, there is a unique Feynman diagram at each order of the loop expansion. The Feynman diagrams depict string world sheets, and therefore they are two-dimensional surfaces. For these four theories the unique L-loop diagram is a closed orientable genus-L Riemann surface, which can be visualized as a sphere with L handles. External (incoming or outgoing) particles are represented by N points (or "punctures") on the Riemann surface. A given diagram represents a well-defined integral of dimension $6L+2N-6$. This integral has no ultraviolet divergences, even though the spectrum contains states of arbitrarily high spin (including a massless graviton). From the viewpoint of point-particle contributions, string and supersymmetry properties are responsible for incredible cancellations. Type I superstrings are unoriented and breakable. As a result, the perturbation expansion is more complicated for this theory, and various world-sheet diagrams at a given order have to be combined properly to cancel divergences and anomalies .

An important discovery that was made between the two superstring revolutions is *T duality.* As we explained earlier, this duality relates two string theories when one spatial dimension forms a circle (denoted S^1). Then the ten-dimensional geometry is $R^9 \times S^1$. T duality identifies this string compactification with one of a second string theory also on $R^9 \times S^1$. If the radii of the circles in the two cases are denoted R_1 and R_2, then

$$R_1 R_2 = \alpha'. \tag{1.93}$$

Here $\alpha' = \ell_s^2$ is the universal Regge slope parameter, and ℓ_s is the fundamental string length scale (for both string theories). Note that T duality implies that shrinking the circle to zero in one theory corresponds to decompactification of the dual theory.

The type IIA and IIB theories are T dual, so compactifying the nonchiral IIA theory on a circle of radius R and letting $R \to 0$ gives the chiral IIB theory in ten dimensions! This means, in particular, that they should not be regarded as distinct theories. The radius R is actually the vacuum value of a scalar field, which arises as an internal component of the 10d metric tensor. Thus the type IIA and type IIB theories in 10d are two limiting points in a continuous moduli space of quantum vacua. The two heterotic theories are also T dual, though there are additional technical details in this case. T duality applied to the type I theory gives a dual description, which is sometimes called type I$'$ or IA.

5.1. M THEORY

In the 1970s and 1980s various supersymmetry and supergravity theories were constructed. In particular, supersymmetry representation theory showed that the largest possible spacetime dimension for a supergravity theory (with spins ≤ 2) is eleven. Eleven-dimensional supergravity, which has 32 conserved supercharges, was constructed in 1978 by Cremmer, Julia, and Scherk [18]. It has three kinds of fields—the graviton field (with 44 polarizations), the gravitino field (with 128 polarizations), and a three-index gauge field $C_{\mu\nu\rho}$ (with 84 polarizations). These massless particles are referred to collectively as the *supergraviton*. 11d supergravity is nonrenormalizable, and thus it cannot be a fundamental theory. However, we now believe that it is a low-energy effective description of M theory, which is a well-defined quantum theory. This means, in particular, that higher-dimension terms in the effective action for the supergravity fields have uniquely determined coefficients within the M theory setting, even though they are formally infinite (and hence undetermined) within the supergravity context.

Intriguing connections between type IIA string theory and 11d supergravity have been known for a long time, but the precise relationship was only explained in 1995. The field equations of 11d supergravity admit a solution that describes a supermembrane. In other words, this solution has the property that the energy density is concentrated on a two-dimensional surface. A 3d world-volume description of the dynamics of this supermembrane, quite analogous to the 2d world volume actions of superstrings (in the GS formalism [19]), was constructed by Bergshoeff, Sezgin, and Townsend in 1987 [20]. The authors suggested that a consistent 11d quantum theory might be defined in terms of this membrane, in analogy to string theories in ten dimensions. (Most experts now believe that M theory cannot be defined as a supermembrane theory.) Another striking result was that a suitable dimensional reduction of this supermembrane gives the (previously known) type IIA superstring world-volume action. For many years these facts remained unexplained curiosities until they were reconsidered by Townsend [21] and by Witten [22]. The conclusion is that type IIA superstring theory really does have a circular 11th dimension in addition to the previously known ten spacetime dimensions. This fact was not recognized earlier because the appearance of the 11th dimension is a nonperturbative phenomenon, not visible in perturbation theory.

To explain the relation between M theory and type IIA string theory, a good approach is to identify the parameters that characterize each of them and to explain how they are related. Eleven-dimensional supergravity (and hence M theory, too) has no dimensionless parameters. The only parameter is the 11d Newton constant, which raised to a suitable power ($-1/9$), gives the 11d Planck mass m_p. When M theory is compactified on a circle (so that the spacetime geometry is $R^{10} \times S^1$) another parameter is the radius R of the circle. Now consider the parameters of type IIA superstring theory. They are the string mass scale m_s, introduced earlier, and the dimensionless string coupling constant g_s.

We can identify compactified M theory with type IIA superstring theory by making the following correspondences:

$$m_s^2 = 2\pi R m_p^3 \tag{1.94}$$

$$g_s = 2\pi R m_s. \tag{1.95}$$

Using these one can derive $g_s = (2\pi R m_p)^{3/2}$ and $m_s = g_s^{1/3} m_p$. The latter implies that the 11d Planck length is shorter than the string length scale at weak coupling by a factor of $(g_s)^{1/3}$.

Conventional string perturbation theory is an expansion in powers of g_s at fixed m_s. Equation (1.95) shows that this is equivalent to an expansion about $R = 0$. In particular, the strong coupling limit of type IIA superstring theory corresponds to decompactification of the eleventh dimension, so in a sense M theory is type IIA string theory at infinite coupling. (The $E_8 \times E_8$ heterotic string theory is also eleven-dimensional at strong coupling.) This explains why the eleventh dimension was not discovered in studies of string perturbation theory.

These relations encode some interesting facts. For one thing, the fundamental IIA string actually *is* an M2-brane of M theory with one of its dimensions wrapped around the circular spatial dimension. Denoting the string and membrane tensions (energy per unit volume) by T_{F1} and T_{M2}, one deduces that

$$T_{F1} = 2\pi R T_{M2}. \tag{1.96}$$

However, $T_{F1} = 2\pi m_s^2$ and $T_{M2} = 2\pi m_p^3$. Combining these relations gives eq. (1.94).

5.2. TYPE II *P*-BRANES

Type II superstring theories contain a variety of p-brane solutions that preserve half of the 32 supersymmetries. These are solutions in which the energy is concentrated on a p-dimensional spatial hypersurface. (The world volume has $p + 1$ dimensions.) The corresponding solutions of supergravity theories were constructed in 1991 by Horowitz and Strominger [23]. A large class of these p-brane excitations are called *D-branes* (or Dp-branes when we want to specify the dimension), whose tensions are given by

$$T_{Dp} = 2\pi m_s^{p+1}/g_s. \tag{1.97}$$

This dependence on the coupling constant is one of the characteristic features of a D-brane. Another characteristic feature of D-branes is that they carry a charge that couples to a gauge field in the RR sector of the theory [24]. The particular RR gauge fields that occur imply that p takes even values in the IIA theory and odd values in the IIB theory.

In particular, the D2-brane of the type IIA theory corresponds to the supermembrane of M theory, but now in a background geometry in which one of the transverse dimensions is a circle. The tensions check, because (using eqs. (1.94) and (1.95))

$$T_{D2} = 2\pi m_s^3/g_s = 2\pi m_p^3 = T_{M2}. \tag{1.98}$$

The mass of the first Kaluza–Klein excitation of the 11d supergraviton is $1/R$. Using eq. (1.95), we see that this can be identified with the D0-brane. More identifications of this type arise when we consider the magnetic dual of the M theory supermembrane, which is a five-brane, called the M5-brane.[1] Its tension is $T_{M5} = 2\pi m_p^6$. Wrapping one of its dimensions around the circle gives the D4-brane, with tension

$$T_{D4} = 2\pi R T_{M5} = 2\pi m_s^5/g_s. \tag{1.99}$$

If, on the other hand, the M5-frame is not wrapped around the circle, one obtains the NS5-brane of the IIA theory with tension

$$T_{NS5} = T_{M5} = 2\pi m_s^6/g_s^2. \tag{1.100}$$

To summarize, type IIA superstring theory is M theory compactified on a circle of radius $R = g_s \ell_s$. M theory is believed to be a well-defined quantum theory in 11d, which is approximated at low energy by 11d supergravity. Its excitations are the massless supergraviton, the M2-brane, and the M5-brane. These account both for the (perturbative) fundamental string of the IIA theory and for many of its nonperturbative excitations. The identities that we have presented here are exact, because they are protected by supersymmetry.

5.3. TYPE IIB SUPERSTRING THEORY

Type IIB superstring theory, which is the other maximally supersymmetric string theory with 32 conserved supercharges, is also 10-dimensional, but unlike the IIA theory its two supercharges have the same handedness. At low-energy, type IIB superstring theory is approximated by type IIB supergravity, just as 11d supergravity approximates M theory. In each case the supergravity theory is only well-defined as a classical field theory, but still it can teach us a lot. For example, it can be used to construct p-brane solutions and compute their tensions. Even though such solutions are only approximate, supersymmetry considerations ensure that the tensions, which are related to the kinds of conserved charges the p-branes carry, are exact. Since the IIB spectrum contains massless chiral fields, one should check whether there are anomalies that break the gauge invariances—general coordinate invariance, local Lorentz invariance, and local supersymmetry. In fact, the UV finiteness of the string theory Feynman diagrams ensures that all anomalies must cancel, as was verified from a field theory viewpoint by Alvarez-Gaumé and Witten [15].

Type IIB superstring theory or supergravity contains two scalar fields, the dilation ϕ and an axion χ, which are conveniently combined in a complex field

$$\rho = \chi + ie^{-\phi}. \tag{1.101}$$

[1] In general, the magnetic dual of a p-brane in d dimensions is a $(d-p-4)$-brane.

The supergravity approximation has an $SL(2, R)$ symmetry that transforms this field nonlinearly:

$$\rho \to \frac{a\rho + b}{c\rho + d}, \tag{1.102}$$

where a, b, c, d are real numbers satisfying $ad - bc = 1$. However, in the quantum string theory this symmetry is broken to the discrete subgroup $SL(2, Z)$ [25], which means that a, b, c, d are restricted to be integers. Defining the vacuum value of the ρ field to be

$$\langle \rho \rangle = \frac{\theta}{2\pi} + \frac{i}{g_s}, \tag{1.103}$$

the $SL(2, Z)$ symmetry transformation $\rho \to \rho + 1$ implies that θ is an angular coordinate. Moreover, in the special case $\theta = 0$, the symmetry transformation $\rho \to -1/\rho$ takes $g_s \to 1/g_s$. This symmetry, called *S duality*, implies that coupling constant g_s is equivalent to coupling constant $1/g_s$, so that, in the case of Type II superstring theory, the weak coupling expansion and the strong coupling expansion are identical! (An analogous S-duality transformation relates the Type I superstring theory to the $SO(32)$ heterotic string theory.)

Recall that the type IIA and type IIB superstring theories are T dual, meaning that if they are compactified on circles of radii R_A and R_B one obtains equivalent theories for the identification $R_A R_B = \ell_s^2$. Moreover, we saw that the type IIA theory is actually M theory compactified on a circle. The latter fact encodes nonperturbative information. It turns out to be very useful to combine these two facts and to consider the duality between M theory compactified on a torus ($R^9 \times T^2$) and type IIB superstring theory compactified on a circle ($R^9 \times S^1$).

A torus can be described as the complex plane modded out by the equivalence relations $z \sim z + w_1$ and $z \sim z + w_2$. Up to conformal equivalence, the periods w_1 and w_2 can be replaced by 1 and τ, with $\mathrm{Im}\,\tau > 0$. In this characterization τ and $\tau' = (a\tau + b)/(c\tau + d)$, where a, b, c, d are integers satisfying $ad - bc = 1$, describe equivalent tori. Thus a torus is characterized by a modular parameter τ and an $SL(2, Z)$ modular group. The natural, and correct, conjecture at this point is that one should identify the modular parameter τ of the M theory torus with the parameter ρ that characterizes the type IIB vacuum [26, 27]. Then the duality of M theory and type IIB superstring theory gives a geometrical explanation of the nonperturbative S duality symmetry of the IIB theory: the transformation $\rho \to -1/\rho$, which sends $g_s \to 1/g_s$ in the IIB theory, corresponds to interchanging the two cycles of the torus in the M theory description. To complete the story, we should relate the area of the M theory torus (A_M) to the radius of the IIB theory circle (R_B). This is a simple consequence of formulas given above

$$m_p^3 A_M = (2\pi R_B)^{-1}. \tag{1.104}$$

Thus the limit $R_B \to 0$, at fixed ρ, corresponds to decompactification of the M theory torus, while preserving its shape. Conversely, the limit $A_M \to 0$ corresponds to decompactification of the IIB theory circle. The duality can be explored further by matching the various p-branes in 9 dimensions that can be obtained from either the M theory or the IIB theory viewpoints. When this is done, one finds that everything matches nicely and that one deduces various relations among tensions [28].

Another interesting fact about the IIB theory is that it contains an infinite family of strings labeled by a pair of integers (p, q) with no common divisor [26]. The $(1, 0)$ string can be identified as the fundamental IIB string, while the $(0, 1)$ string is the D-string. From this viewpoint, a (p, q) string can be regarded as a bound state of p fundamental strings and q D-strings [29]. These strings have a very simple interpretation in the dual M theory description. They correspond to an M2-brane with one of its cycles wrapped around a (p, q) cycle of the torus. The minimal length of such a cycle is proportional to $|p + q\tau|$, and thus (using $\tau = \rho$) one finds that the tension of a (p, q) string is given by

$$T_{p,q} = 2\pi |p + q\rho| m_s^2. \tag{1.105}$$

Imagine that you lived in the 9-dimensional world that is described equivalently as M theory compactified on a torus or as the type IIB superstring theory compactified on a circle. Suppose, moreover, you had very high energy accelerators with which you were going to determine the "true" dimension of spacetime. Would you conclude that 10 or 11 is the correct answer? If either A_M or R_B was very large in Planck units there would be a natural choice, of course. But how could you decide otherwise? The answer is that either viewpoint is equally valid. What determines which choice you make is which of the massless fields you regard as "internal" components of the metric tensor and which ones you regards as matter fields. Fields that are metric components in one description correspond to matter fields in the dual one.

5.4. THE D3-BRANE AND $\mathcal{N} = 4$ GAUGE THEORY

D-branes have a number of special properties, which make them especially interesting. By definition, they are branes on which strings can end—D stands for *Dirichlet* boundary conditions. The end of a string carries a charge, and the D-brane world-volume theory contains a $U(1)$ gauge field that carries the associated flux. When n Dp-branes are coincident, or parallel and nearly coincident, the associated (p+1)-dimensional world-volume theory is a $U(n)$ gauge theory [29]. The n^2 gauge bosons A_μ^{ij} and their supersymmetry partners arise as the ground states of oriented strings running from the ith Dp-brane to the jth Dp-brane. The diagonal elements, belonging to the Cartan subalgebra, are massless. The field A_μ^{ij} with $i \neq j$ has a mass proportional to the separation of the ith and jth branes.

The $U(n)$ gauge theory associated with a stack of n Dp-branes has maximal supersymmetry (16 supercharges). The low-energy effective theory, when the brane separations are small compared to the string scale, is supersymmetric Yang–Mills theory. These theories can be constructed by dimensional reduction of 10d supersymmetric $U(n)$ gauge theory to $p+1$ dimensions. A case of particular interest, which we shall now focus on, is $p = 3$. A stack of n D3-branes in type IIB superstring theory has a decoupled $\mathcal{N} = 4$, $d = 4$ $U(n)$ gauge theory associated to it. This gauge theory has a number of special features. For one thing, due to boson–fermion cancellations, there are no UV divergences at any order of perturbation theory. The beta function $\beta(g)$ is identically zero, which implies that the theory is scale invariant. In fact, $\mathcal{N} = 4$, $d = 4$ gauge theories are conformally invariant. The conformal invariance combines with the supersymmetry to give a superconformal symmetry, which contains 32 fermionic generators. Another important property of $\mathcal{N} = 4$, $d = 4$ gauge theories is an electric-magnetic duality, which extends to an $SL(2, Z)$ group of dualities. Now consider the $\mathcal{N} = 4$ $U(n)$ gauge theory associated to a stack of n D3-branes in type IIB superstring theory. There is an obvious identification, that turns out to be correct. Namely, the $SL(2, Z)$ duality of the gauge theory is induced from that of the ambient type IIB superstring theory. The D3-branes themselves are invariant under $SL(2, Z)$ transformations.

As we have said, a fundamental $(1, 0)$ string can end on a D3-brane. But by applying a suitable $SL(2, Z)$ transformation, this configuration is transformed to one in which a (p, q) string ends on the D3-brane. The charge on the end of this string describes a dyon with electric charge p and magnetic charge q, with respect to the appropriate gauge field. More generally, for a stack of n D3-branes, any pair can be connected by a (p, q) string. The mass is proportional to the length of the string times its tension, which we saw is proportional to $|p + q\rho|$. In this way one sees that the electrically charged particles, described by fundamental fields, belong to infinite $SL(2, Z)$ multiplets. The other states are nonperturbative excitations of the gauge theory. The field configurations that describe them preserve half of the supersymmetry. As a result their masses are given exactly by the considerations described above. An interesting question, whose answer was unknown until recently, is whether $\mathcal{N} = 4$ gauge theories in four dimensions also admit nonperturbative excitations that preserve 1/4 of the supersymmetry. The answer turns out to be that they do, but only if $n \geq 3$. This result has a nice dual description in terms of three-string junctions [30].

5.5. CONCLUSION

In this lecture we have described some of the interesting advances in understanding superstring theory that have taken place in the past few years. The emphasis has been on the nonperturbative appearance of an eleventh dimension in type IIA superstring theory, as well as its implications when combined with superstring T dualities. In particular, we argued that there should be a consistent quantum vacuum, whose low-energy effective description is given by 11d supergravity.

What we have described makes a convincing self-consistent picture, but it does not constitute a complete formulation of M theory. In the past several years there have been some major advances in that direction, which we will briefly mention here. The first, which goes by the name of *Matrix Theory*, bases a formulation of M theory in flat 11d spacetime in terms of the supersymmetric quantum mechanics of N D0-branes in the large N limit. Matrix Theory has passed all tests that have been carried out, some of which are very nontrivial. The construction has a nice generalization to describe compactification of M theory on a torus T^n. However, it does not seem to be useful for $n > 5$, and other compactification manifolds are (at best) awkward to handle. Another shortcoming of this approach is that it treats the eleventh dimension differently from the other ones.

Another proposal relating superstring and M theory backgrounds to large N limits of certain field theories has been put forward by Maldacena in 1997 [31] and made more precise by Gubser, Klebanov, and Polyakov [32], and by Witten [33] in 1998. (For a review of this subject, see [34].) In this approach, there is a conjectured duality (*i.e.*, equivalence) between a conformally invariant field theory (CFT) in d dimensions and type IIB superstring theory or M theory on an Anti-de-Sitter space (AdS) in $d+1$ dimensions. The remaining $9-d$ or $10-d$ dimensions form a compact space, the simplest cases being spheres. Three examples with unbroken supersymmetry are $AdS_5 \times S^5$, $AdS_4 \times S^7$, and $AdS_7 \times S^4$. This approach is sometimes referred to as *AdS/CFT duality*. This is an extremely active and very promising subject. It has already taught us a great deal about the large N behavior of various gauge theories. As usual, the easiest theories to study are ones with a lot of supersymmetry, but it appears that in this approach supersymmetry breaking is more accessible than in previous ones. For example, it might someday be possible to construct the QCD string in terms of a dual AdS gravity theory, and use it to carry out numerical calculations of the hadron spectrum. Indeed, there have already been some preliminary steps in this direction.

Despite all of the successes that have been achieved in advancing our understanding of superstring theory and M theory, there clearly is still a long way to go. In particular, despite much effort and several imaginative proposals, we still do not have a convincing mechanism for ensuring the vanishing (or extreme smallness) of the cosmological constant for nonsupersymmetric vacua. Superstring theory is a field with very ambitious goals. The remarkable fact is that they still seem to be realistic. However, it may take a few more revolutions before they are attained.

References

[1] M.B. Green, J.H. Schwarz, and E. Witten, *Superstring Theory*, in 2 vols., Cambridge Univ. Press, 1987.

[2] J. Polchinski, *String Theory*, in 2 vols., Cambridge Univ. Press, 1998.

[3] J. Scherk and J. H. Schwarz, *Nucl. Phys.* **B81** (1974) 118.

[4] T. Yoneya, *Prog. Theor. Phys.* **51** (1974) 1907.

[5] P. Candelas, G.T. Horowitz, A. Strominger, and E. Witten, *Nucl. Phys.* **B258** (1985) 46.

[6] Y. Nambu, Notes prepared for the Copenhagen High Energy Symposium (1970).

[7] T. Goto, *Prog. Theor. Phys.* **46** (1971) 1560.

[8] A. M. Polyakov, *Phys. Lett.* **103B** (1981) 207.

[9] M. Virasoro, *Phys. Rev.* **D1** (1970) 2933.

[10] G. Veneziano, *Nuovo Cim.* **57A** (1968) 190.

[11] P. Ramond, *Phys. Rev.* **D3** (1971) 2415.

[12] A. Neveu and J. H. Schwarz, *Nucl. Phys.* **B31** (1971) 86.

[13] M. B. Green and J. H. Schwarz, *Nucl. Phys.* **B181** (1981) 502; *Nucl. Phys.* bf B198 (1982) 252; *Phys. Lett.* **109B** (1982) 444.

[14] F. Gliozzi, J. Scherk, and D. Olive, *Phys. Lett.* **65B** (1976) 282.

[15] L. Alvarez-Gaumé and E. Witten, *Nucl. Phys.* **B234** (1983) 269.

[16] M.B. Green and J.H. Schwarz, *Phys. Lett.* **149B** (1984) 117.

[17] D.J. Gross, J.A. Harvey, E. Martinec, and R. Rohm, *Phys. Rev. Lett.* **54** (1985) 502.

[18] E. Cremmer, B. Julia, and J. Scherk, *Phys. Lett.* **76B** (1978) 409.

[19] M.B. Green and J.H. Schwarz, *Phys. Lett.* **136B** (1984) 367.

[20] E. Bergshoeff, E. Sezgin, and P.K. Townsend, *Phys. Lett.* **B189** (1987) 75.

[21] P.K. Townsend, *Phys. Lett.* **B350** (1995) 184, hep-th/9501068.

[22] E. Witten, *Nucl. Phys.* **B443** (1995) 85, hep-th/9503124.

[23] G.T. Horowitz and A. Strominger, *Nucl. Phys.* **B360** (1991) 197.

[24] J. Polchinski, *Phys. Rev. Lett.* **75** (1995) 4724, hep-th/9510017.

[25] C. Hull and P. Townsend, *Nucl. Phys.* **B438** (1995) 109, hep-th/9410167.

[26] J.H. Schwarz, *Phys. Lett.* **B360** (1995) 13, Erratum: *Phys. Lett.* **B364** (1995) 252, hep-th/9508143.

[27] P.S. Aspinwall, *Nucl. Phys. Proc. Suppl.* **46** (1996) 30, hep-th/9508154.

[28] J.H. Schwarz, *Phys. Lett.* **B367** (1996) 97, hep-th/9510086.

[29] E. Witten, *Nucl. Phys.* **B460** (1996) 335, hep-th/9510135.

[30] O. Bergman, *Nucl. Phys.* **B525** (1998) 104, hep-th/9712211.

[31] J. Maldacena, *Adv. Theor. Phys.* **2** (1998) 231, hep-th/9711200.

[32] S.S. Gubser, I.R. Klebanov, and A.M. Polyakov, *Phys. Lett.* **B428** (1998) 105, hep-th/9802109.

[33] E. Witten, *Adv. Theor. Math. Phys.* **2** (1998) 253, hep-th/9802150.

[34] O. Aharony, S.S. Gubser, J. Maldacena, H. Ooguri, and Y. Oz, *Phys. Rept.* **323** (2000) 183.

NEUTRINO MASS AND OSCILLATIONS

Janet Conrad
Columbia University,
Nevis Laboratories,
Irvington, NY 10533
conrad@nevis.columbia.edu

1. INTRODUCTION

We still have much to learn about neutrinos as compared to their charged partners. For example, while we have measured the masses of the quarks and charged leptons, we do not know if neutrinos have mass at all. If neutrinos do have mass, this also has important cosmological consequences. With 10^9 remnant neutrinos from the big bang in every cubic meter of space, even a tiny mass will affect the large-scale galactic structure and the expansion of the universe. If neutrinos do have mass, this has important consequences for the Standard Model of particle physics. The neutrino masses may give us clues to the lepton mass hierarchy. Neutrino mass may lead to the introduction of new particles to the theory. Finally, the charged fermions have right-handed as well as left-handed components, but neutrinos appear to be entirely left-handed. Is there a right-handed neutrino which we have not observed because it is "sterile"?

These are the questions which may be answered through the study of neutrino oscillations. Neutrinos must have mass in order for oscillations to occur. This effect may allow one of the few probes for sterile neutrinos.

The purpose of this paper is to give a general review of neutrinos and neutrino oscillations. This discussion begins with an introduction to neutrinos within the Standard Model; progresses through direct and indirect measurements of neutrino mass; provides an overview of the present experimental picture; considers alternatives for interpreting the results within various theoretical frameworks; and finally, presents the ongoing and future oscillation experiments which will resolve many of the outstanding questions.

H.B. Prosper and M. Danilov (eds.), Techniques and Concepts of High-Energy Physics, 189–249.

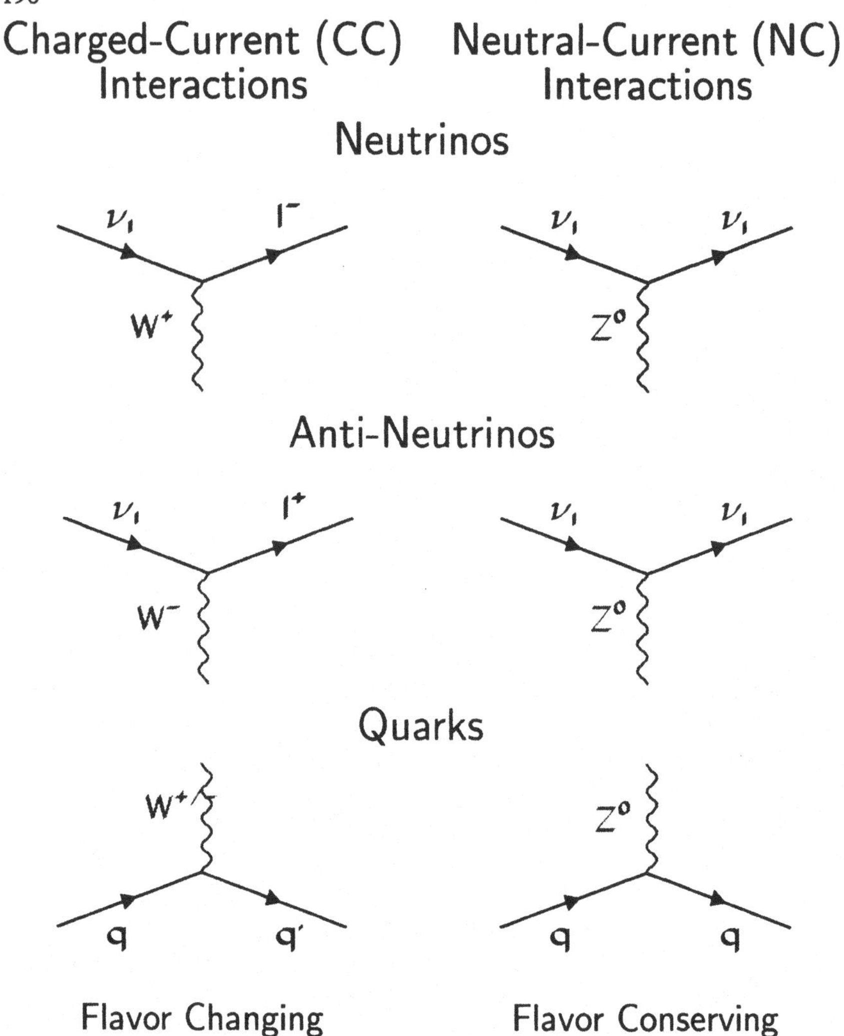

Figure 1 Charged Current (CC) and Neutral Current (NC) neutrino interactions.

2. NEUTRINOS IN THE STANDARD MODEL

Neutrinos are the only Standard Model fermions to interact strictly via the weak interaction. This proceeds through two types of boson exchange, as illustrated in Fig. 1. Exchange of the Z^0 is called the neutral current (NC) interaction, while exchange of $W^{\pm}$ is called the charged current (CC) interaction. When a W is emitted, charge conservation at the vertex requires that a charged lepton exit the interaction. Therefore, this is also called the flavor-changing interaction, since it converts a neutrino to its charged partner. We know the family of an incoming neutrino by the charged partner which exits the CC interaction. For example, a scattered electron tags a ν_e interaction, a μ tags a ν_μ interaction, etc. The neutrino always emits the W^+ and the antineutrino always emits the W^- in the CC interaction. The CC weak interaction is also flavor-changing for the target particle. For example, if a u quark absorbs a W^-, a d will exit the interaction vertex.

In 1989, measurements of the Z^0 width at LEP[1] and SLD[2] determined that there are only three families of light-mass weakly-interacting neutrinos. These are the ν_e, the ν_μ, and the ν_τ. The interactions of the ν_e and ν_μ have been shown to be consistent with the Standard Model weak interaction. Until recently, there has only been indirect evidence for the ν_τ through the decay of the τ meson. However, in July 2000, the DoNuT Experiment (E872) at Fermilab presented the first evidence for ν_τ interactions[3], having observed four candidate events in an emulsion detector. One candidate is shown in Fig. 2 as an example. Each candidate had the signature of a kink in the outgoing lepton track, indicating a scattered τ from a CC interaction which subsequently decayed. The observation is consistent with the Standard Model, and for the discussion below, I will assume that the ν_τ interacts according to the predictions.

Within the Standard Model, neutrinos are massless. This assumption is consistent with direct experimental observation. It is also the simplest way to explain the feature of "handedness" associated with neutrinos. To understand handedness, it is simplest to begin by discussing "helicity," since for massless particles helicity and handedness are identical.

For a spin 1/2 particle, helicity is the projection of a particle's spin ($\mathbf{\Sigma}$) along its direction of motion $\hat{\mathbf{p}}$, with operator $\mathbf{\Sigma} \cdot \hat{\mathbf{p}}$. Helicity has two possible states: spin aligned opposite the direction of motion (negative, or "left helicity") and spin aligned along the direction of motion (positive or "right helicity"). If a particle is massive, then the sign of the helicity of the particle will be frame dependent. When one boosts to a frame where one is moving faster than the particle, the sign of the momentum will change but the spin will not, and therefore the helicity will flip. For massless particles, hence traveling at the speed of light, one cannot boost to a frame where helicity changes sign.

Handedness (or chirality) is the Lorentz invariant (*i.e.* frame-independent) analogue of helicity for both massless and massive particles. There are two states: "left handed" (LH) and "right handed" (RH). For the case

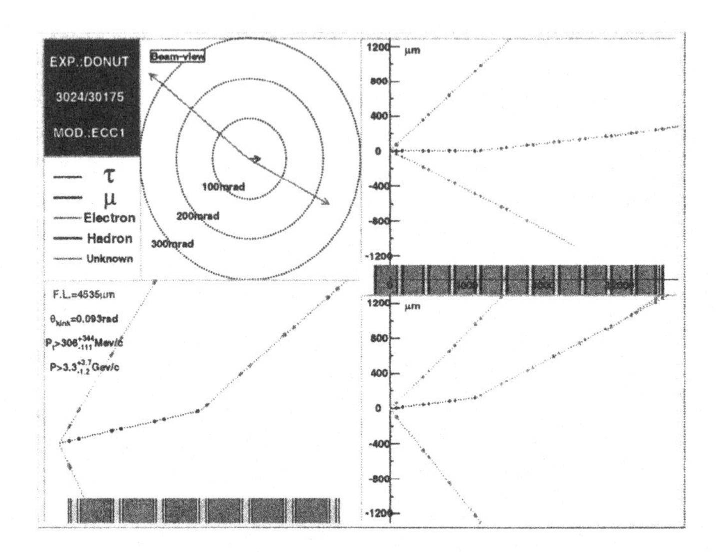

Figure 2 A candidate interaction for ν_τ + nucleon $\rightarrow \tau$ + hadrons seen in emulsion by the DoNuT experiment. Four views of the same event are shown. The neutrino enters into the page for the top left view. For the remaining three, the incoming neutrino (which leaves to track) is from the left. The kinked track is the signature for τ decay. Note the scale. The τ decay will occur within millimeters of the vertex for typical fixed target energies. To date, four candidates have been observed.

of massless particles, including Standard Model neutrinos, helicity and handedness are identical. A massless fermion is either purely LH or RH, and, in principle, can appear in either state. Massive particles have both RH and LH components. A helicity eigenstate for a massive particle is a combination of handedness states. It is only in the high energy limit, where particles are effectively massless, that handedness and helicity coincide for massive fermions. Nevertheless, people tend to use the terms "helicity" and "handedness" interchangeably.

Unlike the electromagnetic and strong interactions, the weak interaction involving neutrinos has a definite preferred handedness. In the late 1950's, it was shown that neutrinos are LH and outgoing antineutrinos are RH [4]. To see how to include this in calculations, consider the following argument using the Weyl, or chiral, spinor representation. A spinor can be written in terms of LH and RH components, $\psi = \psi_L + \psi_R$, and operating with the matrix γ^5 gives: $\gamma^5\psi_{L,R} = \mp\psi_{L,R}$. We can project out the LH or RH portions of the general spinor, ψ through the operation: $\psi_{L,R} = 1/2(1 \mp \gamma^5)\psi$. Hence, we can force the correct handedness in calculations by requiring a factor of $(1 - \gamma^5)/2$ at every weak vertex involving a neutrino. As a result of this factor, which corresponds to the LH projection operator, we often say the charged weak interaction (W exchange) is "left handed."

It should be noted that RH neutrinos (and LH antineutrinos) could, in principle exist, but be undetected because they do not interact. They will not interact via the electromagnetic interactions because they are neutral, or via the strong interaction because they are leptons. RH neutrinos do not couple to the Standard Model W, because this interaction is "left handed," as discussed above. Because they are non-interacting, they are called "sterile neutrinos."

Sterile neutrinos raise obvious theoretical and experimental questions. From a theoretical viewpoint: how do sterile neutrinos come into existence since they cannot interact? This is solved relatively easily if we extend the Standard Model to include, at energy scales well beyond the range of present accelerators, a right-handed W interaction that could produce the RH neutrino. From the experimental viewpoint: if there are sterile neutrinos out there, how do we observe them if they do not interact? As discussed below, the quantum mechanical effect called "neutrino oscillations" can give us a method.

Neutrino interactions in the Standard Model come in four basic types, illustrated by Fig. 3. *Elastic* scattering is a NC interaction where the target does not go in to an excited state or break up. *Quasi-elastic* scattering is the CC analogue to elastic scattering. Exchange of the W causes the incoming lepton and the target to change flavors, but

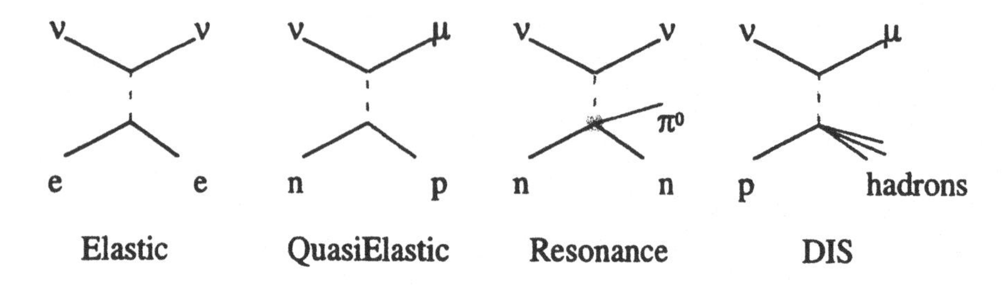

Figure 3 Examples of types of neutrino interactions.

the target does not go into an excited state or break apart. Resonance production may be caused by either NC or CC interactions. The target goes into an excited state, and then emits one or more particles. A case of particular experimental interest is NC resonant production of a π^0, because this forms a background in many oscillation searches. Finally *DIS*, or Deep Inelastic Scattering, is the case where there is large 4-momentum exchange, breaking the nucleon apart. One can have NC or CC deep inelastic scattering.

The total neutrino cross section rises linearly with energy. Thus, from the viewpoint of sheer statistics, one uses the highest energy neutrino beam which is practical for the physics to be addressed. The total cross section can be thought of as the sum of many individual cross sections: quasi-elastic + single pion + two pions + three pions + etc. As the energy increases, each of these cross sections sequentially "turns on" and then becomes constant with energy. Thus the sum, which is the total cross section, increases with energy.

Mass effects produce deviations in the linear dependence for CC scattering. In the CC interaction, you must have enough CM energy to actually produce the outgoing charged lepton. Just above mass threshold, there is very little phase space for producing the lepton, and so production will be highly suppressed. The cross section increases in a non-linear manner until well above threshold. Consider, for example, the ν_τ CC interaction. The mass of the τ is 1.8 GeV. This results in a mass-suppression effect for CC scattering for ν_τ beam energies up to 100 GeV. Fig. 4 [6] illustrates this by comparing the ratio of CC ν_τ interactions to ν_μ interactions for energies up to 100 GeV. If there was no mass-suppression effect for the ν_τ, then both cross sections would rise linearly and the ratio would be constant with energy. However, one can see that even at 100 GeV, which corresponds to a center-of-mass energy

Beware of threshold effects due to charged lepton mass!

- CC interactions may not be possible at a given energy.
- Above threshold, the CC interaction can be surprisingly suppressed for a large energy range!

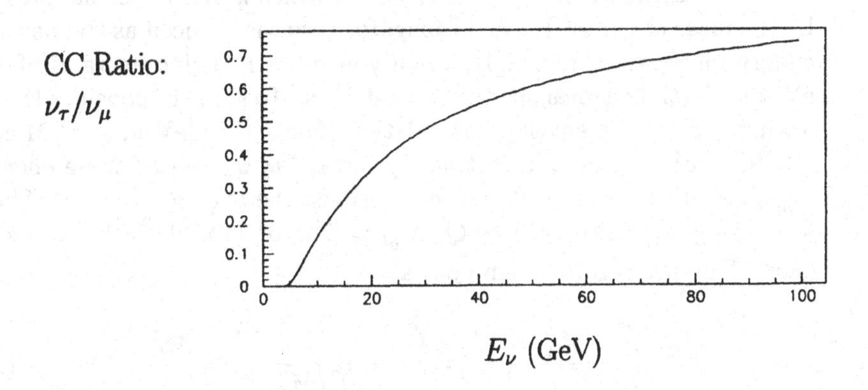

Figure 4 **The ratio of the ν_τ to ν_μ CC cross sections as a function of neutrino energy. (Figure is adapted from [6]).**

of $\sqrt{2ME} \approx 14$ GeV, there is still a 30% reduction in CC interaction rate due to mass suppression.

In discussing neutrino scattering, several kinematic quantities are used to describe events. The squared center of mass energy is represented by the Mandelstam variable, s. The energy transferred by the boson is ν and $y = \nu/E_\nu$ is the fractional energy transfer. The distribution of events as a function of y depends on the helicity. For neutrinos, the y-dependence is flat, but for antineutrinos, the differential cross section is peaked at low y. The variable Q^2 is the negative squared four-momentum transfer. Deep inelastic scattering begins to occur at $Q^2 \sim 1$ GeV^2. M is the target mass. x is the fractional momentum carried by a struck quark in a deep inelastic scatter, and is given by $Q^2/2M\nu$. Elastic and quasieleastic scattering occur at $x = 1$, hence $Q^2 = 2M\nu = sxy$.

This is called the "weak interaction" for good reason. The total cross section for neutrino scattering is small. For 100 GeV ν_μ interactions with electrons is $\sim 10^{-40}$ cm^2 and for nucleons is $\sim 10^{-36}$ cm^2. This is many orders of magnitude less than the strong interaction. For example, for pp scattering, the cross section is $\sim 10^{-25}$ cm^2. The result is that a 100 GeV neutrino will have a mean free path in iron of 3×10^9 meters. Thus most neutrinos which hit the Earth travel through without interacting. With such a small interaction probability, it is clear that large detectors and intense neutrino sources are needed to have high statistics in a neutrino experiment.

The weakness of the weak interaction, which is due to the suppression by the mass of the W in the propagator term, is reduced as the neutrino energy increases. Amazingly, when you reach neutrino energies of 10^{17} eV, the Earth becomes opaque to neutrinos. To see this, consider the following back-of-the-envelope calculation. For a 10^8 GeV ν, $s = 2ME_\nu = 2 \times 10^8$ GeV^2. Most interactions occur at low x; and at these energies $x_{typical} \sim 0.001$. For neutrino interactions, the average y is 0.5. Therefore, using $Q^2 = sxy$, we find $Q^2_{typical} = (2\times10^8)(1\times10^{-3})(0.5) = 1\times10^5$ Gev^2. The propagator term goes as

$$\left(\frac{M_W^2}{M_W^2 + Q^2}\right)^2 = \left(\frac{1}{1 + Q^2/M_W^2}\right)^2 \approx \frac{M_W^4}{Q^4}, \tag{1.1}$$

which is approximately 10^{-3} for our "typical" case. The typical cross section $\sigma_{typical}$ is:
$E \times (\sigma_{tot}/E) \times (\text{prop term}) = 10^8(\text{GeV})(0.6 \times 10^{-38}\text{cm}^2/\text{GeV})10^{-3}$
$= 0.6 \times 10^{-33}\text{cm}^2$. From this we can extract the interaction length, λ_0, on iron, by scaling from hadronic interactions, which tells us that at 30 mb ($= 0.3 \times 10^{-25}$ cm^2), $1\lambda_0 \sim 10$ cm. This implies that λ_0 for our very high energy ν's is $\approx 10^4$ km. However, the Earth is a few $\times 10^4$ km. Thus all of the neutrinos interact; the Earth is opaque to them. The neutrino experiments discussed in detail below are at much lower energies than 10^8 GeV; however this is an effect which must be considered for neutrino astrophysics experiments such as Amanda [7] and Antares [8].

The primary sources of neutrino for interactions observed on Earth are the Sun, cosmic-ray interactions, reactors, and accelerator beams. The solar neutrinos range in energy from approximately 0.1 to about 15 MeV. Reactors produce low energy $\bar{\nu}_e$'s with energy ~ 3 MeV. The atmospheric neutrinos produced by cosmic ray collisions with nucleons have a wide range of energies, with the predominant range used by experiments between 0.1 and 10 GeV. Accelerator neutrino beams presently span from $\sim$10 MeV to 300 GeV.

At high energies, pion decay is the dominate source for production of neutrino beams. Pions preferentially decay to produce ν_μ rather than ν_e. To understand this effect, consider the case of pion decay to a lepton and an antineutrino: $\pi^- \to \ell^- \bar{\nu}_\ell$. The pion has spin zero and so the spins of the outgoing leptons from the decay must be opposite from angular momentum conservation. In the center of mass of the pion, this implies that both the antineutrino and the charged lepton have spin projected along the direction of motion ("right" or "positive" helicity). However, this is a weak decay where the W only couples to the RH antineutrino and the LH component of the charged particle. The amplitude for the LH component to have right-helicity is proportional to m/E. Thus it is very small for an electron compared to the muon, producing a significant suppression for decays to electrons. Calculating the expected branching ratios:

$$R_{theory} = \frac{\Gamma(\pi^\pm \to e^\pm \nu_e)}{\Gamma(\pi^\pm \to \mu^\pm \nu_\mu)} \tag{1.2}$$

$$= \left(\frac{m_e}{m_\mu}\right)^2 \left(\frac{m_\pi{}^2 - m_e{}^2}{m_\pi{}^2 - m_\mu{}^2}\right)^2 \tag{1.3}$$

$$= 1.23 \times 10^{-4}; \tag{1.4}$$

This compares well to the data: $R_{exp} = (1.230 \pm 0.004) \times 10^{-4}$ [5]. Thus, since pions decay almost entirely to ν_μ and since pions are copiously produced from proton-nucleon interactions, one can produce intense, relatively pure ν_μ beams at accelerators.

3. DIRECT MEASUREMENTS OF NEUTRINO MASS

All experimental evidence indicates that, if neutrinos have mass, it is tiny in comparison to the masses of the fermions. For neutrinos, there are no mass measurements, only mass limits. The mass of the neutrino can be measured from decay kinematics, from time of flight from super novae, and from the lifetime for nuclei exhibiting neutrinoless double β decay. Observations of neutrino oscillations, which will be discussed at the end of this paper, are sensitive to the mass differences between neutrinos, not the actual mass of the neutrino. Therefore, they do not fall into the category of a "direct measurement".

The simplest method for measuring neutrino mass is applied to the ν_μ. The mass is obtained from the 2-body decay-at-rest kinematics of $\pi \to \mu\nu_\mu$. One begins in the center of mass with the 4-vector relationship: $p_\pi = p_\mu + p_\nu$. Squaring and solving for neutrino mass gives: $m_\nu^2 = m_\pi^2 + m_\mu^2 - \sqrt{4m_\pi^2(|\mathbf{p}_\mu|^2 + m_\mu^2)}$. From this, one can see that this measurement requires accurate measurement of the muon momentum, $\mathbf{p}_\mu$, as well as

Table 1 Overview of ν_e squared mass measurements.

Experiment	measured m^2 (eV^2)	limit (eV), 95% C.L.	Year
Troitsk [11]	-1.0 $\pm$ 3.0$\pm$ 2.1	2.5	2000
Mainz[12]	-3.7 $\pm$ 5.3 $\pm$ 2.1	2.8	2000
LLNL [13]	- 130 $\pm$ 20$\pm$ 15	7.0	1995
CIAE [14]	- 31 $\pm$ 75$\pm$ 48	12.4	1995
Zurich [15]	-24 $\pm$ 48$\pm$ 61	11.7	1992
Tokyo INS [16]	- 65 $\pm$ 85$\pm$ 65	13.1	1991
Los Alamos [17]	- 147 $\pm$ 68$\pm$ 41	9.3	1991

the masses of the muon, m_μ and the pion, m_π. In fact, the uncertainty on the mass of the pion is what dominates the ν_μ mass measurement. As a result, a limit is set at $m_{\nu_\mu} < 170$ keV [9].

The mass for the ν_τ is obtained from the kinematics of τ decays. the τ typically decays to many hadrons. However, the four vectors for each of the hadrons can be summed. Then the decay can be treated as a two-body problem with the neutrino as one 4-vector and the sum of the hadrons as the other vector. At this point, the same method described for the ν_μ can be applied. Measurements are again error-limited, so a limit on the mass is placed. The best limit, which is $m_{\nu_\tau} < 18.2$ MeV, comes from fits to $\tau^- \rightarrow 2\pi^-\pi^+\nu_\tau$ and $\tau^- \rightarrow 3\pi^-2\pi^+(\pi^0)\nu_\tau$ decays observed by the ALEPH experiment [10].

The experimental situation for the ν_e mass measurement is more complicated. The endpoint of the electron energy spectrum from tritium β decay is used to determine the mass. Just as in the case of the ν_μ and ν_τ, the experiments measure a value of m^2. The problem is that the measurements have been systematically negative. A review of measurements, as a function of time, is given in Tab. 1. Recent measurements at Troitsk [11] and Mainz[12] are negative, but in agreement with zero. Based on these results, one can extract a limit of approximately < 3 eV for the mass of the ν_e.

Another method for measuring neutrino mass from simple kinematics is to use time of flight for neutrinos from supernovae. Neutrinos carry away $\sim$ 99% of the energy from a supernova. The mass limit is obtained from the spread in the propagation times of the neutrinos. The propagation time for a single neutrino is given by

$$t_{obs} - t_{emit} = t_0\left(1 + \frac{m^2}{2E^2}\right) \tag{1.5}$$

where t_0 is the time required for light to reach Earth from the supernova. Because the neutrinos escape from a supernova before the photons, we do not know t_{emit}. But we can obtain the time difference between 2 events:

$$\Delta t_{obs} - \Delta t_{emit} \approx \frac{t_0 m^2}{2}\left(\frac{1}{E_1^2} - \frac{1}{E_2^2}\right). \tag{1.6}$$

using the assumption that all neutrinos are emitted at the same time, one can obtain a mass limit of ~ 30 eV from the ~ 25 events observed from SN1987a at 2 sites.[19, 20]

This is actually an oversimplified argument. The models for neutrino emission are actually quite complicated. The pulse of neutrinos has a prompt peak followed by a broader secondary peak with a long tail distributed over an interval which can be 4 s or more. The prompt peak is from "neutronization" and is mainly ν_e, while all three neutrino flavors populate the secondary peak. However, the rate of ν_e escape is slower compared to ν_μ and ν_τ produced at the same time, because the ν_e's can experience CC interactions, while the kinematic suppression from the charged lepton mass prevents this for the other flavors. However, when all of the aspects of the modeling are put together, the bottom line remains the same: it will be possible to set stringent mass limits if we observe neutrinos from nearby supernovae.

The final method for directly accessing neutrino mass is through neutrinoless double β decay: $(Z, A) \to (Z + 2, A) + (e^- e^-)$. This is a beyond-the-standard model analogue to double β decay: $(Z, A) \to (Z + 2, A) + (e^- e^- \bar{\nu}_e \bar{\nu}_e)$. Double β decay is a standard nuclear process with an absurdly low rate (proportional to $(G_F \cos\theta_C)^4$). Therefore, if the weak decay is possible, single β decay $((Z, A) \to (Z+1, A) + e^- + \bar{\nu}_e)$ will dominate in most cases. However, there are 13 nuclei, including (^{136}Xe $\to$ ^{136}Ba and ^{76}Ge $\to^{76}$ Se, for which single β decay is energetically disallowed. In these cases double β decay with two neutrinos has been observed [21]. If the neutrino were its own antiparticle, then the neutrinos produces in the double β decay process could annihilate, yielding neutrinoless double β decay (see Fig. 5).

Another way to say this is that neutrinoless double β ($0\nu\beta\beta$) decay demands that neutrinos be "Majorana" rather than "Dirac" particles. In the Standard Model, neutrinos are Dirac particles, meaning that the ν and the $\bar{\nu}$ are distinct particles, just as the elec-

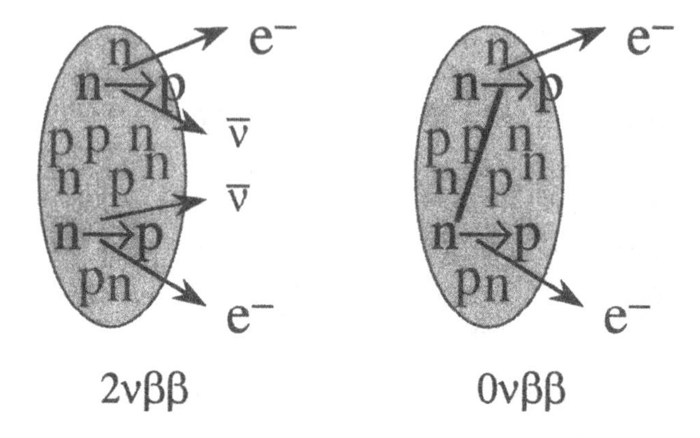

Figure 5 Illustration of two-neutrino double β decay and neutrinoless double β decay.

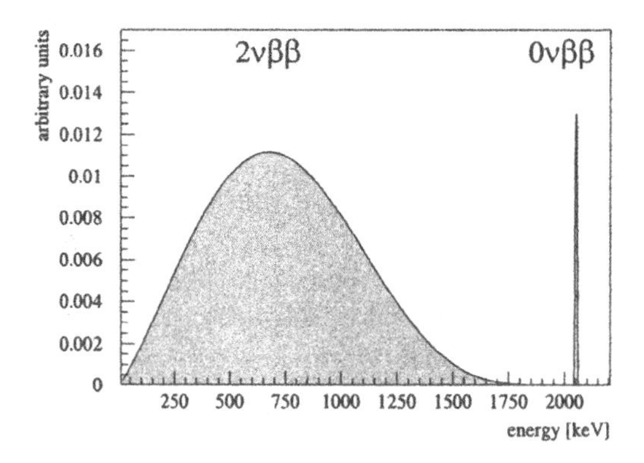

Figure 6 Spectrum for two-neutrino double β decay and expected peak for neutrinoless double β decay.

tron and positron are distinct. The particle, ν has lepton number +1 and the antiparticle, $\bar{\nu}$ has lepton number −1. Lepton number is conserved in an interaction. Thus, using the muon family as an example, ν's ($L = +1$) must produce μ^- ($L = +1$) and $\bar{\nu}$'s ($L = -1$) must produce μ^+ ($L = -1$). The alternative viewpoint, first put forth by Majorana, is that the ν and $\bar{\nu}$ are two helicity states of the same particle, which I will call the ν_{maj}. The π^+ decay produces the left-handed ν^{maj} and the π^- decay produces the right-handed ν^{maj}. This explains all of the data without invoking lepton number. This viewpoint has the nice feature of economy of total particles and quantum numbers, but it renders the neutrino different from the other Standard Model fermions.

How can we experimentally tell the difference between the Dirac (ν, $\bar{\nu}$) and Majorana (ν^{maj}) scenarios? One can imagine a straightforward experiment. First, produce left handed neutrinos in π^+ decays. These may be ν's or they may be ν_{LH}^{maj}'s. Next, run the neutrino through a magic helicity-flipping device. If the neutrinos are Majorana, then what comes out of the flipping-device will be ν_{RH}^{maj}. These particles will behave like antineutrinos when they interact, showing the expected RH y-dependence for the cross section. But if the initial neutrino beam is Dirac, then what comes out of the flipping-device will be right-handed ν's, which are sterile. They do not interact at all. Such a helicity-flipping experiment is presently essentially impossible to implement. If neutrinos do have mass, then they may have an extremely tiny magnetic moment and a very intense magnetic field could flip their helicity. But the design requirements of such an experiment are far beyond our capability at the moment. Therefore, at the moment, we do not know if neutrinos are Majorana or Dirac in nature.

If there are Majorana neutrinos, then the half-life for $0\nu\beta\beta$ is proportional to the squared neutrino mass, m_{maj}^2. The number of $0\nu\beta\beta$ events which occur can be determined above the standard 2 neutrino double β ($2\nu\beta\beta$) decay background through simple kinematics cuts. The two-body nature of $0\nu\beta\beta$ decay will cause a peak at the endpoint of the $2\nu\beta\beta$ decay (4-body) spectrum, as shown in Fig. 6. No experiment has observed $0\nu\beta\beta$ decay, and so limits are set on m_{maj}^2. Presently the best limit, < 0.2 to 0.6 eV (depending on nuclear model) comes from the Heidelberg-Moscow experiment, which uses ^{76}Ge [22].

4. MOTIVATING NEUTRINO MASS AND STERILE NEUTRINOS IN THE THEORY

There is no lack of creative ideas for adding neutrino masses to the theory. The simplest method is to introduce a Dirac mass, by analogy with the electron. This allows us to introduce a small neutrino mass, but it is entirely without motivation, so theorists have looked for other approaches. Among the oldest ideas is the "see-saw" mechanism, which fits well with Grand Unified Theories. There are also ideas based on extra dimensions, where neutrinos live outside of $3+1$-space. Alternatively, one can modify the Higgs sector to accommodate neutrino mass. In this section, I briefly examine the more common schemes. It should be noted that most of the options introduce at least one EW isosinglet, which is a right-handed partner of the left-handed ν. This particle doesn't couple to the left-handed W or Z, and hence it is the sterile neutrino. There is debate over whether it is light (< 1 eV) or heavy ($> 10^{16}$ eV).

In general, the mass term in the Lagrangian will be of the form $m\bar{\psi}\psi$. Recall that in the Weyl, or chiral, representation, where $\gamma^5\psi_{L,R} = \mp\psi_{L,R}$, we can project out the LH or RH portions of the general spinor, ψ through the operation: $\psi_{L,R} = 1/2(1 \mp \gamma^5)\psi$. Using $\bar{\psi} = \psi^\dagger\gamma^0$, we can also write: $\bar{\psi}_{L,R} = 1/2\bar{\psi}(1 \pm \gamma^5)$. From this we find:

$$\bar{\psi}\psi = \bar{\psi}\left[\frac{1+\gamma^5}{2} + \frac{1-\gamma^5}{2}\right]\left[\frac{1+\gamma^5}{2} + \frac{1-\gamma^5}{2}\right]\psi = \bar{\psi}_L\psi_R + \bar{\psi}_R\psi_L.$$

Thus the scalar "mass" term mixes the RH and LH states of the fermion. If the fermion has only one chirality, then the mass term will automatically vanish. For this reason, a standard mass term for the neutrino will require the RH neutrino and LH antineutrino states.

To motivate the mass term, the most straightforward approach is to use the Higgs mechanism, as was done for the electron in the Standard Model. In the case of the electron, when we introduce a spin-0 Higgs doublet ,(h^0, h^+), into the Lagrangian, we find terms like: $g_e\bar{\psi}_e(\psi_\nu(h^+)^\dagger + \psi_e(h^0)^\dagger)$, where g_e is the coupling constant. The second piece of this term can be identified with the Dirac mass term, $m_e\bar{\psi}_e\psi_e$. If we let $\langle h^0\rangle = v/\sqrt{2}$, so that we obtain $g\langle h^0\rangle\bar{\psi}_e\psi_e$ and $m_e = g_e v/\sqrt{2}$. This is the Standard Model method for conveniently converting the *ad hoc* electron mass, m_e, into an *ad hoc* coupling to the Higgs, g_e and a vacuum expectation value (VEV) for the Higgs, v. Following the same procedure for neutrinos allows us to identify the Dirac mass term with $m_\nu = g_\nu v/\sqrt{2}$. The VEV, v, has to be the same as for all other leptons. Therefore, the small mass must come from a very small coupling, g_ν. This implies that $g_e > 5 \times 10^4 g_\nu$. The troublesome feature of this procedure is that it gives no physical insight into what is occurring. While it does introduce neutrino mass, it has simply shifted the arbitrary magnitude of that mass into an arbitrary coupling. This approach also does nothing to answer the question: why would the relative couplings be so different?

An alternative is to consider Majorana mass terms as well as Dirac mass terms. This pops out naturally if we consider the ν and $\bar{\nu}$ to be different helicity states of the same particle. This is the case where the neutrino is its own charge conjugate, $\psi^c = \psi$. The operators which appear in the Lagrangian for the neutrino in this case are the set $(\psi_L, \psi_R, \psi_L^c, \psi_R^c)$ and $(\bar{\psi}_L, \bar{\psi}_R, \bar{\psi^c}_L, \bar{\psi^c}_R)$. Certain bilinear combinations of these in the Lagrangian can be identified as Dirac masses (*i.e.* $m(\bar{\psi}_L \psi_R + ...)$). However, we also get a set of terms terms of the form: $(M_L/2)(\bar{\psi^c}_L \psi_L) + (M_R/2)(\bar{\psi^c}_R \psi_R) + ...$ These are the "Majorana mass terms," which mix the pair of charge-conjugate states of the fermion. If the particle is not its own charge conjugate, then these terms automatically vanish. The notation is more compact if we define: $\phi = (\psi_L^c + \psi_L)/\sqrt{2}$ and $\Phi = (\psi_R^c + \psi_R)/\sqrt{2}$. In this case, the mass terms of the Lagrangian can be written in matrix form:

$$(\bar{\phi} \;\; \bar{\Phi}) \begin{pmatrix} M_L & m \\ m & M_R \end{pmatrix} \begin{pmatrix} \phi \\ \Phi \end{pmatrix}, \tag{1.7}$$

where ϕ and Φ are not describing states with definite mass. The Dirac mass, m, is on the off-diagonal elements, while the Majorana mass constants, M_L, M_R are on the diagonal. To obtain the physical masses, you have to diagonalize the matrix.

Despite appearances, introducing the idea of Majorana neutrinos is actually a theoretical improvement because one can invoke "see-saw models" which motivate small observable neutrino masses. It turns out that left-right symmetric GUT's motivate mass matrices that look like [23]:

$$\begin{pmatrix} 0 & m_\nu \\ m_\nu & M \end{pmatrix}. \tag{1.8}$$

When you diagonalize this matrix to obtain the physical masses, this results in two states which can be measured experimentally:

▷ $m_{light} \approx m_\nu^2/M$

▷ $m_{heavy} \approx M$

Grand Unified Theories favor very large masses for the "heavy neutrino" (often called a "neutral heavy lepton"). If the Dirac mass terms for the leptons are approximately equal (that is, $m_\nu \approx m_e$), then $m_{light} < 1$ eV. In this theory neutrinos have only approximate handedness, where the light neutrino is mostly LH with a very small admixture of RH and the neutral heavy lepton is essentially RH. Thus we have a LH neutrino which is light, which matches observations, and a RH neutrino which is not yet observed because it is too massive.

However, the picture is not perfectly rosy. In the simplest GUT's, protons decay at a rate which is much higher than the experimental limits. (At present, the best limit for proton decay comes from $p \to \pi^+ e^-$, which is greater than 10^{33} years [24].) In order to stabilize the proton, very elaborate GUT's must be constructed, and this has made this solution less fashionable than other Beyond-the-Standard-Model theories. A lesser evil is that this extension to the Standard Model produces at least 12 extra, arbitrary parameters. So it is not clearly a step toward simplicity and deeper meaning.

We would like a tidier package, and so we keep trying. We want to motivate why the neutrino mass is light, why all masses are $\ll M_{planck}$, why protons are stable, and why gravity is weak. Recently, there has been a lot of excitement about theories with extra dimensions, because these can provide explanations for the above list of problems. The idea is that the universe has many dimensions, but we are "localized" to only 3+1 dimensions within the multidimensional "bulk". Only electroweak singlet states can propagate through the bulk. One example is the graviton, and another could be the light-mass sterile neutrino [25]. Thus, these theories represent one way that light mass sterile neutrinos can be motivated.

Another method for motivating sterile neutrinos explains dark matter by introducing a "mirror world" which is a parity-flipped duplication of our world in every way except that its electroweak scale is larger by some factor [26]. The mirror world spatially overlaps our world, but can interact with our world only through gravity. The big bang relic neutrinos in the mirror world could provide the necessary dark matter, if the factor multiplying the electroweak scale in the mirror world is approximately 30. In this case, dark matter would not be a mixture of hot dark matter (light standard model neutrinos) and cold dark matter (wimps, neutralinos, etc.); instead it is "warm dark matter" (mirror world neutrinos). Planck-scale effects can induce neutrinos in our world to convert, or oscillate, into mirror world neutrinos. Since these neutrinos no longer interact in our world, they are sterile. There are three species of sterile neutrino in this model.

In summary, there are many ways to introduce neutrino mass into the theory, and the picture is confusing. Therefore, it is important for experiments to cover the widest possible set of models. For the remaining discussion, which will focus on oscillation experiments, we will assume:

- ▷ the ν_e, ν_μ, and ν_τ exist and have "Standard Model" interactions, and
- ▷ at least one light-mass sterile neutrino, ν_s, may exist.

All of these neutrinos are assumed to have mass ~few eV or less.

5. NEUTRINO OSCILLATION FORMALISM

If neutrinos have mass, it is likely that the mass eigenstates are different from the weak interaction eigenstates. In this case, the weak eigenstates can be written as mixtures of the mass eigenstates, for example:

$$\begin{aligned} \nu_e &= \cos\theta\ \nu_1 + \sin\theta\ \nu_2 \\ \nu_\mu &= -\sin\theta\ \nu_1 + \cos\theta\ \nu_2 \end{aligned}$$

where θ is the "mixing angle." In this case, a pure flavor (weak) eigenstate born through a weak decay can oscillate into another

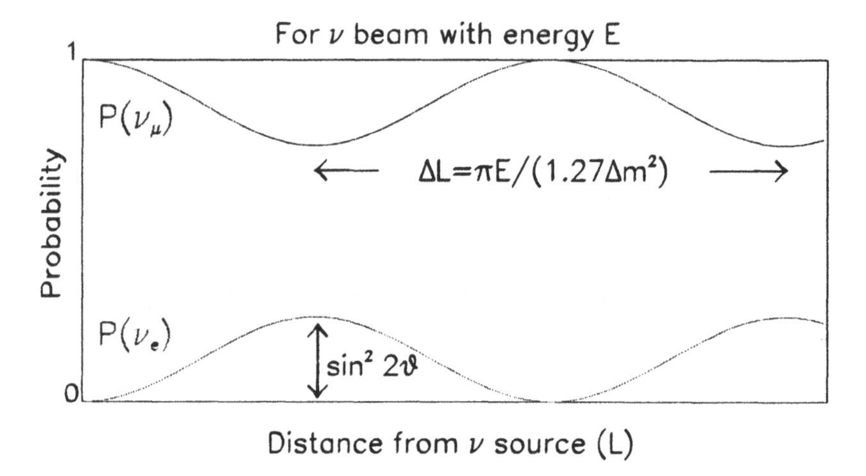

Figure 7 Example of neutrino oscillations as a function of distance from the source, L. The wavelength depends upon the experimental parameters L and E (neutrino energy) and the fundamental parameter Δm^2. The amplitude of the oscillation is constrained by the mixing term, $\sin^2 2\theta$.

flavor as the state propagates in space. This oscillation is due to the fact that each of the mass eigenstate components propagates with different frequencies if the masses are different, $\Delta m^2 = |m_2^2 - m_1^2| > 0$. In such a two-component model, the oscillation probability for $\nu_\mu \to \nu_e$ oscillations is then given by:

$$\text{Prob}\,(\nu_\mu \to \nu_e) = \sin^2 2\theta\ \sin^2\left(\frac{1.27\ \Delta m^2\left(\text{eV}^2\right)\ L\,(\text{km})}{E\,(\text{GeV})}\right). \quad (1.9)$$

where L is the distance from the source, and E is the neutrino energy. As shown in Fig. 7, the oscillation wavelength will depend upon L, E, and Δm^2. The amplitude will depend upon $\sin^2 2\theta$.

Most neutrino oscillation analyses consider only two-generation mixing scenarios, but the more general case includes oscillations between all three neutrino species. This can be expressed as:

$$\begin{pmatrix} \nu_e \\ \nu_\mu \\ \nu_\tau \end{pmatrix} = \begin{pmatrix} U_{e1} & U_{e2} & U_{e3} \\ U_{\mu 1} & U_{\mu 2} & U_{\mu 3} \\ U_{\tau 1} & U_{\tau 2} & U_{\tau 3} \end{pmatrix} \begin{pmatrix} \nu_1 \\ \nu_2 \\ \nu_3 \end{pmatrix}.$$

This formalism is analogous to the quark sector, where strong and weak eigenstates are not identical and the resultant mixing is described conventionally by a unitary mixing matrix. The oscillation probability is then:

$$\text{Prob}\,(\nu_\alpha \to \nu_\beta) \;= \delta_{\alpha\beta} - 4\sum_{j>i} U_{\alpha i}U_{\beta i}U^*_{\alpha\, j}U^*_{\beta\, j}\sin^2\left(\frac{1.27\,\Delta m^2_{ij}\,L}{E}\right) \quad (1.10)$$

where $\Delta m^2_{ij} = \left|m_i^2 - m_j^2\right|$. Note that there are three different Δm^2 (although only two are independent) and three different mixing angles.

Although in general there will be mixing among all three flavors of neutrinos, two-generation mixing is often assumed for simplicity. If the mass scales are quite different ($m_3 >> m_2 >> m_1$ for example), then the oscillation phenomena tend to decouple and the two-generation mixing model is a good approximation in limited regions. In this case, each transition can be described by a two-generation mixing equation. However, it is possible that experimental results interpreted within the two-generation mixing formalism may indicate very different Δm^2 scales with quite different apparent strengths for the same oscillation. This is because, as is evident from equation 1.10, multiple terms involving different mixing strengths and Δm^2 values contribute to the transition probability for $\nu_\alpha \to \nu_\beta$.

From equation 1.9, one can see that three important issues confront the designer of the ideal neutrino experiment. First, if one is searching for oscillations in the very small Δm^2 region, then large L/E must be chosen in order to enhance the $\sin^2(1.27\Delta m^2 L/E)$ term. However if L/E is too large in comparison to Δm^2, then oscillations occur rapidly. Because experiments have finite resolution on L and E, and a spread in beam energies, the $\sin^2(1.27\Delta m^2 L/E)$ averages to 1/2 when $\Delta m^2 \gg L/E$ and one loses sensitivity to Δm^2. This point is illustrated in Fig. 8. Finally, because the probability is directly proportional to $\sin^2 2\theta$, if the mixing angle is small, then high statistics are required to observe an oscillation signal.

There are two types of oscillation searches: "disappearance" and "appearance." To be simplistic, consider a pure source of neutrinos of type x. In a disappearance experiment, one looks for a deficit in the expected flux of ν_x. Appearance experiments search for $\nu_x \to \nu_y$ by directly observing interactions of neutrinos of type y. The case for oscillations is most persuasive if the deficit or excess has the

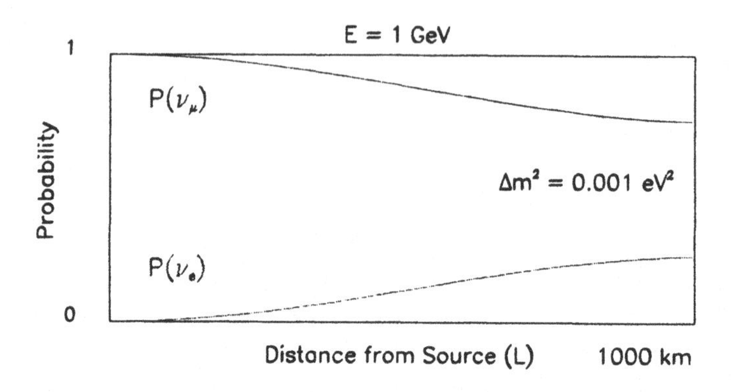

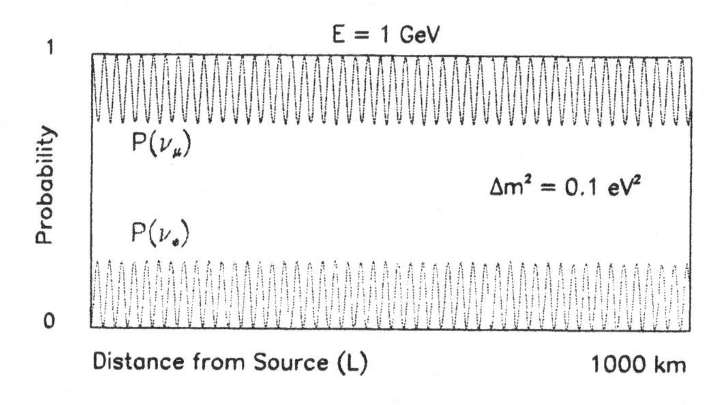

Figure 8 If Δm^2 is small, then one must design an experiment with large L/E to observe a large oscillation probability (top). However if $\Delta m^2 \gg L/E$, then rapid oscillations are smeared by statistics, variations in beam energy and detector effects, causing an experiment to be insensitive to the value of Δm^2 (bottom).

(L/E) dependence predicted by the neutrino oscillation formula (equation 1.9).

Let us say that a hypothetical perfect neutrino oscillation experiment sees no oscillation signal, based on N events. The experimenters can rule out the probability for oscillations at some confidence level. A typical choice of confidence level is 90%, so in this case, the limiting probability is $P = 1.28\sqrt{N}/N$. There is only one measurement and there are two unknowns, so this translates to an excluded region within $\Delta m^2 - \sin^2 2\theta$ space. As shown in Fig. 9, this is indicated by a solid line, with the excluded region on the right. At high Δm^2, the limit on $\sin^2 2\theta$ is driven by the experimental statistics. The L and E of the experiment drive the low Δm^2 limit. The imperfections of a real experiment affect the limits which can be set. Systematic uncertainties in the efficiencies and backgrounds reduce the sensitivity of a given experiment. Background sources introduce multiple flavors of neutrinos in the beam. Misidentification of the interacting neutrino flavor in the detector can mimic oscillation signatures. In addition, systematic uncertainties in the relative acceptance versus distance and energy need to be understood and included in the analysis of the data. These systematics are included in the 90% CL excluded regions presented by the experiments in this paper. The "sensitivity" of an experiment is defined as the average expected limit if the experiment were performed many times with no true signal (only background).

Indications of neutrino oscillations appear as allowed regions, indicated by shaded areas (see example in Fig. 9), on plots of Δm^2 *vs.* $\sin^2 2\theta$. The most convincing signature for oscillations is a statistically and systematically significant signal (that is, appearance as opposed to disappearance) with the dependence on L and E as predicted for oscillations. This has not yet been observed in any experiment. Deficits have been observed in the expected rate of two neutrino sources: solar and atmospheric. A signal has been been observed by the LSND experiment, but it is not at 5σ significance and the L and E dependence is not clearly demonstrated.

Care must be taken when comparing excluded and allowed regions near the boundaries. First, one should remember that a 90% CL exclusion limit means that if a signal were in this region, 10% of the time it would not be seen. Second, there is disagreement about how to handle data in the case of background plus a small expected

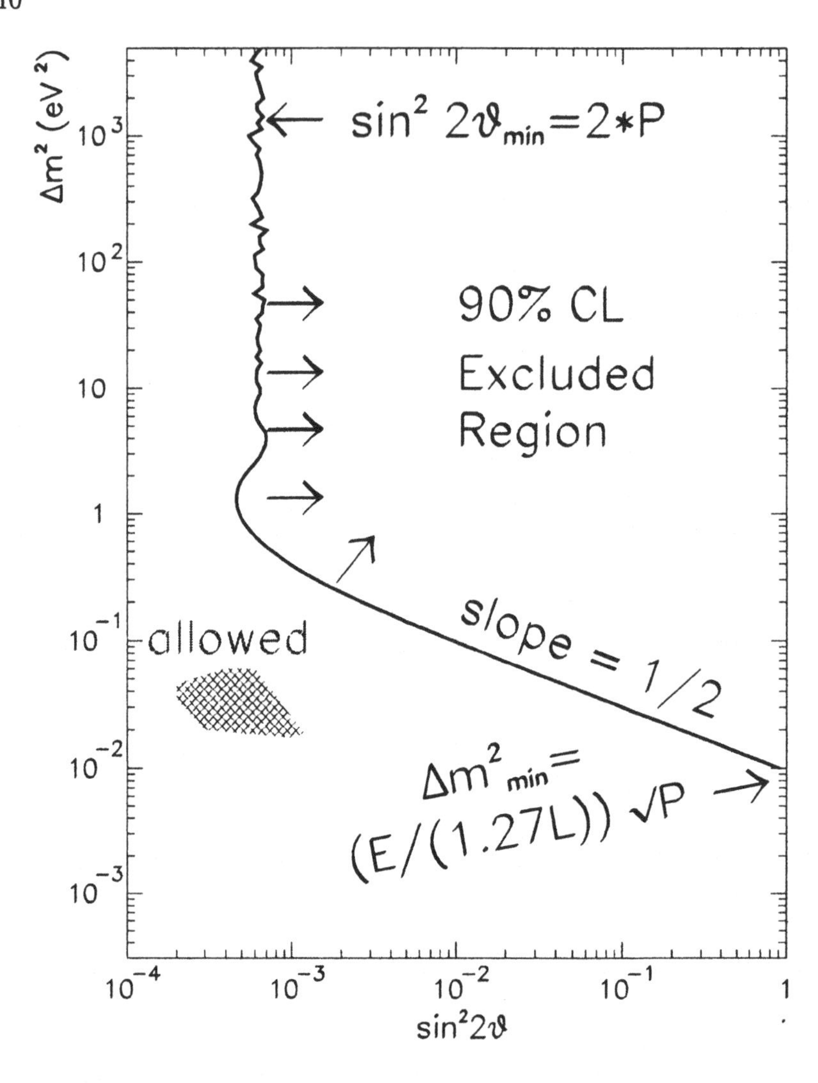

Figure 9 Generic example of a neutrino oscillation plot. The region to the right of the solid line is excluded at 90% CL. The shaded blob represents an "allowed" region.

signal. If an experiment sees the expected background, then the differences in the statistical methods are relatively small, but when the background fluctuates low there can be significant differences in limits. To address this, when an experiment sets a significantly better limit than the sensitivity, the experiment should also indicate the sensitivity on the plot. This allows readers to draw conclusions based upon their own opinion of what is acceptable.

6. EXPERIMENTAL SIGNALS FOR OSCILLATIONS

Three different sources of neutrinos have shown deviations from the expectation, consistent with oscillations. The first, called the "Solar Neutrino Deficit," is a low rate of observed ν_e's from the Sun. The data are consistent with $\Delta m^2 \sim 10^{-10}\mathrm{eV}^2$ or $\Delta m^2 \sim 10^{-5}\mathrm{eV}^2$, depending on the theoretical interpretation. The second, called the "Atmospheric Neutrino Deficit," refers to neutrinos produced by decays of mesons from cosmic ray interactions in the atmosphere. An observed anomalously low ratio of ν_μ/ν_e can be interpreted as oscillations with $\Delta m^2 \sim 10^{-3}\mathrm{eV}^2$. The third observation is an excess of $\overline{\nu}_e$ events in a $\overline{\nu}_\mu$ beam by the LSND experiment, with $\Delta m^2 \sim 10^{-1} eV^2$. Fig. 10 summarizes the allowed regions from these results.

6.1. THE SOLAR NEUTRINO DEFICIT

The first indication of neutrino oscillations came from the solar neutrino experiments. These experiments observe a deficit of solar neutrinos compared to the Standard Solar Model (SSM) [27] as shown in Fig. 11. The first observation of this ν_e deficit was made using Cl target in the Homestake [28] mine by Davis and collaborators. Four additional experiments have confirmed these observations. The GALLEX and SAGE experiments search for electron neutrino interactions in a Ga target [29]. The Kamiokande and Super Kamiokande ("Super K") experiments observe $\nu_e + e \rightarrow \nu_e + e$ reactions in water.

Each type of solar experiment has a different energy threshold for observing ν_e interactions, and thus is sensitive to different reactions producing neutrinos in the Sun. The characteristic range of solar ν energies from each production mechanism is shown in Fig. 12.

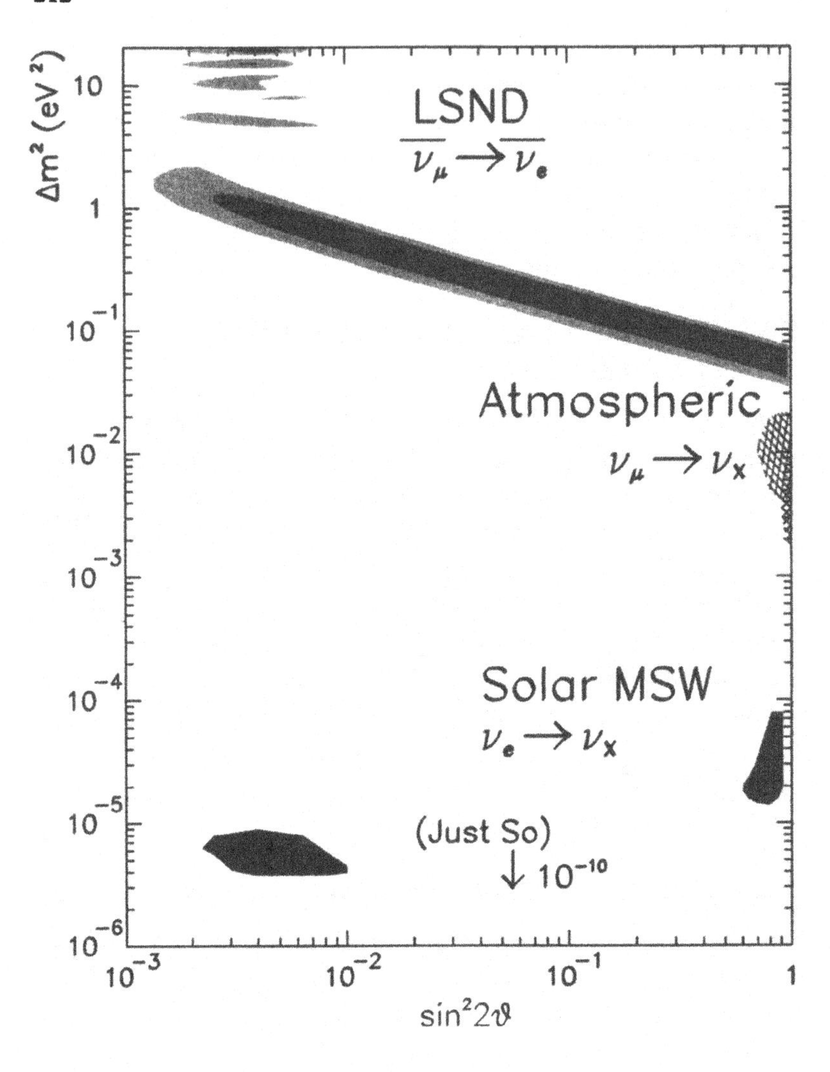

Figure 10 Allowed regions for the three possible oscillation signals.

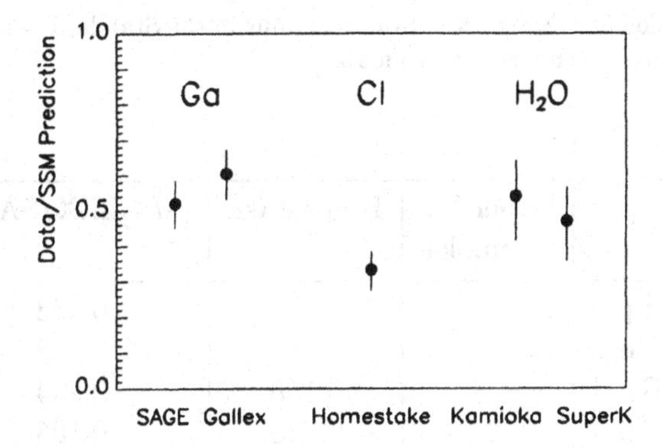

Figure 11 Ratio of measurements from five solar neutrino experiments, to the SSM (BP98) prediction.

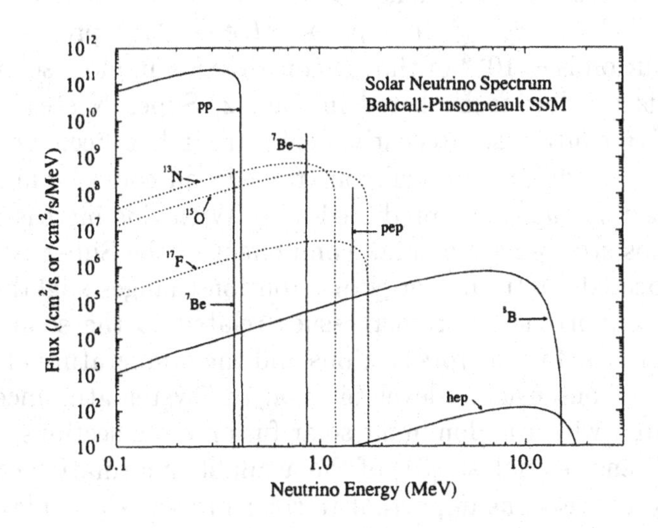

Figure 12 Neutrino fluxes as a function of energy from the Sun.

Table 2 Fraction of ν_e's expected from reactions in the Standard Solar Model for the three types of solar neutrino experiments.

	Super K/ Kamioka	Homestake	GALLEX/SAGE
pp			0.538
^{7}Be I			0.009
^{7}Be II		0.150	0.264
^{8}B	1	0.775	0.105
pep		0.025	0.024
^{1}3N		0.013	0.023
^{1}5O		0.038	0.037

Major sources of solar neutrinos for each experiment are listed in Tab. 2. The "*hep*" ($^3He + p \rightarrow\ ^4He + e^+\nu_e$) process neutrino contribution is $\sim 10^{-4}$ of the ^8B contribution in most solar models, which is too low to be listed in Tab. 2; Super K sees effectively 100% ^8B neutrinos. Recently, however, it has been pointed out that the *hep* flux is not well constrained and could be much larger than past models have predicted [30]. With this increase, the *hep* neutrinos still remain a small component of the Super K flux, but the expected neutrino energy distribution changes slightly.

Two important theoretical issues related to the solar neutrino fluxes are the fusion cross sections and the temperature of the solar interior. Consensus is developing on the systematic uncertainties associated with the dominant solar fusion cross sections. A recent comprehensive analysis [31] of the available information on nuclear fusion cross sections important to solar processes provides the best values along with estimated uncertainties. These are included in the uncertainty in the SSM value for the ratio of data to prediction shown in Fig. 11. The flux of neutrinos from certain processes, particularly ^8B, depends strongly upon temperature. Recent results in helioseismology have provided an important test of the SSM [32]. The Sun is a resonant cavity, with oscillation frequencies dependent upon $U = P/\rho$, the ratio of pressure to density. Helioseismological data confirm the SSM prediction of U to better than 0.1% [33].

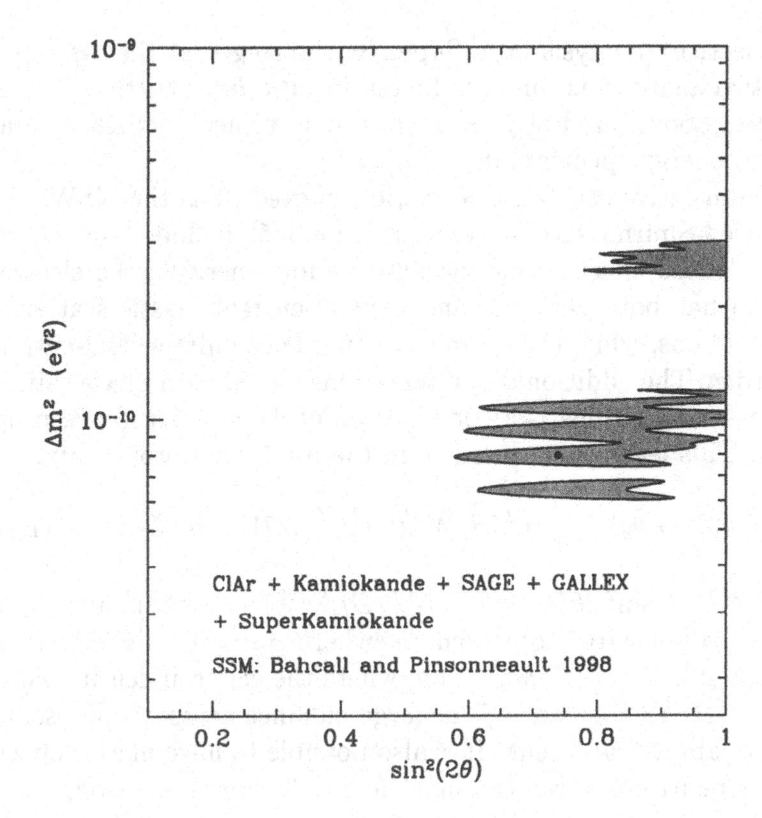

Figure 13 Allowed regions in the Δm^2 vs $\sin^2 2\theta$ parameter space from combining the data from four solar neutrino experiments assuming vacuum oscillations.

Helioseismological data are not included in the data used to determine the SSM, making this an independent cross-check of the model. The systematic error associated with temperature dependence is included in the theoretical error on the SSM prediction.

Interpreting the deficit of solar neutrinos as a signal for oscillations, one calculates the vacuum oscillation probability using equation 1.9, as shown in Fig. 13. Vacuum oscillations are often referred to as the "Just So Hypothesis" because this theory assumes that the distance of the Earth from the Sun is an oscillation maximum. The energy of the neutrinos (only a few MeV) combined with the long path length from the Sun to the Earth ($\sim 10^{11}$m) results in allowed regions of Δm^2 which are very low ($\Delta m^2 \sim 10^{-10}$eV^2). In Fig. 13, an analysis of the overall fluxes [34] is compared to the recent energy spectrum analysis from Super K, showing that the overlap of these two analyses is limited. Increasing the *hep* neutrino flux, as discussed above, modestly improves the agreement between overall flux and energy spectrum data [30].

An alternative oscillation scenario, referred to as the MSW (Mikheyev-Smirnov-Wolfenstein) solution [35], includes "matter effects." These effects occur because at low energies, the electron neutrino has both charged- and neutral-current elastic scattering with electrons, while the ν_μ and ν_τ experience only neutral-current scattering. The additional ν_e interactions introduce a phase shift as the mass state, which is a combination of flavor eigenstates, propagates. This leads to an increase in the oscillation probability:

$$\text{Prob}\,(\nu_e \to \nu_\mu) = \left(\sin^2 2\theta / W^2\right)\, \sin^2\left(1.27 W \Delta m^2 L/E\right) \qquad (1.11)$$

where $W^2 = \sin^2 2\theta + (\sqrt{2} G_F N_e (2E/\Delta m^2) - \cos 2\theta)^2$ and N_e is the electron density. In a vacuum, where $N_e = 0$, this reduces to equation 1.9. But within the Sun, where the electron density varies rapidly, "MSW resonances," or large enhancements of the oscillation probability can occur. It is also possible to have matter effects occur as neutrinos travel through the Earth. For this reason, Super K has searched for a "day-night effect" – a difference in interaction rate at night due to the MSW effect in the Earth's core.

Results of a combined analysis of the solar neutrino data within the MSW framework are shown in Fig. 14. Allowed regions are indicated by the two solid areas, which are referred to as the small mixing angle (SMA) solution and large mixing angle (LMA) solution. Super K has seen no evidence of a day-night effect. As a result, a region indicated by the hatched area can be excluded at the 90% CL.

6.2. THE ATMOSPHERIC NEUTRINO DEFICIT

Neutrinos may be produced in the upper atmosphere. They result from cosmic rays colliding with atmospheric nucleons, producing mostly pions which then decay into muons and muon-neutrinos. The resulting muons may also decay, producing both a muon- and an electron-flavored neutrino. Thus, this decay chain is expected to produce a two-to-one ratio for ν_μ to ν_e.

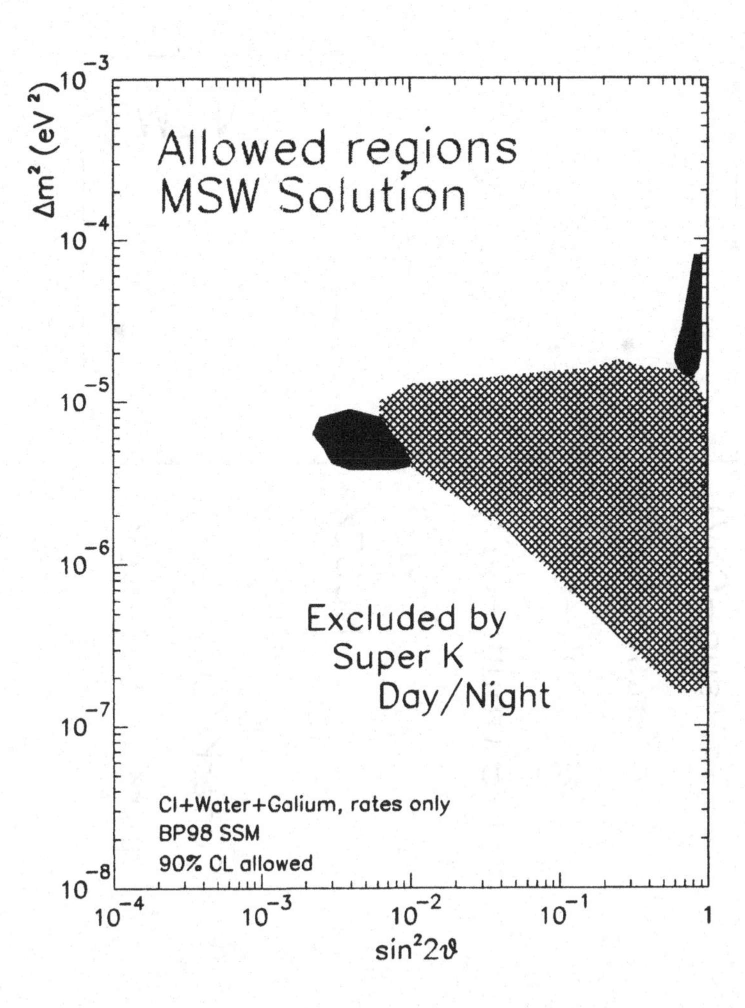

Figure 14 Solid: Allowed regions from the solar neutrino experiments, including the MSW effect. Hatched: Excluded (90% CL) region due to no day-night effect.

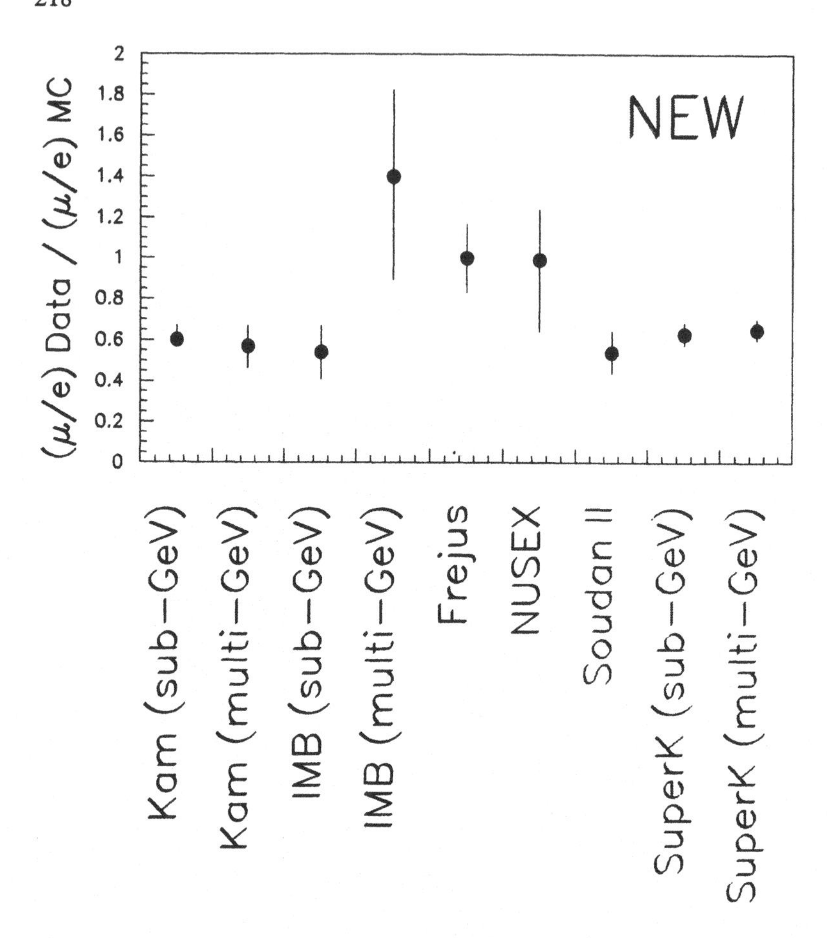

Figure 15 $(\nu_\mu/\nu_e)_{data}/(\nu_\mu/\nu_e)_{MC}$ for atmospheric experiments. The older experiments appeared to show an effect for water-based Cerenkov detectors (Kamioka and IMB) but not for iron-based calorimeter detectors (Frejux and Nusex). However, the three newest results indicate a deficit with small error bars. Super K is a water-based Cerenkov detector while Soudan II is an iron calorimeter. Based on these results, the deficit does not appear to be due to a detector effect.

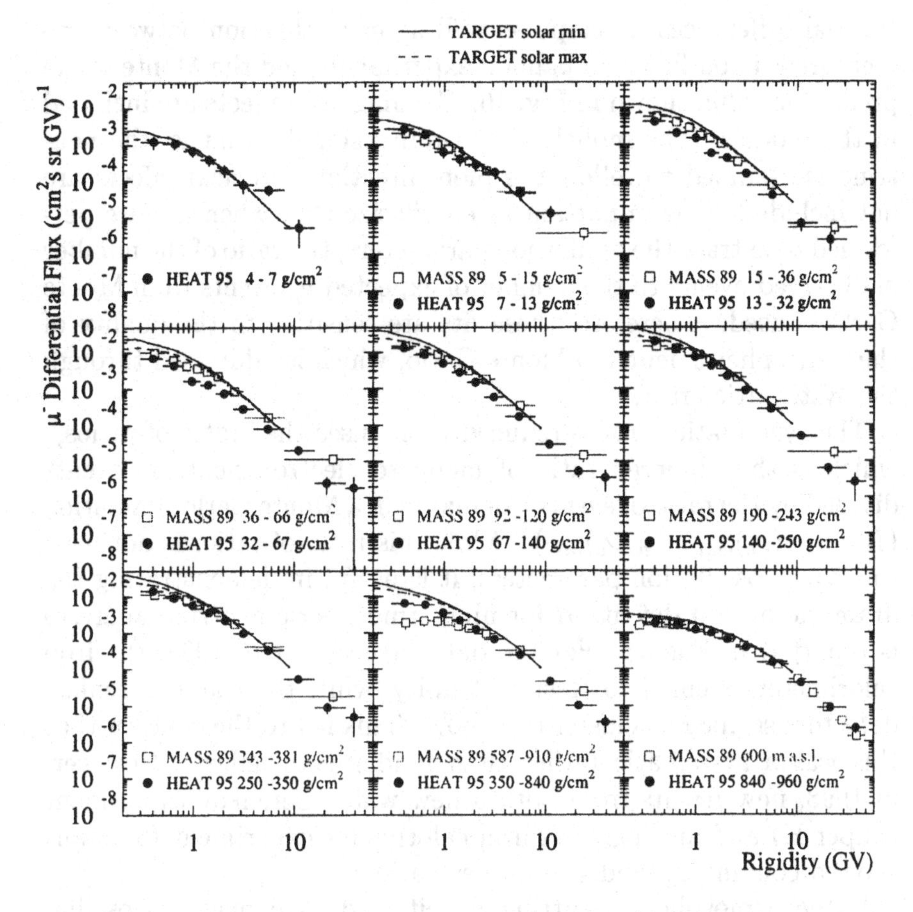

Figure 16 Comparison of the atmospheric Monte Carlo "TARGET" to atmospheric μ data.

Atmospheric neutrinos have been detected through their charged–current interactions in detectors on the Earth's surface. The sensitivity of these detectors to electrons and muons varies over the observed energy range. For example, electron events are mostly contained in the detector, while muon events have longer range and escape the detector at higher energies. Therefore, the results are often divided into sub-GeV (contained) and multi-GeV (partially-contained) samples.

The absolute flux of atmospheric neutrinos is only known to $\sim 20\%$. The inputs to the Monte Carlo simulation of the neutrino flux come from atmospheric balloon-based experiments and accelerator-based experiments. One should note that there are substantial differences in shape as well as normalization between the measured muon fluxes in balloon experiments and the Monte Carlo predictions [36], shown in Fig. 16. Geomagnetic effects are included in the models. The models are "1-dimensional," that is, the neutrino is assumed to follow the pion direction. Nuclear effects are not included. It is important to emphasize that when fits are performed to extract the oscillation parameters, the ratio of the number of observed events to the number of expected ν_μ events from Monte Carlo is used. Therefore, these fits are sensitive to the quality of the atmospheric neutrino Monte Carlo, which is addressed through the systematic error.

The systematic error is reduced if one uses the "ratio-of-ratios," which is the observed ratio of muon to electron neutrino events divided by the ratio of events calculated in a Monte Carlo simulation ($R = (\nu_\mu/\nu_e)_{data}/(\nu_\mu/\nu_e)_{MC}$). While this ratio-of-ratios is not used to extract oscillation parameters, it is useful in demonstrating the dramatic overall deficit. R for nine atmospheric neutrino analyses is reported in Fig. 15. For the data taken prior to 1998, the iron calorimeters seemed to cluster at unity, while the water Cerenkov detectors seemed to cluster at $\sim 60\%$. This led to the concern that this was a systematic rather than fundamental effect. However, in 1998, new results from both a new water Cerenkov experiment (Super K) and an upgraded iron calorimeter experiment (Soudan) showed convincing evidence for oscillations.

If the atmospheric neutrino deficit is due to neutrino oscillations, then one would expect a change in the number of neutrinos of a specific flavor as a function of neutrino path length. Neutrinos which are produced in interactions directly above the detector, called "downward-going," traverse $L \sim 10$ km. Neutrinos which are produced on the opposite side of the Earth, called "upward-going," travel $L \sim 10,000$ km. Traditionally, the path of the neutrino is described by $\cos\theta_z$, where θ_z is the zenith angle measured from directly above the detector. It should be noted that the actual path length, L, is rapidly changing with $\cos\theta_z$. As an example, the fraction of non-oscillating ("surviving") neutrinos *vs.* $\cos\theta_z$ for oscillations of 1 GeV ν_μ's is shown in Fig. 17 for low, medium, and high Δm^2 values. At very low Δm^2, the probability for oscillation is low, even for very long path lengths, so the fraction of surviving neutrinos will be consistent with one. At very high Δm^2, even the

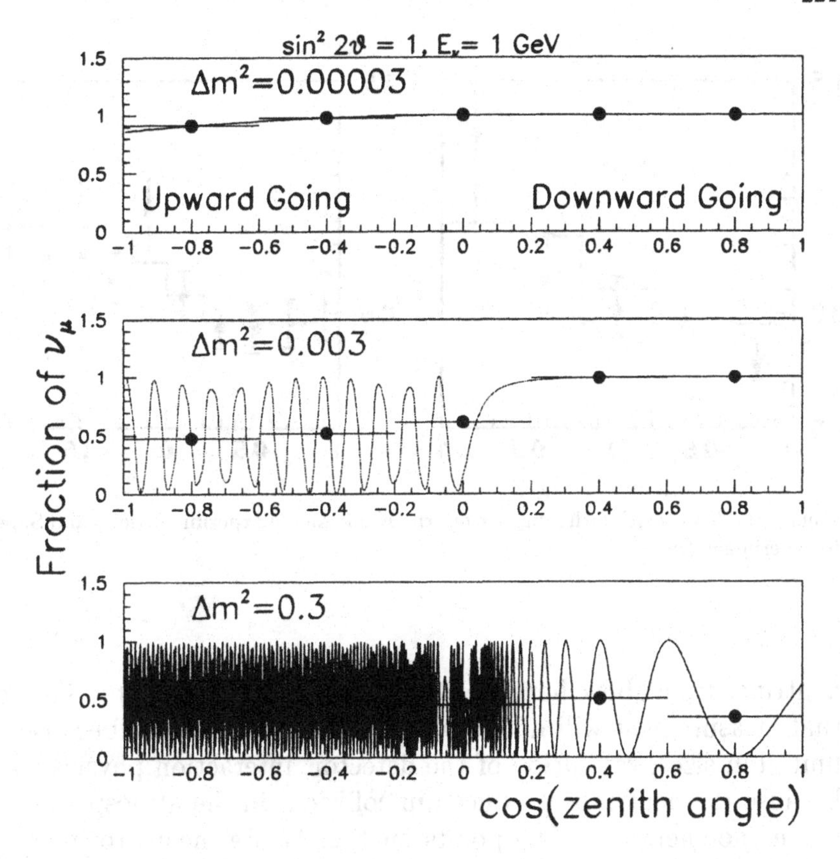

Figure 17 Fraction of surviving ν_μ as a function of cosine of zenith angle for three regions of Δm^2. Resolution of the experiments and limitations of statistics smooth rapid oscillations such that the data will tend to look like the points.

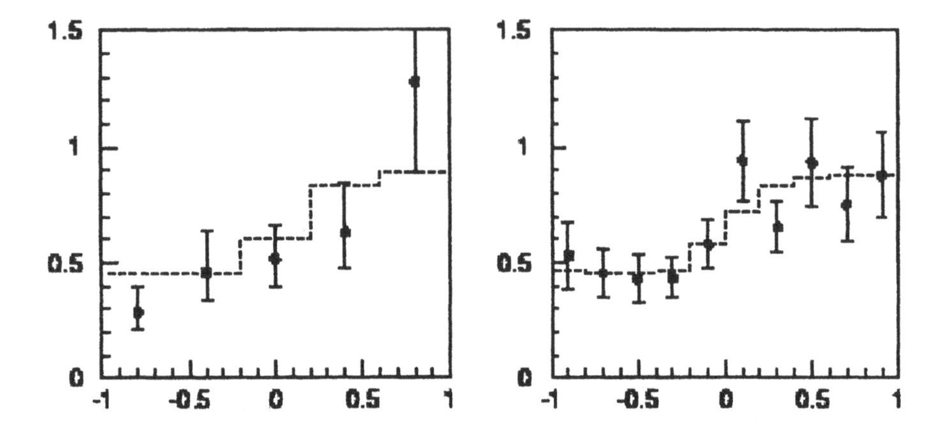

Figure 18 R vs cos(Zenith Angle) from the Kamiokande experiment (left) and Super K experiment (right).

neutrinos from above have had the opportunity to oscillate. The actual measurement will not resolve the rapid oscillations because of finite bin sizes, resolution of the detector, interaction physics, and because the neutrinos produced in collisions in the atmosphere are not mono-energetic. The points on Fig. 17 are meant to indicate what might be measured for the three Δm^2 cases. A typical experiment would be unable to resolve rapid oscillations. This makes interpretation of the data, particularly in the moderate Δm^2 region, difficult.

In 1994, the Kamiokande group reported a zenith angle dependence of R [37]. The data appear to be consistent with expectations for moderate Δm^2, as seen on Fig. 18 (left). The best fit for $\nu_\mu \to \nu_\tau$ oscillations is $\Delta m^2 = 1.6 \times 10^{-2}$ eV^2. The zenith angle dependence has been confirmed by recent data from Super K [18], as shown in Fig. 18 (right). The Soudan Experiment has observed a similar zenith angle dependence, providing important corroboration from an iron-calorimeter detector.

Neutrinos which travel through the Earth may interact in the matter under the detectors. MACRO, Super K, and Kamiokande [39] have measured the upward-going muons from these neutrino interactions. All of these experiments see rates below that which is expected and an angular dependence which is consistent with oscillations. However, in these studies, systematics are a larger issue because one is not able to use a ratio-of-ratios technique. In order to address this, Super K has used a ratio of horizontal-to-upward-going muons through the detector. However, this still leaves several concerns. The energy of neutrinos producing these events is the range of 10 to several 100 GeV. Concerns include: how well is the E-dependence of this flux known? Do uncertainties in flux due to differences in decay length between horizontal and vertical showers affect the ratio? Are nuclear effects important, considering these energies correspond, on average, to low x events? While all of these effects are likely to be small, interpretation of the data can vary significantly with 10% variations in these systematics.

Fig. 19 summarizes the analyses of these experiments under a two-generation $\nu_\mu \rightarrow \nu_\tau$ oscillation hypothesis. As will be noted again in the section on Theoretical Interpretation, the best fits with no $\sin^2 2\theta < 1$ constraint extend into the unphysical region of $\sin^2 2\theta > 1$. Taken as a whole, the results are consistent with oscillations $\Delta m^2 \sim 2.5 \times 10^{-3}\mathrm{eV}^2$ and $\sin^2 2\theta \sim 1$.

6.3. THE LSND SIGNAL

The LSND hint for neutrino oscillations is the only indication for oscillations which is a *signal*, as opposed to a deficit. Evidence has been seen for both $\bar{\nu}_\mu \rightarrow \bar{\nu}_e$ and $\nu_\mu \rightarrow \nu_e$ oscillations. In 1995 the LSND experiment published data showing candidate events that were consistent with $\bar{\nu}_\mu \rightarrow \bar{\nu}_e$ oscillations [40]. Additional event excesses were published in 1996 and 1998 for both $\bar{\nu}_\mu \rightarrow \bar{\nu}_e$ ("DAR") oscillations [41] and $\nu_\mu \rightarrow \nu_e$ ("DIF") oscillations. [42] The two oscillation searches are complementary, having different backgrounds and systematics, yet yielding consistent results. The experiment has now been completed; final results were presented at conferences in summer, 2000, and a final paper is expected to be submitted soon.

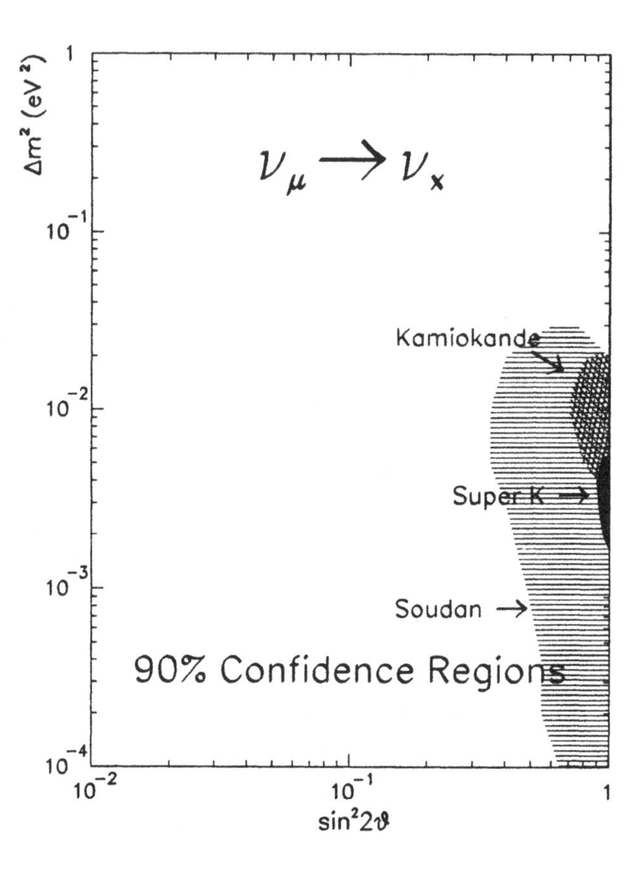

Figure 19 A summary of 90% CL allowed regions for Soudan II (horizontal lines), Kamiokande (cross-hatch) and Super K (solid).

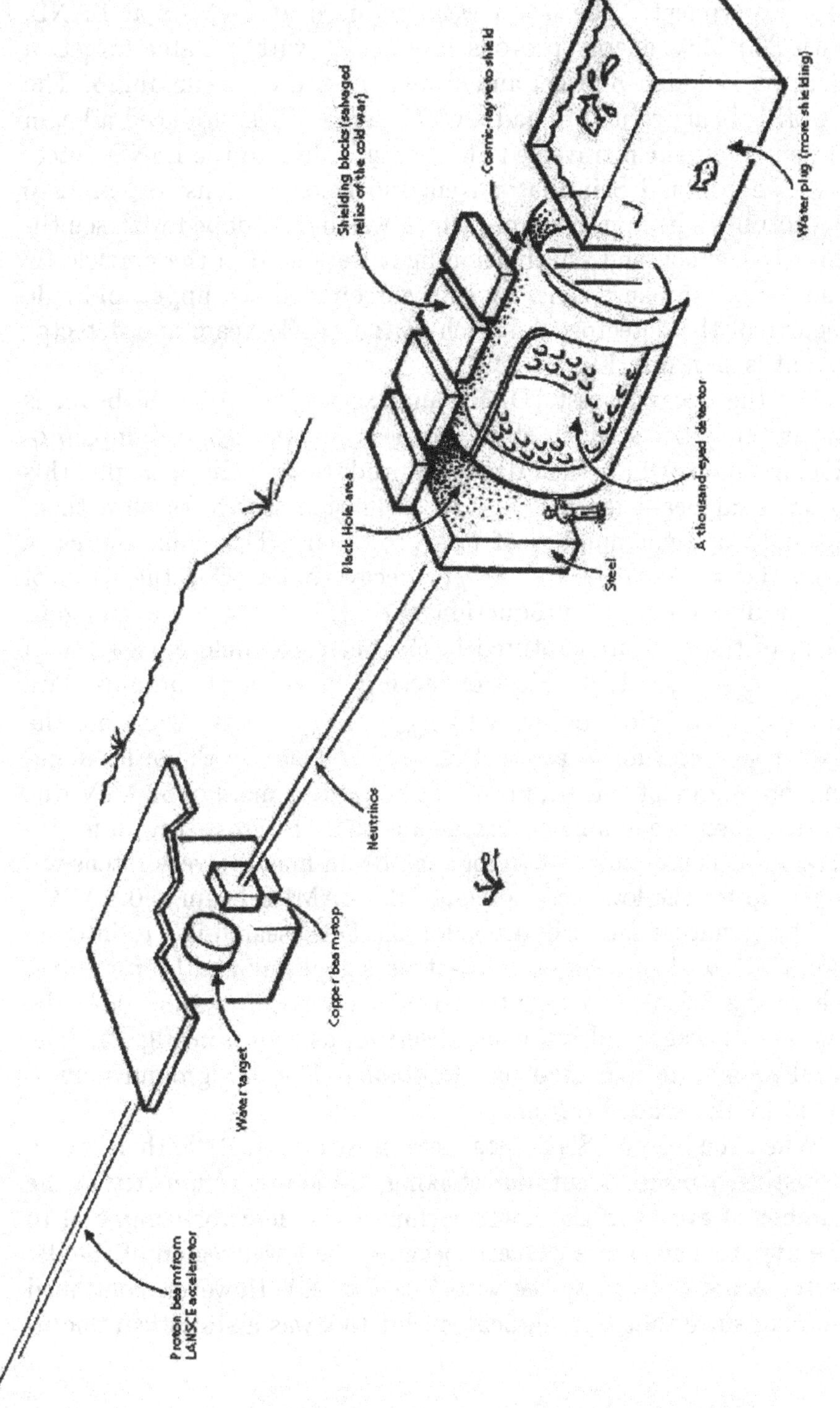

Figure 20 Illustration of the LSND beam and detector design.

This is the only oscillation signature observed from an accelerator experiment. The beam was produced at LANCE at LANL, with 800 MeV energy protons interacting with a water target, a close-packed high-Z target and a water-cooled Cu beam dump. The LAMPF beam structure had a 600 μs spill. This produced a beam of neutrinos which traversed 30 m of shielding to the LSND detector. This Liquid Scintillator Neutrino Detector consisted of 1220 phototubes within mineral oil, which was lightly doped with scintillator. Cerenkov and scintillation light were used in the particle ID and energy measurement. A veto surrounded the upper and side regions of the detector. A sketch of the LSND beam and detector layout is shown in Fig. 20.

For the decay-at-rest (DAR) analysis ($\overline{\nu}_\mu \rightarrow \overline{\nu}_e$), the beam is produced by π^+'s which stop and decay in the beam dump, producing muons which then decay to produce $\overline{\nu}_\mu$'s. In principle, this is an ideal beam for a $\bar{\nu}_\mu \rightarrow \bar{\nu}_e$ oscillation search because there are only a small number of $\bar{\nu}_e$'s produced. The main source is from the $\pi^- \rightarrow \bar{\nu}_\mu \mu^- \rightarrow e^- \nu_\mu \bar{\nu}_e$ decay chain. But the ratio of π^+ production to π^- production is $\sim 1/7$ at these low energies. Most of the π^-'s are captured in the beamstop before decay, with $\pi^-_{decay}/\pi^-_{produced} \sim 1/20$. Most of those μ^-'s which are produced are also captured before decay, with $\mu^-_{decay}/\mu^-_{total} \sim 1/8$. Therefore the reduction factor for $\bar{\nu}_e$ production is $> 1/1000$. On the other hand, the low energy of this beam, which has a maximum of 52 MeV due to the muon decay kinematics, means that the cross section for interactions is very low. Therefore the beam must be very intense to make up for the low cross section. The LAMPF beam is 0.8 MW.

The signature for oscillations for the DAR search is a $\bar{\nu}_e$ interaction ($\bar{\nu}_e p \rightarrow e^+ n$), yielding a positron signal followed by $np \rightarrow d\gamma$, where the 2.2 MeV γ is detected. An excess of events above the expected background has been observed, as shown in Fig. 21. The total events are indicated by the points. The background is indicated by the shaded region.

When the initial LSND data were presented in 1996, there was a statistically insignificant, but striking-to-the-eye asymmetry in the number of events in the lower region of the detector compared to the upper. This was a concern because the lower region of the detector is not covered by the veto (see Fig. 20). However, continued running since that time indicates that this was a statistical fluctu-

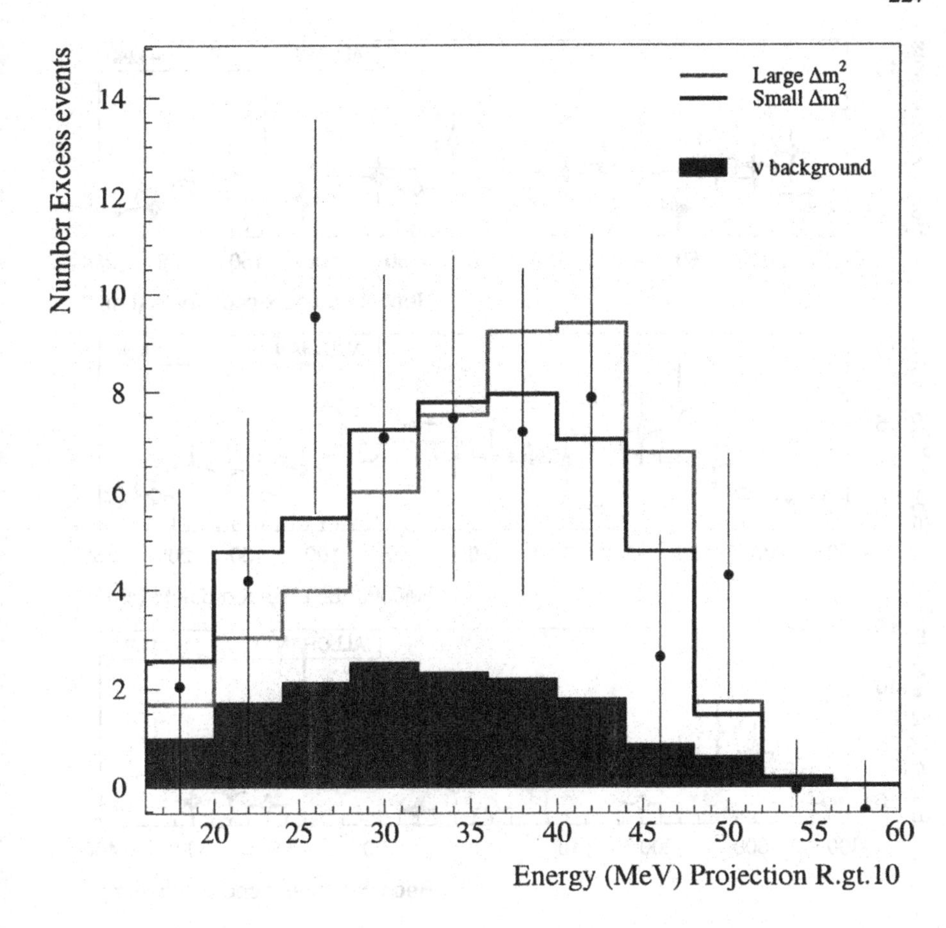

Figure 21 **Observed events from the DAR analysis as a function of energy. The background is shown by the shaded region; the histograms indicate two possible oscillation scenarios.**

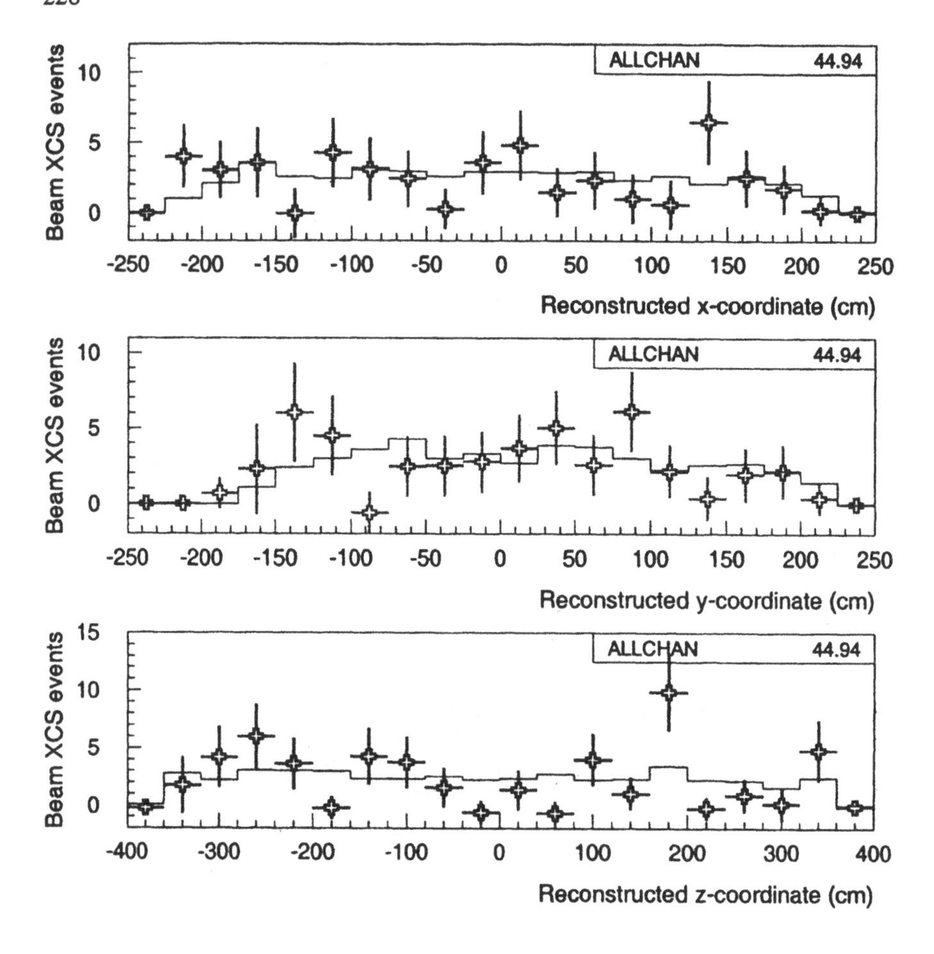

Figure 22 Final x, y and z distributions for the LSND signal. The $\bar{\nu}_e$ signal events are indicated by the points. To show the expectation, the histograms indicate the distribution for ν_eC interactions in the detector.

ation. The final distributions in x, y and z are shown in Fig. 22. The points indicate the distribution for the oscillation candidates.

The histograms in Fig. 22 indicate the expected spatial distribution if the excess is due to neutrino interactions. The events in the histograms are from $\nu_e + \mathrm{C} \to e^- + {}^{12}\mathrm{N}_{\text{ground state}}$ where the β decay of the ^{12}N in coincidence with the e^- from the neutrino interaction provides a clear signature in the detector. The humped shape of the x and y distributions occurs because the detector is cylindrical. The candidates (points) appear to be in agreement with the expectation (histogram). In particular, there is nothing striking about the y-coordinate distribution in the final data set.

Can the excess events be neutrons in coincidence with an e-like interaction? Neutrons produced in the beam will sometimes capture, producing a single photon, which is the event signature. However, much more often, the neutrons will knock into the nuclei and produce a signal of multiple γ's in the detector. Thus, if there is a large excess of neutrons causing a fake signal, the "smoking gun" will be an excess of multiple γ events in the detector. LSND has observed 49.2 ± 9 events with one associated γ (signal for oscillations). However, they have observed -2.8 ± 1.7 events with > 1 associated γ (signal for neutron background). From this, the estimated background from neutrons is < 2 events. Therefore, neutrons are apparently not the source of the excess.

Can these events be $\bar{\nu}_\mu + p \to \mu^+ + n$? The neutrino must have sufficient energy in order to have a CC interaction which knocks out a neutron and overcomes the mass threshold for producing an outgoing muon. The source of high energy $\bar{\nu}_\mu$ is π^- decay in flight. However, DIF π^-'s tend to have a relatively soft spectrum. Furthermore, as discussed above, π^- production is highly suppressed. Finally, the event must have a misidentified muon. The absolute rate of mis-identification can be constrained using the the μ^- events which are identified. In total, the estimated background from $\bar{\nu}_\mu$ nucleon interactions is < 5 events, substantially lower then the signal.

If the DAR signal is interpreted as oscillations, then the allowed region is shown in Fig. 23 as the inner, medium shaded region (90% CL) and middle, light shaded region (99% CL). To get a sense of the variations as a function of energy for a signal, Fig. 21 has two histograms superimposed on the data. The histogram which peaks at lower energy is a low Δm^2 oscillation signal and the other is a

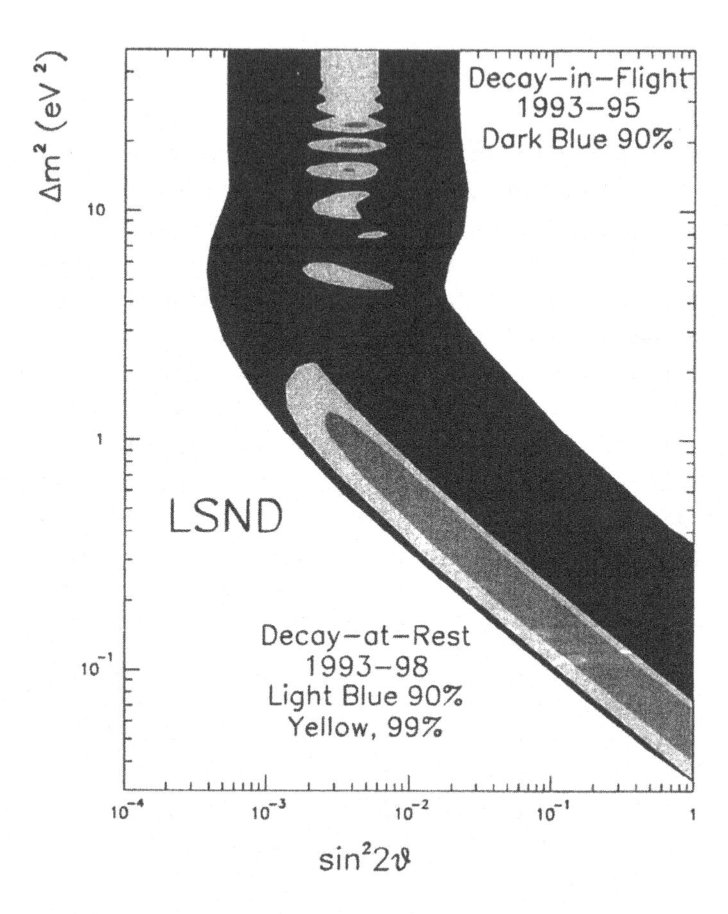

Figure 23 Interpreting the LSND DAR signal as oscillations yields the medium shaded region at 90% CL and the light shaded region at 99% CL. Interpreting the DIF excess as oscillations yields the dark shaded allowed region at 90% CL. The regions agree.

high Δm^2 solution. Clearly one cannot resolve the two cases from this data.

An important check on the LSND DAR signal comes from the search for oscillations from the ν_μ's in the beam which come from pion decay-in-flight (DIF). The DIF neutrinos are produced by targets upstream of the beamstop and are high energy compared to DAR. Therefore, the L and the E of this search differs from the DAR analysis, but the ratio, L/E, is nearly the same. The signature for $\nu_\mu \to \nu_e$ oscillations is an electron from the reaction $\nu_e C \to e^- X$ in the energy range $60 < E_e < 200$ MeV. This signal is substantially different from the DAR data. While the cross section is higher due to the increased energy, the intensity of the beam is substantially lower because of the DIF requirement, making this analysis statistics limited. An excess of events consistent with $\nu_\mu \to \nu_e$ oscillations is observed. This is indicated by the dark shaded region in Fig. 23, which is in agreement with the DAR $\bar{\nu}_\mu \to \bar{\nu}_e$ signal.

7. EXPERIMENTS WHICH SET LIMITS ON OSCILLATIONS

Many experiments have searched for neutrino oscillations and seen no signal. Tab. 3 provides a list of the exclusion experiments whose results appear in this section.

7.1. LIMITS ON $\nu_\mu \leftrightarrow \nu_E$ OSCILLATIONS

Several completed experiments have excluded the high Δm^2 solution to the LSND excess, including CCFR[43] (shown on Fig. 24). NOMAD has a preliminary limit which extends somewhat lower in Δm^2 than CCFR because the L/E of the neutrino experiment is larger [44]. BNL 776 extends still lower in Δm^2 [45]. The low Δm^2 regions with high mixing angle are ruled out by the Bugey [46] reactor experiment.

The only ongoing experiment which can test the LSND signal is KARMEN. The KARMEN experiment [47] is complementary to LSND, running at the ISIS facility in Rutherford, England, with a neutrino source produced by pion decay-at-rest. The detector is

Table 3 Some of the recent experiment which have set limits on oscillations.

Experiment	Source/Beam	$\sim E_\nu$	$\sim L$	Detector
CCFR/NuTeV	accel. ν_μ, $\bar{\nu}_\mu$	100 GeV	1 km	Iron/scint cal and muon spect
NOMAD	accel. ν_μ	26 GeV	1 km	DC targ. w/i mag., EM cal, TRD
CHORUS	accel. ν_μ	26 GeV	1 km	Emuls. targ. w/ scint fiber, tracking, cal
CDHS	accel. ν_μ	1 GeV	1 km	Iron/scint cal and muon spect
BNL E776	accel. ν_μ	1.4 GeV	1 km	Concrete/DC cal and muon spect
KARMEN I, II	π DAR $\bar{\nu}_\mu$	20-60 MeV	17 m	Liquid Scint Detector
CHOOZ	reactor $\bar{\nu}_e$	3 MeV	1 km	Gd-doped scintillator oil
Bugey	reactor $\bar{\nu}_e$	3 MeV	15, 40, 95m	Gd-doped scintillator oil

smaller and closer than LSND and the beam is less intense than at LAMPF, resulting in lower statistics than LSND by $\sim \times 10$. The experiment has had two data runs. The KARMEN I experiment took data through 1995. The KARMEN II experiment is ongoing. They have reported on data taken from 1997 to 2000. This experiment observes 11 events and expects a total background of 12.29 ± 0.63. Using a frequentist approach, Karmen II can set an upper limit at 90% CL on the number of events in their data set which could come from an oscillation signal in the LSND region. They find 90% CL limits of 3.1 events at $\Delta m^2 > 20$ eV^2 (which is a highly unlikely range based on previous experimental results, see above) and 3.8 events at $\Delta m^2 < 1$ eV^2 [48] For those who prefer Bayesian statistics, the corresponding limit is 6.3 events. The low-Δm^2-low-$\sin^2 2\theta$ region has not been addressed.

7.2. LIMITS ON $\nu_\mu \leftrightarrow \nu_\tau$ OSCILLATIONS

The two most recent $\nu_\mu \rightarrow \nu_\tau$ oscillation searches were performed by the CHORUS and NOMAD experiments, which share a high-intensity ν_μ beam produced at CERN. The neutrino energies range from 10 to 40 GeV. The $\bar{\nu}_\mu$ contamination in this beam is only $\sim 5\%$. The prompt ν_τ contamination is $(3 \pm 4) \times 10^{-6}$.

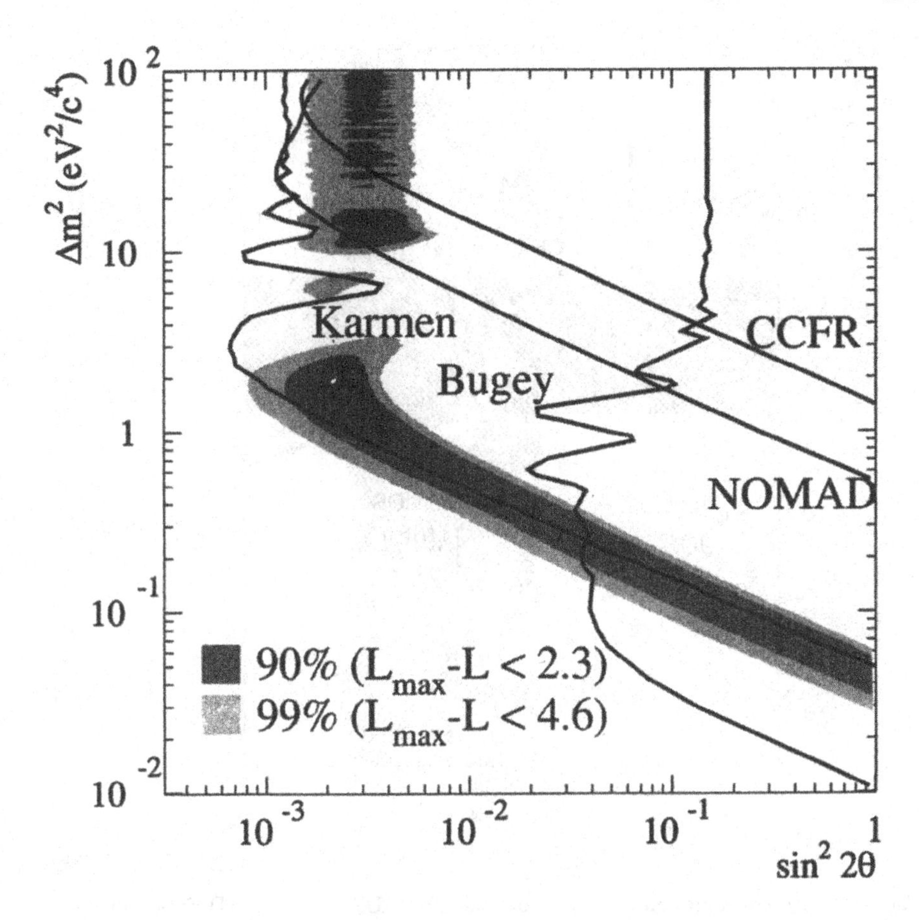

Figure 24 **Limits from other oscillation searches compared to the LSND signal.**

The CHORUS experiment uses an 800 kg emulsion target which provides $< 1\mu$m spatial resolution. This experiment can identify ν_τ charged current interactions by seeing the τ decay in the emulsion after a few tenths of a millimeter, producing a kink in the track. Automatic emulsion scanning systems have been developed to handle the large quantities of data ($\sim 300,000$ events). The emulsion target is followed by a magnetic spectrometer, calorimeter and muon spectrometer allowing momentum reconstruction and particle identification in each event.

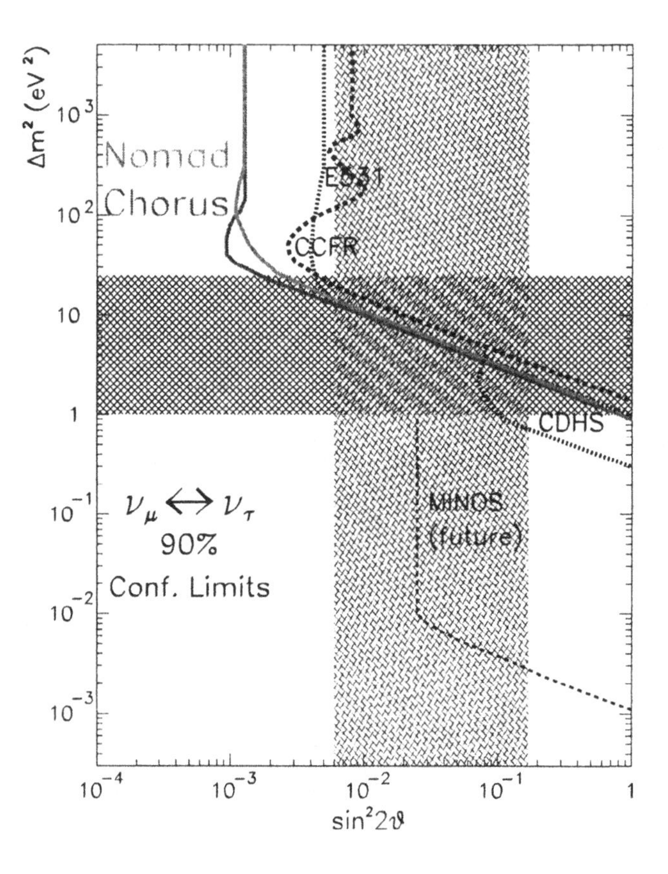

Figure 25 Recent limits on $\nu_\mu \to \nu_\tau$ from the CHORUS and NOMAD Experiments. Also shown is a limit from a previous experiment, CCFR. The sensitivity of a future experiment, Minos, is indicated. The horizontal hatched region indicates the allowed band for neutrino masses from dark matter studies, assuming one neutrino mass dominates. The vertical hatched region indicates the range of mixing angles of interest, if one uses an analogy to the quark sector.

NOMAD is a fine-grained electronic detector composed of a large aperture dipole magnet (3 m $\times$ 3 m $\times$ 7 m with B = 0.4 T) filled with drift chambers that act as both the target and tracking medium. The experiment uses kinematic cuts associated with the missing energy from outgoing ν's in the τ decay to separate statistically a possible oscillation signal.

CHORUS and NOMAD experiments have sensitivity to oscillations with $\Delta m^2 > 1$ eV^2, as does an older experiment, CCFR. Preliminary negative search results are reported, as shown in Fig. 25. This addresses the region where one may expect a signature from neutrinos which contribute to "dark matter" in the universe. Recent astrophysical data has indicated that some of the dark matter may be "hot." One candidate for hot dark matter is massive neutrinos. If we assume that one neutrino is substantially more massive than the others, by analogy with the variation in masses between generations for the charged fermions, the allowed region for masses from dark matter models may be translated into a range in Δm^2. Present data can accommodate masses indicated by the horizontal hatched region. As a toy model, one might expect the near-generation mixing to dominate and to be of a similar size to the mixing in the quark sector. The allowed range for this model is indicated by the vertical hatched band in $\sin^2 2\theta$. As shown on this plot, the entire region of interest for this model, where the two hatched regions overlap, is not completely covered by these experiments. An upcoming experiment, Minos, with sensitivity indicated on the figure, will be able to address an unexplored portion of this region in the near future.

Among the hints for oscillations discussed previously, only the atmospheric neutrino deficit may result from $\nu_\mu \leftrightarrow \nu_\tau$ oscillations. The Δm^2 reach of the present experiments does not cover the atmospheric allowed region. The lowest limit, $\Delta m^2 \sim 0.3$ eV^2, is from the CDHS experiment.[49]

7.3. LIMITS ON $\nu_E \leftrightarrow \nu_\tau$ OSCILLATIONS

Naively, $\nu_e \rightarrow \nu_\tau$ oscillations appear least likely, because this skips the second generation. However, this oscillation does appear in some models. The one hint which can be interpreted as such an oscillation is the solar neutrino deficit. In addition, one or more of the neutrinos may represent a fraction of the dark matter in the universe. If there is a mass difference between the neutrinos, this might manifest itself through $\nu_e \rightarrow \nu_\tau$ oscillations at relatively high Δm^2.

Recent searches from NOMAD and CCFR have addressed high Δm^2's, while the reactor experiments have explored down to 10^{-3}eV^2. The terrestrial experiments remain a few orders of magnitude away from addressing the solar $\nu_e \rightarrow \nu_\tau$ hypothesis.

8. THEORETICAL INTERPRETATION OF THE DATA

When comparing the evidence for oscillations with the excluded regions, we are faced with theoretical problems with both the suggested Δm^2 regions and the mixing angles. There are apparently three distinct Δm^2 regions:

$$\begin{aligned}
\Delta m^2_{solar} &= 10^{-5} \text{ or } 10^{-10} \text{ eV}^2 \\
\Delta m^2_{atmos} &= (10^{-2} \text{ to } 10^{-4}) \text{ eV}^2 \\
\Delta m^2_{LSND} &= (0.2 \text{ to } 2) \text{ eV}^2
\end{aligned}$$

This cannot be accommodated in simplistic 2-generation mixing models.

However, in a straightforward three-generation mixing model, one can accommodate three Δm^2 values if one assumes that the atmospheric deficit is a mixture of $\nu_\mu \rightarrow \nu_\tau$ and $\nu_\mu \rightarrow \nu_e$. As a result, Δm^2_{atmos} extracted in a 2-generation model is a convolution of Δm^2_{solar} and Δm^2_{LSND} [50]. If the upper region of the LMA solar solution is the correct, then this theory is plausible because the L/E of the atmospheric experiments is of the correct magnitude to have a contribution from $\nu_\mu \rightarrow \nu_e$. Fitting a three-generation oscillation signal using a two-generation oscillation model can lead to a solution for $\sin^2 2\theta$ which is unphysical. This may explain why the atmospheric data best fit extends into the region of $\sin^2 2\theta > 1$. The Super K data for 70.5 kiloton-year has been fit with a three-generation scenario, and preliminary results for the best-fit oscillation parameters are available [51]. The model correctly predicts the ratio of data to Monte Carlo for the upward-going ν_e's and ν_μ's and for the downward going ν_e's. The main conflict between the data and predictions of this scheme are for downward-going ν_μ interactions, where the central predictions are $\sim 2\sigma$ away from the measurement.

The scenarios become much richer as soon as one adds the extra degree of freedom of the sterile neutrino. Obviously sterile neutrinos can only be invoked for cases where a deficit (as opposed to a signal) are observed. Sterile neutrinos may provide the explanation for the atmospheric or solar deficits, but LSND is required to be $\nu_\mu \to \nu_e$.

The first option is that the atmospheric neutrino deficit is caused entirely or in part by $\nu_\mu \to \nu_s$. In a simple two-generation analysis, Super K has found that their data favor $\nu_\mu \to \nu_\tau$ rather than $\nu_\mu \to \nu_s$ [53], although one should note that this analysis relies heavily on a precise understanding of the systematics of the upward, through-going muons. However, in a full four-generation oscillation analysis [52], the atmospheric data are shown to accommodate sterile neutrinos with the parameters $\Delta m^2_{34} = 2.0 \times 10^{-3}$ eV^2, $(\theta_{24}, \theta_{34}, \theta_{23}) = (45°, -30°$, and $20°)$. In this model the LSND solution is $\Delta m^2_{23} = 0.3$ eV^2 and the LMA solar solution is chosen for the fit.

The second option within the sterile neutrino solution is to assume the atmospheric result is entirely due to $\nu_\mu \to \nu_\tau$. In this case, four-generation fits can describe all of the data in a scheme where one mass eigenstate dominates the others. In this case the sterile neutrino has a small mixing with all of the active flavors. The small mixing angle of the LSND result is thus explained by a small mixing of ν_e to ν_μ via the sterile neutrino [54]. If this is the case, the solar experiments will not be able to observe the sterile contribution to oscillations.

If one allows for some of the data to be "slightly off" then the three and four-generation models expand dramatically. For example, the only solar neutrino experiment which measures less than half the expected number of neutrinos is the Homestake chlorine-based experiment. If, for some reason, this experiment were systematically low, then one can get good three generation fits, as well as alternative mass hierarchies for the four generation case.

The final possibility is that one of the experiments is not observing oscillations. However, "not oscillations" does not necessarily mean "not interesting". Other beyond-the-Standard-Model solutions to the three oscillation signals have been put forward. As a few representative examples:

▷ The solar neutrino deficit could be due to a large neutrino magnetic moment ($\mu_\nu \sim \mathcal{O}(10^{-12}\mu_B)$ [55]. If neutrinos travel through a high magnetic field within the Sun, they may experience a spin-flip and be rendered RH, hence sterile. This is disfavored because it is difficult to create a solar model with sufficiently high fields, but it is still a possibility. Alternatively, another solution to the solar neutrino deficit which fits all of the current data introduces a beyond-the-Standard-Model Flavor Changing Neutral Current interaction [56]. This would also have the consequence of producing flavor violating τ decays. However, these are at a rate just below the present upper bound. This model can be tested by searching for these flavor violating decays at the B factories.

▷ The atmospheric neutrino deficit could result from neutrino decay. However, the upward-going muon data strongly disfavor this scenario compared to oscillations [57].

▷ The LSND signal may be due to lepton-flavor violating decays, where $\mu^+ \rightarrow e^+\nu_\mu\bar{\nu}_e$ at a very small level. This is as unlikely explanation because KARMEN does not see an effect.

9. THE FUTURE (NEAR AND FAR)

The existing indications of neutrino oscillations raise many questions for future experiments to address. Many new experiments are proposed or are running which address the issues that have been raised by the present data. This section provides an overview of some of the exciting results which can be expected in the near future.

9.1. FUTURE TESTS OF SOLAR NEUTRINO OSCILLATIONS

The issues related to the solar neutrino deficit are:

▷ Can we see the L/E dependence which will clearly demonstrate neutrino oscillations?

▷ Is this $\nu_e \rightarrow \nu_\mu$, ν_τ, or ν_s (or some combination)?

▷ What is the Δm^2? Is MSW or Just-So the right explanation?

▷ If the solution is MSW, is it the small or large angle solution?

▷ Is there any room for doubting the Standard Solar Model?

In order to address the L/E dependence, a wide range of experiments with varying energy thresholds is needed. Tab. 4 provides a

Table 4 Some future Solar Neutrino Experiments.

Experiment	Detector	Search	Source	Approx. Range
SNO	Deuterium	$\nu_x + d \to p + n + \nu_x$ types) (NC $x = e, \mu, \tau$) $\nu_e + d \to p + p + e^-$ (CC ν_e only)	Sun	> 5 MeV
BOREXINO	Liquid Scint.	ν_e elastic scatters	Sun	$0.5 < E < 1.0$ MeV
GNO	Gallium	ν_e capture in Ga	Sun	> 0.2 MeV
HELLAZ	Helium TPC	ν_e elastic scatters	Sun	> 0.05 MeV
KamLAND	Liquid Scint.	ν_e elastic scatters	Reactors	> 1 MeV

summary of the upcoming solar experiments. As can be seen from the Tab. 4 these and other proposed future experiments will cover energies ranging upward from 0.05 MeV, permitting tests of the L/E dependence.

The SNO experiment [58] has already taken data for over a year. First oscillation search results are expected in spring, 2001. This experiment will test the hypothesis for sterile neutrino solar oscillations. This experiment observes neutral current (NC) interactions for all three standard neutrinos and charged current (CC) ν_e interactions. Sterile neutrinos will not have neutral current interactions. Therefore the ratio of NC to CC interaction rates will be lower than predicted if solar oscillations are $\nu_e \to \nu_s$.

The BOREXINO experiment [59] is sensitive to neutrinos from the ${}^7Be + e^- \to {}^7 Li + \nu_e$ interaction, which is a delta function in the flux distribution, as shown on Fig. 12. Therefore, this experiment will be highly sensitive to seasonal variations in L, the Earth-Sun distance, if the "Just So" solution is correct.

As the only terrestrial experiment which can address the solar neutrino question, the KamLAND experiment [60] is not affected by theoretical errors from the Standard Solar Models. This experiment will be located in the Kamiokande cavern and will make use of neutrinos from five reactor sites, resulting in $L \sim 160$ km. This experiment is sensitive only to the LMA MSW solution.

In the far future, an interesting test of LMA MSW $\nu_e \rightarrow \nu_\tau$ and $\nu_e \rightarrow \nu_\mu$ oscillations may be made by the first stage of the muon collider. A muon storage ring would provide an intense beam of ν_e's (and ν_μ's) from muon decays. Using a 50 GeV storage ring and beams directed from the US to Italy and Japan, the single event sensitivity covers the upper region of the LMA solution [61].

9.2. FUTURE TESTS OF ATMOSPHERIC NEUTRINO OSCILLATIONS

The issues related to the atmospheric neutrino deficit are:

▷ Can we see an effect in the controlled environment of an accelerator experiment?

▷ Is this mainly $\nu_\mu \rightarrow \nu_\tau$ or ν_s (or some combination)?

▷ Is there any $\nu_\mu \rightarrow \nu_e$ component?

▷ What is the Δm^2?

It is possible to be sensitive to the moderate Δm^2's indicated by the atmospheric neutrino deficit through "long baseline" experiments. In the near future, beams will be built at accelerator facilities with $E_\nu \sim 1 - 10$ GeV which point to detectors at distances of hundreds of kilometers. This opens a new era of tests of neutrino oscillations in the atmospheric region, with entirely different systematics from the previous experiments.

The sensitivities of two long baseline experiments which are approved to run in the near future are shown in Fig. 26. The K2K experiment, which has already begun running [62], uses a 250 km baseline from KEK to the Super K detector. The beam is produced using 12 GeV protons on target, with an average energy of 1.4 GeV. K2K has two "near detectors" located 0.3 km downstream of the meson decay region to monitor the beam flux before possible oscillations. The signal would be a deficit of events in the far detector (Super K) compared to the prediction based on the near detector measurements. As of summer, 2000, K2K reported 27 observed events. A total of 40.3 events were expected if no oscillations occurred. The energy dependence of the events has not

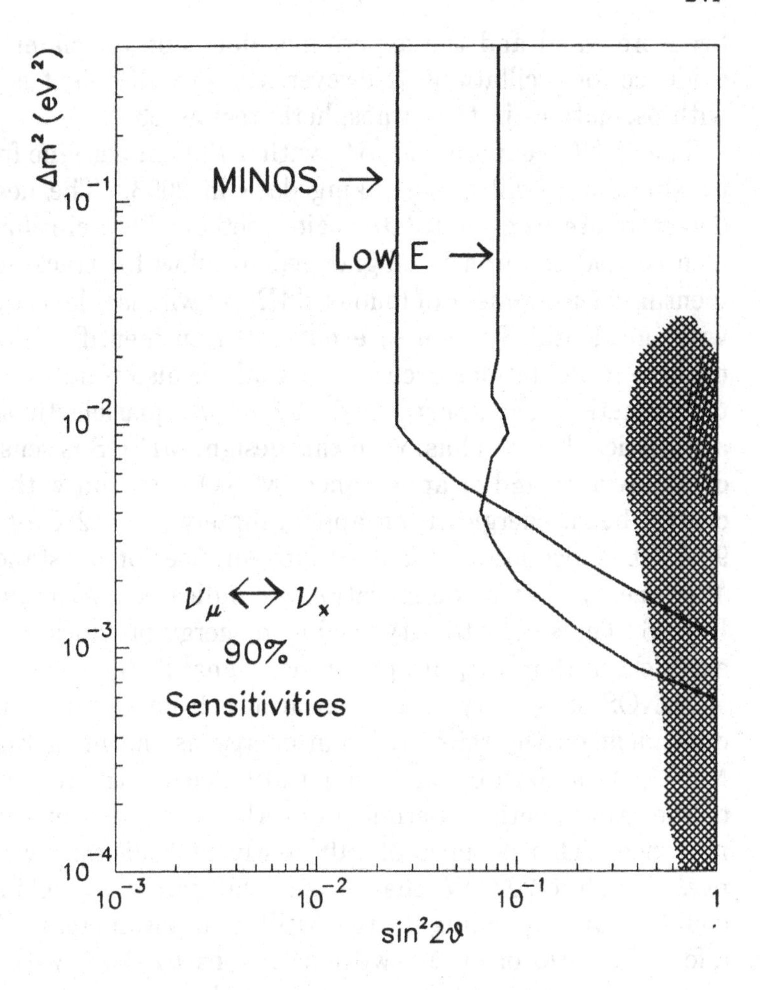

Figure 26 **Sensitivities of the MINOS experiment with two possible beam designs.**

been presented and the experiment does not yet claim to observe evidence for oscillations. However, the overall deficit is consistent with oscillations in the atmospheric region [63].

The MINOS experiment [64], with a 730 km baseline from FNAL to Minnesota, will begin taking data in 2003. The near and far detectors are iron-scintillator calorimeters. The scintillator is segmented and the iron is magnetized to allow for tracking and momentum measurement of muons. MINOS will use the ratio of events with no identified muon to events with an identified muon to determine if oscillations occur. The analysis also isolates highly electromagnetic events, thereby identifying ν_e quasielastic scatters on a statistical basis. Thus, with this design, MINOS is sensitive to ν_μ disappearance and ν_e appearance. MINOS can run with a range of central beam energies from approximately 2 to 12 GeV [65]. The 90% CL expectations for ν_μ disappearance for a "standard" and "low energy" beam configuration are indicated by the two lines on Fig. 26. The ability to vary the beam energy provides an important extra check that a ν_μ disappearance signal is due to oscillations.

MINOS sensitivity to ν_τ appearance from $\nu_\mu \rightarrow \nu_\tau$ is strongly dependent on the selected beam energy as shown in Fig. 4. The MINOS "standard beam" configuration covers the region where all of the atmospheric experiments overlap with greater than 5σ significance. This beam is of sufficiently high energy, with a broad peak from 5 to 20 GeV, that a significant rate of ν_τ CC interactions could occur, depending on the oscillation parameters. This would affect the ratio of events without muons to those with muons in a characteristic manner, since 17% of the τ decays are to μ while the remaining are electromagnetic or, mostly, hadronic. Full coverage of the Super K region is obtained with the "low energy" beam configuration, which peaks at about 2 GeV. In this configuration, however, ν_τ interactions will be almost entirely in NC mode due to the mass suppression of the 1.8 GeV τ. MINOS plans to begin its initial running in the low energy beam mode.

CERN is also in the process of planning an extensive long baseline program using a beam directed to the Gran Sasso facility. The baseline from CERN to the Gran Sasso is 740 km. The detectors for this program are designed to search for $\nu_\mu \rightarrow \nu_\tau$ appearance. The designers of this program must balance competing goals related to the beam. One would like to use a low energy beam in order to maximize L/E and, therefore, sensitivity to the lowest Δm^2 of the Super K allowed region. On the other hand, the ν_τ CC cross section increases with beam energy, as shown by Fig. 4. Thus high beam energy allows experiments to obtain sufficient statistics on a short timescale. Compromising between these conflicting needs, the 1999 "reference beam design" has a peak energy of about 20 GeV [66]. To characterize the expectation for this program, one can use the Opera experiment, which is a proposed 2000 ton hybrid emulsion detector [67]. Assuming 5 years of running, with an oscillation signal at $\Delta m^2 \times 10^{-3}$ (the Super K most-likely value), Opera will obtain 18.3 events with a background expectation of 0.6 events.

9.3. FUTURE TESTS OF THE LSND SIGNAL

The issues related to the LSND signal are:

- ▷ Is the signal due to oscillations?
- ▷ What is the Δm^2?
- ▷ What is the $\sin^2 2\theta$?

At this point enticing signals have been seen in three types of experiments exploring this channel (see section 2). What is required at this point is an experiment which definitively covers the entire LSND allowed region at $> 5\sigma$ and which can accurately measure the oscillation parameters.

BooNE (Booster Neutrino Experiment), which has been approved at FNAL, will be capable of observing both $\nu_\mu \rightarrow \nu_e$ appearance and ν_μ disappearance. The first phase, MiniBooNE, is a single detector experiment designed to obtain $\sim$ 1000 events per year if the LSND signal is due to $\nu_\mu \rightarrow \nu_e$ oscillations. This establishes the oscillation signal at the $\sim 8\sigma$ level. The second phase of the experiment introduces a second detector, with the goal to accurately measure the Δm^2 and $\sin^2 2\theta$ parameters of observed oscillations.

The MiniBooNE experiment [68] (phase 1 of BooNE) will begin taking data in 2001. The detector will consist of a spherical tank 6 m in radius. An inner structure at 5.5 m radius will support 1280 8-inch phototubes (10% coverage) pointed inward and optically isolated from the outer region of the tank. The vessel will be filled with 769 tons of mineral oil, resulting in a 445 ton fiducial volume. The outer volume will serve as a veto shield for identifying

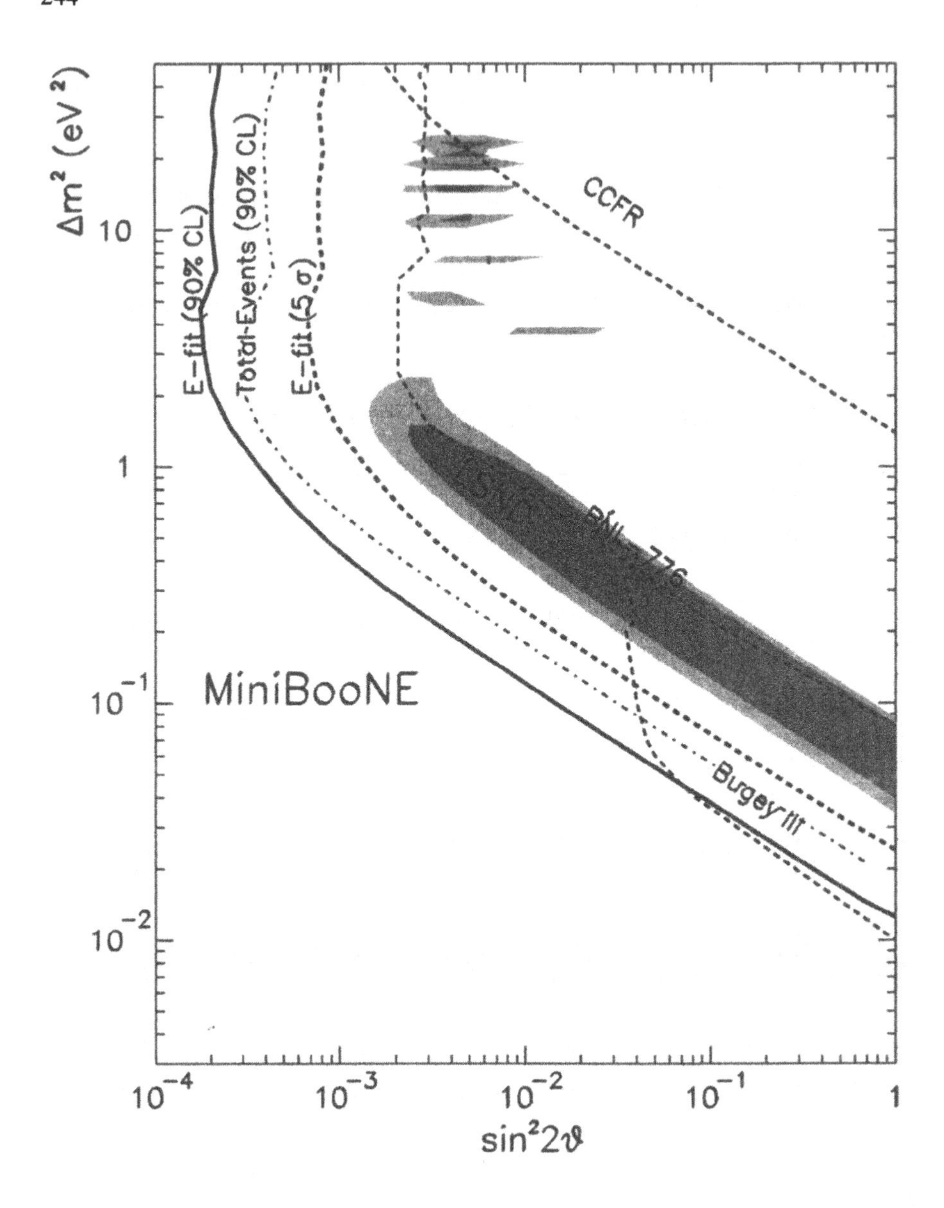

Figure 27 Sensitivity regions for the MiniBooNE experiment with 5×10^{20} protons on target (1 year). The solid (dashed) curve is the 90% CL (5σ) region using the energy fit method and the dashed-dot curve is the 90% CL region using the total event method.

particles both entering and leaving the detector, with 292 phototubes mounted on the support structure facing outwards. The first detector will be located 500 m from the Booster neutrino source. The neutrino beam, constructed using the 8 GeV proton Booster at FNAL, will have an average beam energy of approximately 0.75 GeV.

The sensitivity to oscillations can be calculated by summing over energy or by including energy dependence in the fit. As shown in Fig. 27, both the summed analysis and the energy-dependent analysis extend well beyond the LSND allowed region at 90% CL. Also shown is the region where MiniBooNE will see a 5σ or greater signal above background and make a conclusive measurement, which again extends well beyond the LSND signal region.

9.4. AND BEYOND...

Depending on what we learn from the upcoming experiments described above, the next step may be a "neutrino factory." This would make use of an intense, highly collimated beam of neutrinos produced via muon decay from a muon storage ring. The initial design goal is to produce a beam of $> 10^{19}$ neutrinos per year. This is two orders of magnitude higher than the planned MINOS beam and four orders of magnitude higher than the CCFR or NOMAD/CHORUS beams. With such high intensities, baselines as long as a few thousand kilometers would be realistic for neutrino oscillation experiments. Because the beams are produced by muon decay, this accelerator will also provide the first ν_e beam.

An inital design study was completed at Fermilab last summer [69]. The assumption was that in 10 years time:

- ▷ LSND will be confirmed/refuted. If confirmed, we could know Δm^2_{12} to 0.1eV^2, and $\sin^2 2\theta_{12}$ to 10%
- ▷ K2K & MINOS will confirm/refute atmospheric neutrino deficit;
- ▷ SuperK & MINOS may rule out $\nu_\mu \rightarrow \nu_{\text{sterile}}$. The assumption is that these experiments will then measure δm^2_{23} to 30%, $\sin^2 2\theta_{23}$ to 20%; and
- ▷ SNO & Borexino will determine if solar is $\nu_e \rightarrow \nu_{\text{sterile}}$ and if SMA, LMA or Just-So is the correct solution.

If these experiments, demonstrate oscillations, then the remaining questions will be:

- ▷ Is 3-generation mixing matrix unitary?
- ▷ What is the sign of δm^2?
- ▷ Is there CP violation?
- ▷ Can we exploit matter effects?
- ▷ If sterile neutrinos: how many?

The program which is envisioned is very rich, with great potential for new physics.

10. CONCLUSIONS

In the past few years three exciting indications for neutrino oscillations have been observed. However, these results open new questions, including: Are all signals attributable to oscillations? What are the correct Δm^2 and $\sin^2 2\theta$? In the case of the solar and atmospheric deficits, oscillations to which ν flavor? It is too early to draw firm conclusions. To answer these questions, we must wait for the results a new generation of oscillation experiments. But one thing is without question: these are exciting times for neutrino physics!

References

[1] D. Decamp, *et al.*, CERN-EP/89-169, Phys.Lett.B235:399, 1990.

[2] H. Band, *et al.*, SLAC-PUB-4990, published in the Proceedings of the Fourth Family of Quarks ,and Leptons, Santa Monica, CA, Feb 23-25, 1989.

[3] http://fn872.fnal.gov/

[4] C.S. Wu, *et al.* Phys. Rev. 105, 1413, 1957.

[5] D.E. Groom et al, The European Physical Journal C15: 1, 2000.

[6] H.M. Gallagher and M. Goodman, NuMI-112, available at: http://www.hep.anl.gov/ndk/hypertext/numi_notes.html

[7] http://amanda.berkeley.edu/

[8] http://antares.in2p3.fr/

[9] K. Assamagan, *et al.*, Phys.Rev.D53:6065, 1996.

[10] Barate, *et al*, EPJ C2 395, 1998.

[11] V. M. Lobashev, *et al.*, Phys.Atom.Nucl.63: 962, 2000.

[12] J. Bonn, *et al.*, Phys. Atom. Nucl.63:969, 2000.

[13] W. Stoeffl, *et al.*, PRL 75: 3237, 1995.

[14] C. Ching, *et al.*, Int. Journ Mod. Phys. A10: 2841, 1995

[15] E. Holzschuh, *et al.*, Phys. Lett., B287: 381, 1992.

[16] H. Kawakami, *et al.*, Phys. Lett., B256: 105, 1991.

[17] H. Robertson, *et al.*, Phys.Rev.Lett.67: 957, 1991; H. Robertson, PR D33: R6, 1991.

[18] K. Sato and H. Suzuki, PRL 58: 2722, 1987.

[19] R. Bionta, *et al.*, PRL 58: 1494, 1987.

[20] K. Hirata, *et al.*, PRL 58: 1490, 1987.

[21] A. Alessandrello, *et al.*, Phys. Lett. B486: 13, 2000; L. DeBraekelee, *et al.*, Phys. Atom Nucl 63:1214, 2000; R. Arnold, *et al.*, Nucl. Phys. A678:341, 2000; M. Alston-Garnjost, *et al.*, PR C55: 474, 1997; A. DeSilva, *et al.*, PR C56:2451, 1997; M. Gunther, *et al.*, PR D55:54, 1997; R. Arnold, *et al.*, Z. Phys. C72: 239, 1996; A. Balysh, *et al.*, PRL 77:5186, 1996.

[22] H. Klapdor-Kleingrothaus, Proceedings of Neutrino 96, Helsinki,June, 1996; H. Klapdor-Kleingrothaus, hep-ex/9802007.

[23] A good discussion of GUT motivation for the sea-saw model appears in Kayser, Gibrat-Debu, and Perrier, *The Physics of Massive Neutrinos,* World Scientific Lecture Notes in Physics, 25, World Scientific Publishing, 1989.

[24] Y. Hayato, *et al.*, Phys.Rev.Lett. 83: 1529, 1999.

[25] E. Ma (UC, Riverside), G. Rajasekaran, U. Sarkar, hep-ph/0006340

[26] Z. Berezhiani and R. Mohapatra, Phys.Rev.D 52: 6607, 1995.

[27] J.N. Bahcall, S. Basu, and M. H. Pinsonneault, Phys. Lett. B **433** 1 (1998).

[28] R. Davis, Prog. Part. Nucl. Phys. **32**, 13 (1994).

[29] J. N. Abdurashitov *et al.* , Phys. Lett. B **328**, 234 (1994); P. Anselmanni *et al.* , Phys. Lett. B **328**, 377 (1994).

[30] J. N. Bahcall and P. I. Krastev, Phys. Lett. B **436** (243) 1998.

[31] E. G. Adelberger *et al.* , "Solar Fusion Cross Sections," To be published in Rev. Mod. Phys., Oct. 1998, astro-ph/9805121.

[32] For example, J. Christensen-Dalsgaard *et al.* , Science **272** 1286 (1996).

[33] For example, Castellani *et al.* , Nucl. Phys. Proc. Suppl. **70** 301 (1998).

[34] N. Hata and P. Langacker, Phys. Rev. D **56** 6107 (1997); N. Hata and P. Langacker, Phys. Rev. D **50**, 632 (1994).

[35] L. Wolfenstein, Phys. Rev. D17, 2369 (1978); D20, 2634 (1979); S. P. Mikheyev and A. Yu. Smirnov, Yad. Fiz. 42, 1441 (1985) [Sov. J. Nucl. Phys. 42, 913 (1986)]; Nuovo Cimento 9C, 17 (1986).

[36] S. Coutu, Proceedings of ICHEP'98, Vancouver 23-29 July 1998, p.666.

[37] Y. Hirata *et al.* , Phys. Lett. B**335**, 237 (1994).

[38] Y. Totsuka, Nucl. Phys. A663, 218 (2000); Y.Fukuda, *et al.* Phys. Rev. Lett. 81, 1562 (1998)

[39] S. Hatakeyama *et al.* , Phys. Rev. Lett **81** 2016 (1998).

[40] C. Athanassopoulos *et al.* , Phys. Rev. Lett. **75**, 2650 (1995);

[41] C. Athanassopoulos *et al.* , Phys. Rev. Lett. **77**, 3082 (1996); C. Athanassopoulos *et al.* , Phys. Rev. C. **54**, 2685 (1996).

[42] C. Athanassopoulos *et al.* , LA-UR-97-1998, submitted to Phys. Rev. C.

[43] A. Romosan *et al.* , Phys. Rev. Lett. **78** 2912 (1997).

[44] D. Autiero, *et al.*, Proceedings of the 29th International Conference on High-Energy Physics (ICHEP 98), Vancouver, British Columbia, Canada, 23-29 Jul 1998, 626.

[45] L. Borodovsky *et al.* , Phys. Rev. Lett. **68**, 274 (1992).

[46] B. Achkar *et al.* , Nucl. Phys. **B434**, 503 (1995).

[47] B. Bodmann *et al.* , Phys. Lett. B **267**, 321 (1991); B. Bodmann *et al.* , Phys. Lett. B **280**, 198 (1992); B. Zeitnitz *et al.* , Prog. Part. Nucl. Phys., **32** 351 (1994). K. Eitel, hep-ex/9706023.

[48] K. Eitel, To be published in the Proceedings of 19th International Conference on Neutrino Physics and Astrophysics - Neutrino 2000, Sudbury, Ontario, Canada, 16-21 Jun 2000. hep-ex/0008002.

[49] F. Dydak *et al.* , Phys. Lett. B **134**, 281 (1984).

[50] G. Barenboim and F. Scheck, Phys. Lett., B 440, 211, 1995.

[51] G. Barenboim, private communication.

[52] O. Yasuda, Proceedings of the 30th International Conference on High-Energy Physics (ICHEP 2000), Osaka, Japan, 2000, hep-ph/0008256.

[53] S. Fukuda, *et al.* , submitted to Phys. Rev. Lett., hep-ex/0009001.

[54] V. Barger, B. Kayser, J. Learned, T. Weiler, K. Whisnant, Phys. Lett. B489: 345, 2000.

[55] J. Pulido, FISIST-00-00-CFIF, Sep 2000, To be published in the Proceedings of Europhysics Neutrino Oscillation Workshop (NOW 2000), Conca Specchiulla, Otranto, Lecce, Ita, 9-16 Sep 2000, hep-ph/0012059.

[56] S. Bergmann, M.M. Guzzo, P.C. de Holanda, P.I. Krastev, H. Nunokawa, Phys. Rev. D62: 073001, 2000.

[57] G.L. Fogli, E. Lisi, A. Marrone and G. Scioscia, Nuc. Phys. B85: 159, 2000.

[58] See http://www.sno.phy.queensu.ca/

[59] See http://almime.mi.infn.it/

[60] Y. F. Wang, STANFORD-HEP-98-04 (1998). P. Alivisatos *et al.*, STANFORD-HEP-98-03 (1998).

[61] S. Geer, Phys. Rev. D **57** 6989 (1998); S. Geer, FERMILAB-CONF-97-417.

[62] http://neutrino.kek.jp/

[63] Results presented at Neutrino 2000: http://ichep2000.hep.sci.osaka-u.ac.jp/scan/0801/pl/nishikawa/index.html.

[64] http://www.hep.anl.gov/ndk/hypertext/numi.html

[65] See http://www.hep.anl.gov/NDK/
HyperText/minos_tdr.html. P875, "A Long Baseline Neutrino Oscillation Experiment at Fermilab, February, 1995; Dave Ayres for the MINOS collaboration, "Summary of the MINOS proposal, A Long Baseline Neutrino Oscillation Experiment at Fermilab", March 1995; NuMI note: NuMI-L-71, see:
http://www.hep.anl.gov/NDK/Hypertext/
numi_notes.html

[66] http://ngs.web.cern.ch/NGS/.

[67] see http://opera.web.cern.ch/opera/doctab.html, M. Guler, *et al.*, CERN/SPSC 2000-028, SPSC/P318, LNGS P25/2000.

[68] See: http://www.neutrino.lanl.gov/BooNE/
E. Church *et al.* FERMILAB-P-0898 (1997).

[69] see http://www.fnal.gov/projects/muon_collider/

NEW DEVELOPMENTS IN CHARGED PARTICLE TRACKING

Thinner - Faster - Larger - Harder

Andreas S. Schwarz
DESY Hamburg
Notkestr. 85
22603 Hamburg
Germany
andreas.schwarz@desy.de

Abstract Prompted by the quest for ever higher energies and larger luminosities at both commissioned and planned colliders, new developments in charged particle tracking had to be made over the last decade. The state of the art in both the area of gaseous tracking detectors and semiconductor tracking detectors is described with special emphasis on issues of radiation hardness.

1. INTRODUCTION

In the following lectures, an attempt is made to give an overview of the recent developments in charged particle tracking that have occured over the past several years.

The lectures are structured as follows. After a sketch of the expected experimental environment at new accelerators, the principles of gaseous tracking detectors are outlined. This is followed by a description of the basic elements of semiconductor tracking detectors. The next two chapters concentrate on the important issue of radiation hardness, both for gaseous detectors and for semiconductor detectors. Finally, the lectures are concluded with a brief description of two examples of tracking detector systems of the new generation: the ATLAS Semiconductor Tracker and the HERA-B large area gaseous tracking chambers.

Owing to the limited time available, several important developments and areas could not be covered. I chose to omit the development in emulsions, now gaining importance for new neutrino oscillation experiments, the application of scintillating fibres in tracking, as well as new develop-

H.B. Prosper and M. Danilov (eds.), Techniques and Concepts of High-Energy Physics, 251–311.

ments in applications 'outside' high energy physics, namely in biology, medicine and astroparticle physics. Diamond and GaAs as materials for applications in charged particle tracking will not be discussed and the issue of readout electronics is only peripherally covered where absolutely needed. I apologize for all the groups whose contributions I thus will not cover adequately.

There exists a vast literature covering the different areas touched upon in these lectures and the reader is referred to them for further reading. I would like to specifically point out the series of proceedings of the NATO ASI schools [1], the seminal article about gaseous detectors from Sauli [2], the excellent book on gaseous charged particle detectors by Blum and Rolandi [3] and the new book on semiconductor particle detectors by Lutz [4]. The interested reader is further referred to the proceedings of the Vienna Wire Chamber Conferences [5], of the London Conference on Position Sensitive Devices [6], the International Workshop on Vertex Detectors [7] and the International Workshops on Semiconductor Physics [8]. A summary of the status of vertex detectors and their application in the study of the physics of heavy flavours has been given by the author in Ref. [9].

The topic of radiation hardness in gaseous detectors and in semiconductor detectors has inspired much interest recently but to my knowledge the former has been the topic of an international workshop only once [10].

2. EXPERIMENTAL ENVIRONMENT – NEW CHALLENGES

The development in charged particle tracking over the last 10 years has once again been driven by research interests. They can be divided up as follows:

- CP violation in the bottom meson system, both at e^+e^- machines (BaBar, BELLE) as well as at hadron machines (HERA-B, CDF, D0, LHCb)
- The 'High Energy Frontier': the search for the quark-gluon plasma at RHIC (STAR, PHENIX) and at the Large Hadron Collider (LHC) at CERN (ALICE) and the search for Supersymmetry and the origin of the Higgs mechanism (ATLAS and CMS at the LHC).

Before discussing the different requirements on charged particle tracking of these experiments, in somewhat more detail, I give in Table 1 an overview of the event rates and charged particle multiplicities measured/expected at different facilities.

Table 1 Charged track multiplicities and number of events per bunch crossing for various High Energy Physics facilities.

Facility	Production	Startup Year	# Charged tracks	Crossing Interval [ns]	Events/ crossing
PETRA	e^+e^-	1978	10–15	3,600	Rare
LEP	e^+e^-	1989	20	22,000	Rare
HERA	ep	1992	20	96	Rare
HERA-B	pX	2000	80–150	96	3–5
BaBar	e^+e^-	1999	12	4	Rare
TeVatron	$p\bar{p}$	2001	200	132	2
LHC	pp	2005	800	25	23
LHC/H.I.	AA	2005	10,000	125	Rare

One can see two distinct groups according to the number of interactions per crossing of a particle bunch: the e^+e^- machines (even the newly built ones) and the Heavy Ion programs have to cope with relatively rare interaction rates ($\sim 10^{-3}$ interactions/bunch crossing) whereas at the hadron machines between 2 and $\sim$20 interactions per bunch crossing are needed in order to get access to the tiny cross sections for processes of interest.

2.1. E^+E^- B FACTORIES – BELLE AND BABAR

In the Standard Model, large CP violating effects are predicted for the bottom meson system. This has led to the development of a strong branch in high energy physics research over the last years: the B meson factories. Two different approaches have been followed: the copious production of B mesons at dedicated e^+e^- colliders and the production at hadron machines.

Two e^+e^- storage rings have been successfully commissioned in the last year: the KEK B Factory with the experiment BELLE [11] and the PEP-II collider with the experiment BaBar at SLAC [12]. They aim at luminosities of $(3 - 10)\text{x}10^{33}\text{cm}^{-2}\text{sec}^{-1}$ at a center of mass energy of 10.58 GeV, corresponding to the mass of the $\Upsilon(4S)$ resonance. The beams collide asymmetrically such that the $\Upsilon(4S)$ resonance is boosted in the direction of the high energy beam ($\beta\gamma \approx 0.56$). This allows a separation of the two B meson decay vertices of a given event and thus a determination of which B meson decayed first, a necessary ingredient in order to study the subtle CP violation effects in the decays of the neutral bottom mesons on the $\Upsilon(4S)$ resonance.

The events have low multiplicities and low momenta: the average momentum of particles in the 'golden' decay $B^0 \rightarrow J/\Psi\pi^+\pi^-$ is ~1.8 GeV/c^2 and Kaon/Pion separation has to be effective down to momenta of ~0.3 GeV/c^2.

In order to reconstruct a large fraction of the events fully, a large part of the solid angle has to be covered. The boost of the $\Upsilon(4S)$ resonance results in an average separation of the B meson decay vertices of ~250μm. The CP asymmetry peaks at ~550μm thus putting stringent requirements on the spatial resolution of the vertex detector system.

The requirements for tracking devices at the new e^+e^- B meson factories can be summarized as follows:

- High precision vertex detectors with very low material thickness, located as close to the interaction point as possible. Both experiments have chosen silicon detectors for this task.

- Continuous and low mass tracking at larger distances to the beam line. The detector of choice is the drift chamber, using thin gases (e.g. He(80%)-C_4H_{10}(20%)) and goldcoated Aluminum wires.

- Medium radiation hardness. At the worst position closest to the beam line the silicon detectors have to survive yearly doses of the order of 200 krad. The drift chamber has to sustain rates of 5 kHz/cell and an integrated accumulated charge per cm of wire of O(0.025)C.

These requirements represent a fairly 'low risk' extrapolation from existing tracking detector designs.

2.2. HEAVY ION PHYSICS – ALICE AT THE LHC

The experiments now starting at the RHIC collider and planned for the LHC are designed to study the transformation of nuclear matter at high nuclear densities and high temperatures. To this end, beams of heavy nuclei are bombarded onto each other with the hope of finding signs of the transition to the so-called 'quark-gluon' plasma.

Typical luminosities at the LHC heavy ion collider option are 10^{27} cm^{-2} sec^{-1}, the rate of central collisions is O(100Hz) and irradiation problems are not a big issue at these experiments. The big challenge lies in the fact that per central collision more than 10,000 charged particles are produced and the charged particle tracking has to approximate true three-dimensional reconstruction capabilities at all radii in order to disentangle the subtle effects expected (charm suppression, strangeness production, particle-particle correlations etc.).

The ALICE experiment at the LHC [13] has chosen to combine a silicon vertex detector system at low radius from the beam with a Time Projection Chamber at larger radii.

The vertex detector is primarily intended to allow the reconstruction of secondary vertices (Charm, Hyperons). Since the momenta of the particles can be very low (less than 0.02% of the charged particles have transverse momenta >5GeV/c) great care has to be exercised in limiting the material thickness of the device. The system consists of, in total, 6 layers of silicon detector devices equally spaced between 3.9 and 45cm from the beam. The first 2 layers are pixel detectors providing very high granularity and true 3D reconstruction. Layers 3 and 4 are built up of silicon drift chambers, still providing true 3D reconstruction capabilities with worse resolution but better cost/area ratio. Finally, for the last 2 layers silicon strip detectors are used. At these radii their good spatial resolution is sufficient to cope with the particle densities even if true 3D space point measurements are not provided.

At larger radii (from 88cm to 250cm radius and over a length of 2x2.5m) the charged particle tracking is performed with the Time Projection Chamber. This technology is very well suited for the environmental requirements at a Heavy Ion Collider as it provides large volume tracking with high resolution and excellent 2-track resolution at a very low cost in material thickness. Because of the low rate of central collisions the relatively slow response of the device is not a problem.

2.3. HADRONIC B FACTORIES – HERA-B

Another approach to get access to copious B meson production is hadron-hadron interactions both in a fixed target mode and at colliders. As the issues at hadron-hadron colliders will be discussed in more detail in the next section ('The High Energy Frontier'), I will concentrate here on the specific case of a dedicated fixed target experiment: the HERA-B experiment at DESY [14].

Here the B mesons are produced by colliding the protons of the HERA 920 GeV/c proton beam with the nuclei of a set of wire targets that are inserted into the halo of the proton beam. Since the ratio of the bottom cross section and the inelastic cross section is $O(10^{-6})$, several interactions per bunch crossing (every 96 ns) have to occur in order to get a sufficient number of B meson decays to be able to study the CP violating effects. At 4 interactions per bunch crossing the bottom particle rate is O(40 Hz) and the total number of Bs produced in a canonical running year of 10^7s is $\sim 10^9$.

Fig. 1 shows a Monte Carlo simulation of one such event with 4 interactions superimposed. The charged particles have relatively high momenta (typically 50 to 100 GeV/c). The average decay length of a B meson is $\sim$ 10mm (10,000μm). The occupancy of charged particles falls rapidly as $\propto 1/R^2$ where R is the distance from the beam line. Between 100 and 200 charged particles are generated per bunch crossing.

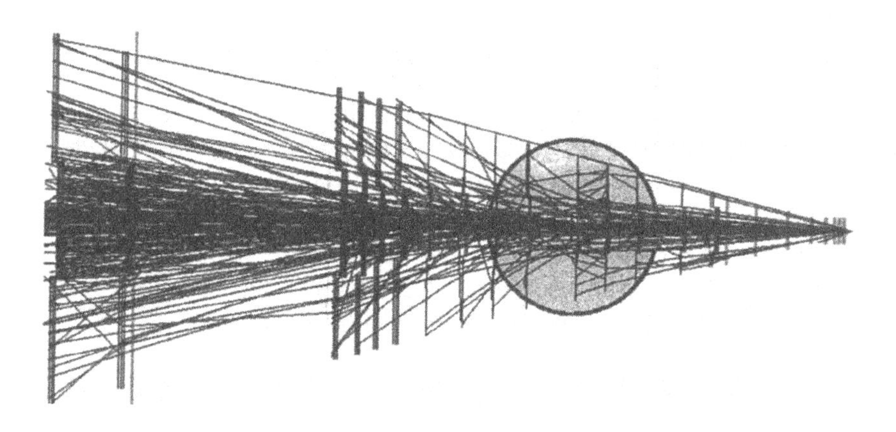

Figure 1 The Monte Carlo simulation of an event with 5 interactions of HERA protons with the target wires as recorded by the HERA-B detector.

The high charged particle flux results in very serious requirements imposed on the radiation hardness of the detectors: at 1cm distance from the beam the silicon detectors will accumulate O(5Mrad) per running year and the integrated charge accumulated on the wires of the drift chamber system approaches O(0.500C) in the same time at a radius of 30cm.

Fig. 2 shows an overview of the detector layout of the HERA-B experiment. The detectors cover more than 90% of the solid angle in the center of mass system, corresponding to angles between 10 and 160 mrad to the beam line in the nonbending view. At very low radii, both vertexing and tracking is done with a silicon detector system. The region at intermediate radii between 15 and 30cm is complicated as the requirements of small cell sizes (to avoid high occupancies), low material, low cost and large area ($\approx 16\text{m}^2$) are hard to fulfill with existing and well-proven technologies. In the end, Micro Strip Gas Chambers with GEM amplification have been chosen.

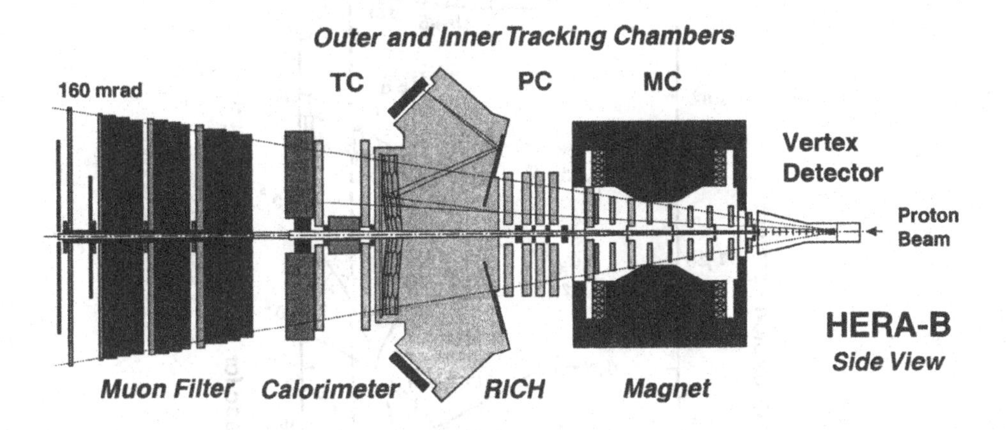

Figure 2 A schematical sideview of the HERA-B experiment at DESY. The protons enter from the right and the particles produced in the interactions with the wire target first traverse a vertex detector, followed by tracking in a dipole magnet. Particle identification is performed using a RICH, an electromagnetic calorimeter and a Muon filter system. Charged particle tracking is performed in a set of 13 superlayers from the magnet (MC) to the calorimeter (PC and TC).

Finally, at larger distances from the beam (30 to $\sim$ 220cm) gaseous detectors (5 and 10mm diameter honeycomb driftcells) are used.

2.4. THE HIGH ENERGY FRONTIER – ATLAS AND CMS AT THE LHC

Fig. 3 shows the cross sections of various physics channels of interest at the center of mass attainable at the LHC (14 TeV) and other machines. Channels of particular interest such as, e.g., the production of the elusive Higgs boson are suppressed by many orders of magnitude compared to the inelastic total proton-proton cross section and other similarly mundane processes. In order to get a reasonable production rate of new physics, at the design luminosity of 10^{34}cm^{-2} sec^{-1}, many (20 or more) interactions per bunch crossing ($\sim$ 25ns) have to be accepted. Given that a typical inelastic collision has O(20) charged tracks and an interesting event has O(200) charged tracks, one expects of the order of 600 charged tracks at each bunch crossing!

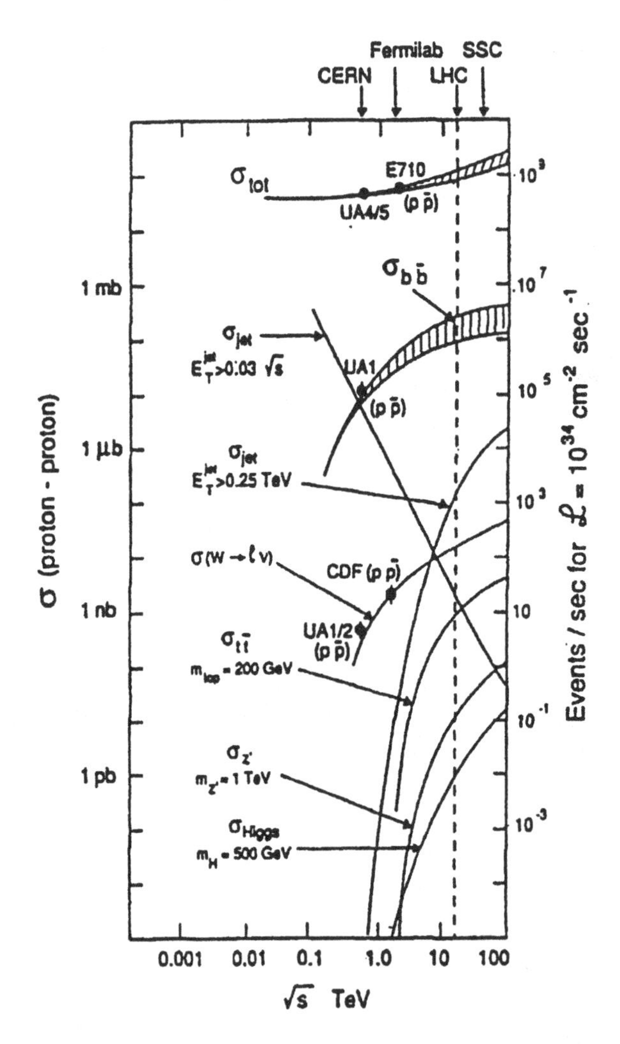

Figure 3 A compilation of cross sections for various physics channels of interest as a function of center of mass energy. The LHC value is indicated with the dashed vertical line. Also shown is the total proton-proton cross section [15].

Owing to the high energies involved, the requirements on the charged particle tracking are very tough. A high magnetic field strength, a long track length with many measurement points per track and a high precision point measurement are needed in order to measure a typical transverse momentum of a muon from the decay of a heavy Higgs particle ($H^0 \rightarrow Z^0 Z^0 \rightarrow \mu^+\mu^-\mu^+\mu^-$) of between 0.5 and 1 TeV/c, with sufficient precision. Hence the tracking detector system has to combine very high precision with large volume coverage.

Because of the high particle flux the radiation load on the detectors is similar to (and in some regions even harsher than) those at the HERA-B experiment. As an example the total fluence expected in the inner layers of the ATLAS Inner Detector exceeds $10^{14}\mathrm{cm}^{-2}\mathrm{yr}^{-1}$.

As a consequence, at radii between 10 and 30 cm silicon pixel detectors with sensor element size $10^{-4}\mathrm{cm}^2$ are planned. In the region between 30 and 60cm silicon strip detectors are used. The lower radius is determined by the requirement that the occupancy not exceed 1%, resulting in a sensor element size of $\approx 1/10\mathrm{cm}^2$. At radii exceeding 60cm, the ATLAS experiment has chosen to use gaseous detectors (4mm straw tubes). CMS instead has opted for an all-silicon solution to the tracking [16, 17].

It is interesting to note that for these new detectors silicon assumes the role of both vertexing and tracking. As a consequence, the silicon detector based systems are large (typically covering an area of $O(50)\mathrm{m}^2$), representing a very nontrivial extrapolation of existing systems at e.g. LEP or the Tevatron.

3. CHARGED PARTICLE TRACKING WITH GASEOUS DETECTORS

3.1. IONIZATION OF GASES BY CHARGED PARTICLES

When a charged particle traverses a layer of gas, incoherent Coulomb interactions between the electromagnetic field of the particle and the gas result in both excitation and ionization of the gas molecules. Contributions from bremsstrahlung, Cherenkov radiation and transition radiation to the total energy loss are negligible in gaseous detectors.

The total ionization of a given thickness of gas can be expressed for a minimum ionizing particle as

$$n_t = \frac{1}{W_i}\frac{dE}{dx}, \tag{1.1}$$

where dE/dx is the total energy loss of the (minimum ionizing) particle in the gas volume considered and W_i is the effective average energy needed to produce one ion-electron pair. Table 2 gives a brief overview of some values for often used gases.

Table 2 Primary (n_p) and total (n_t) ionization for commonly used gases (from Ref. [2, 3, 18]). Z and A denote the nuclear and mass number of the gas, respectively.

Gas	Z	A	W_i [eV]	dE/dx [keV/cm]	n_p [/cm]	n_t [/cm]
He	2	4	41	0.32	5.9	7.8
Ar	18	39.9	26	2.44	29.4	94
Xe	54	131.3	22	6.76	44	307
CO_2	22	44	33	3.01	34	91
CH_4	10	16	28	1.48	16	53
CF_4	42	88	54	5.40	51	100

For mixtures of gases (with fractions f_i), the composition law

$$n_t = \sum_i \frac{1}{W_i} \frac{dE}{dx} f_i, \tag{1.2}$$

holds to a good approximation. As an example, for the gas mixture Ar(65%)CF_4(30%)CO_2(5%), n_p=36/cm and n_t=99/cm. For this gas mixture the average distance between primary interactions is 10,000 μm/36 = 280μm. Each primary interaction produces on the average about 2.8 secondary interactions.

Since the typical threshold for high sensitivity readout electronics lies between 2,500 and 5,000 electrons, the ionization signal in gaseous detectors usually has to be amplified.

3.2. DRIFT AND DIFFUSION

The application of an electric field leads to a net drift of the ions and the electrons. The ions move in the direction of the electric field with a drift velocity v_d^+, which is found experimentally to be proportional to the 'reduced' electric field E/p

$$v_d^+ = \mu^+ \frac{E}{p} p_0, \tag{1.3}$$

where the pressure is measured in units of p_0=760Torr and μ^+ is the net mobility of a given ion in the gas. For gas mixtures, the composition law

$$\frac{1}{\mu_i^+} = \sum_{k=1}^{n} \frac{c_k}{\mu_{ik}^+}, \tag{1.4}$$

holds, where c_k is the volume concentration of gas k and μ_{ik} is the mobility of ion i in gas k. Examples of the mobilities of various ions in different carrier gases are listed in Table 3.

Table 3 Ion mobilities of various ions in various different carrier gases (from Ref. [3]).

Gas	Ion	μ^+ [$cm^2V^{-1}s^{-1}$]
He	He^+	10.2
Ar	Ar^+	1.7
Ar	CH_4^+	2.26
Ar	CO_2^+	1.72
CO_2	CO_2^+	1.09

For electrons the situation is more complicated. Fig. 4 shows the fraction of energy of an electron that goes into elastic collisions, excitations (leading to photon emissions) and ionization of the carrier gas as a function of the reduced electric field and for three different gases [18].

Because of their small mass, electrons can substantially increase their energy between collisions with the gas molecules. At higher fields, inelastic collisions or even further ionization can occur. As a consequence, except at very low field, the electron mobility is not constant but becomes a function of the energy loss associated with such an excitation, which in itself becomes highly dependent on the electric field and the gas composition.

One example is the so-called Penning effect (see Fig. 4 for Ne and Ne+0.1%Ar). When adding a very small amount of Ar to a Ne carrier gas, the excitations from a molecule of one of the components can be transformed to the molecules of the other gas type via the process $A^* + B \rightarrow A + B^+ + e^-$. This results in the ionization of the other gas requiring less energy per electron-ion pair than in the gas without the Ar admixture.

Fig. 5 shows the mean electron energy ϵ_k as a function of the electric field in a variety of gases. One can see that the mean electron energy in Ar is much larger than in e.g. CH_4 or CO_2. This can be explained by the fact that the first excitation level in Argon with 11.6 eV is much

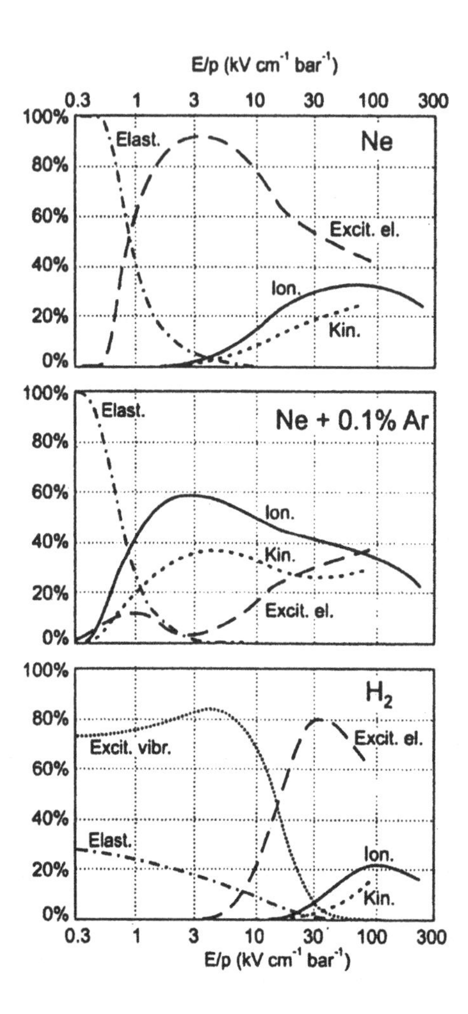

Figure 4 The fraction of energy of an electron going into elastic collisions, excitations (leading to photon emissions) and ionization of the carrier gas as a function of the reduced electric field, and for three different gases (from Ref. [18]).

higher than the first excitation level of CH_4 which is at 0.03 eV due to the existence of molecular rotational and vibrational degrees of freedom. The electrons have to acquire much more energy before they are able to ionize/excite the Ar gas than that required for the CO_2 molecules. Argon is therefore called a 'hot' gas, whereas CH_4 and CO_2 are called 'cold' gases.

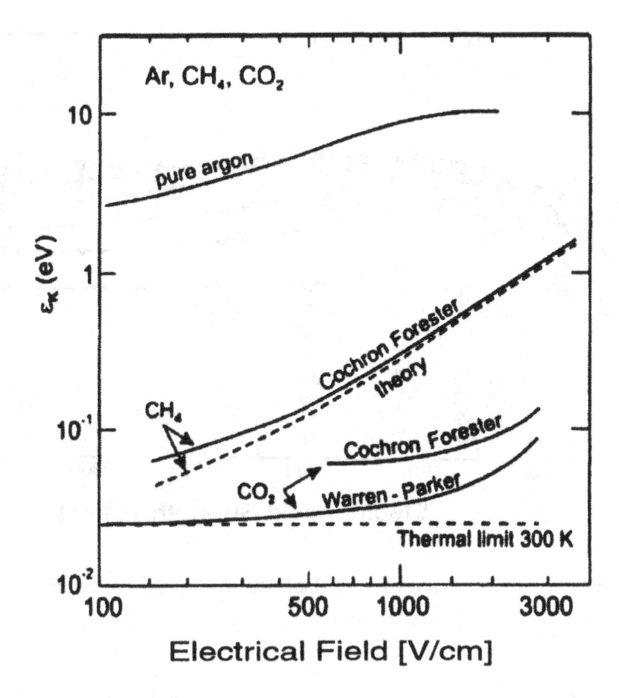

Figure 5 The mean electron energy ϵ_k as a function of the electric field in a variety of gases (from Ref. [18]).

In conclusion, the addition of even small amounts of cold gas to a hot gas can alter the drift properties of a gas mixture drastically. This is experimentally verified in Fig. 6, which shows the electron drift velocity for Ar-Isobutane mixtures of varying composition.

Note that the electron drift velocity is much larger than the ion drift velocity. Therefore the electron drift is generally used to determine the particle trajectory.

Another, rather unwanted effect of admixtures to a carrier gas is the attachment of drifting electrons to molecules that have electron affinity. These electrons are lost for the signal formation and can reduce the efficiency of signal detection. The typical attachment probability is negligible in noble gases and N_2, appreciable in Cl and very large for H_2O and O_2. As an example, about 1% pollution of air in Ar will remove about 33% of the migrating electrons per cm of drift! This fact poses serious constraints on the design and gas tightness of the gas volume of drift chambers as air (and hence H_2O and O_2) is abundant in the environment surrounding the detector cells.

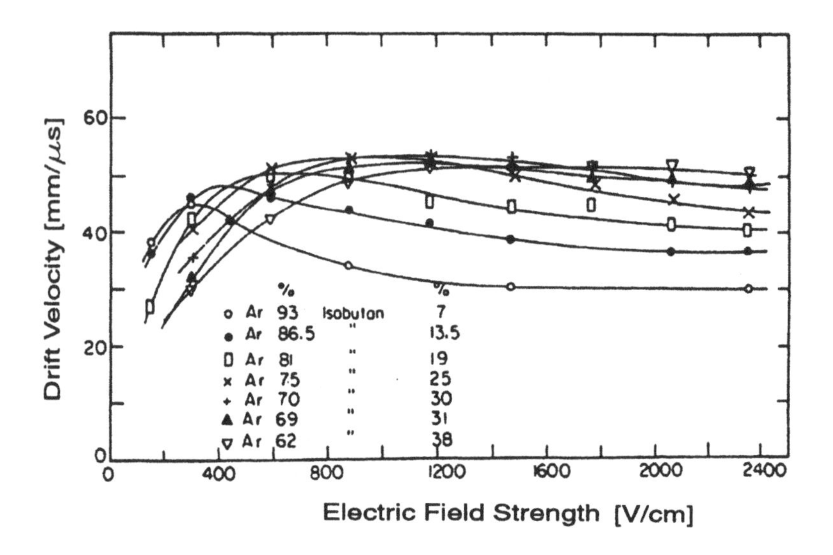

Figure 6 The electron drift velocity for Ar-Isobutane mixtures of varying composition (from Ref. [19]).

During the drift of the electron cloud to the field electrode, the electrons will collide with the molecules of the gas and scatter in random directions resulting in a longitudinal and transverse broadening. For example a point charge at $x = 0$ exhibits a gaussian distribution at position x with a width

$$\sigma_x = \sqrt{\frac{2Dx}{v_d}} \equiv \zeta\sqrt{x} \tag{1.5}$$

If one assumes isotropic scattering,

$$\zeta = \sqrt{\frac{2\epsilon_k}{eE}} \tag{1.6}$$

where ϵ_k is the average electron energy and E is the electric field strength.

This diffusion reduces the accuracy of the position determination. The functional dependence, Eq. 1.6, indicates that hot gases lead to much larger values of the diffusion broadening than cold gases, which therefore are preferred, especially at high values of the electric field.

3.3. GAS AMPLIFICATION

The simplest gaseous tracking detector is shown in Fig. 7. If one choses the electrical field $E = U_0/d$ such that the charged particles created by the track passing at z_0 are collected completely on the capacitor plates, without secondary ionization processes occuring, this is called an ionization chamber. The primary ionization can be measured.

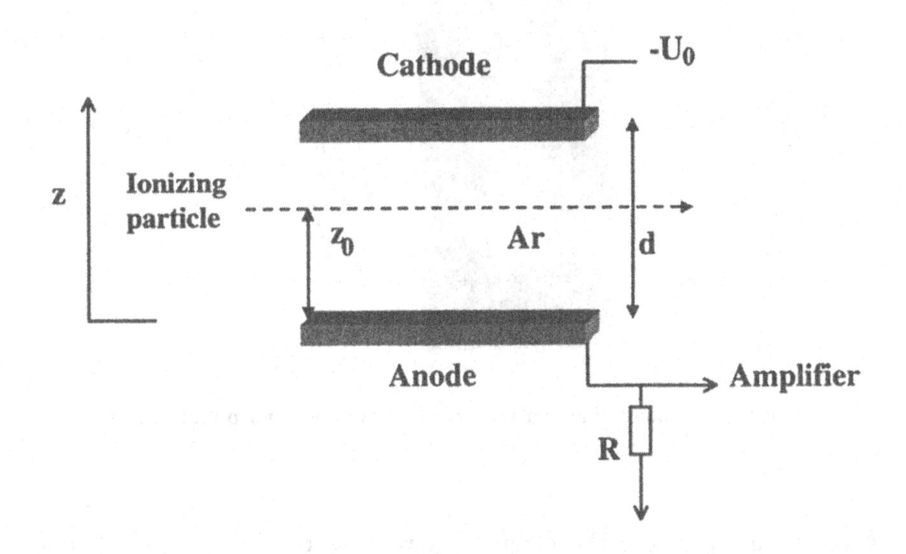

Figure 7 A simple gaseous tracking detector: the parallel plate chamber.

The movement of the charges (only O(100) per cm are produced) in the electric field induces charges on both capacitor plates. If the charge is collected at the electrode, the signal will be $V = Ne/C$. With typical values of $N = 100$ and $C = 10$pF this gives a signal size of $V \approx 2\mu$V.

When boosting up the voltage, more and more electrons receive enough energy between collisions to produce inelastic phenomena, excitations and ionization. Approximately one half of the electrons/ions are at any given time in the front part of the avalanche as they have just been produced in the last mean free path between collisions. Because of the big difference in drift velocity, the positive ions are left behind the much faster electron front. A cloud chamber picture of this avalanche formation is shown in Fig. 8 [20].

Such a parallel plate counter has some disadvantages: the absolute value of the detected signal depends on the avalanche length and hence

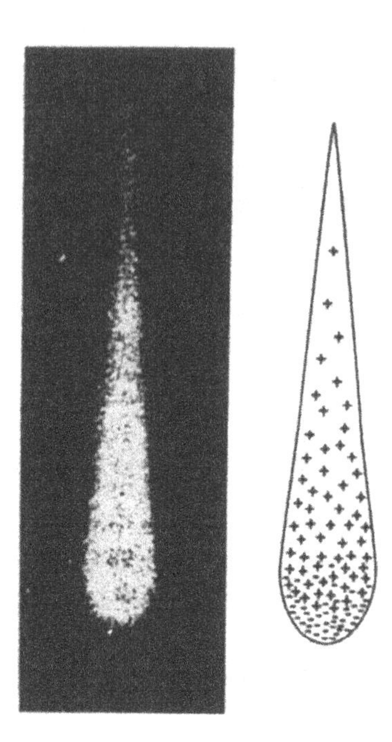

Figure 8 A cloud chamber picture of an avalanche formation [20].

on the point z_0 where the original charge has been produced. There is hence no proportionality between the deposited energy and the detected signal. In addition, only moderate gains can be obtained before breakdown occurs as extremely large voltages between the electrodes are required for a signal to be detectable with standard electronics.

This led to the development of the proportional counter, depicted schematically in Fig. 9. The electric field is generated by applying a voltage between the inner and outer electrode and depends rather sensitively on the choice of the radii of the inner (anode) and outer (cathode) electrode. Typical values are $2r_i$=25μm (a wire), $2r_a$=10mm (a tube) and U_0=1500-2000V (depending on the gas mixture used).

As has been shown in Fig. 4, in standard gases, ionization only occurs at rather high fields of E/p=10-20kVcm^{-1}bar^{-1}. These field strengths can usually only be met very close (few tens of μm) to the anode wire and hence in most of the region the electric field just makes the e^- and ions (A^+) drift towards the electrodes. Only very close to the anode (typically a few wire radii) does multiplication and avalanche formation occur. In these counters the absolute value of the detected charge is

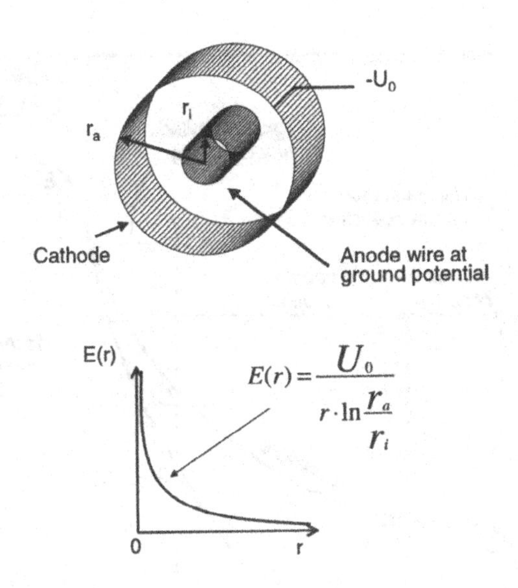

Figure 9 The proportional chamber. r_i and r_a are the radii of the anode wire and the cathode, respectively. E is the electrical field given by the voltage U_0.

hence almost independent of the distance of the ionizing particle from the anode and 'proportional' to the deposited energy.

Fig. 10 from Ref. [21] shows the gain-voltage characteristic for such a proportional counter. At very low voltages charges begin to be collected but recombination is still dominant. At higher voltages (B) full collection of deposited charge begins, amplification does still not occur and the counter is called an ionization chamber. For even higher voltages (C) the electric field is high enough for avalanche multiplication to occur. Here the detected charge is 'proportional' to the deposited charge $Q_{detected} = G \times Q_{deposited}$ where G is the gas gain. The gas gain depends exponentially on the applied voltage and is a strong function of the gas composition, the temperature and pressure. Typically a 1% change in voltage gives a 15% change in gain.

When the voltage is increased further (D), this proportionality is gradually lost due to the large space charge built up around the anode wire thus effectively screening the field for electrons that are generated be-

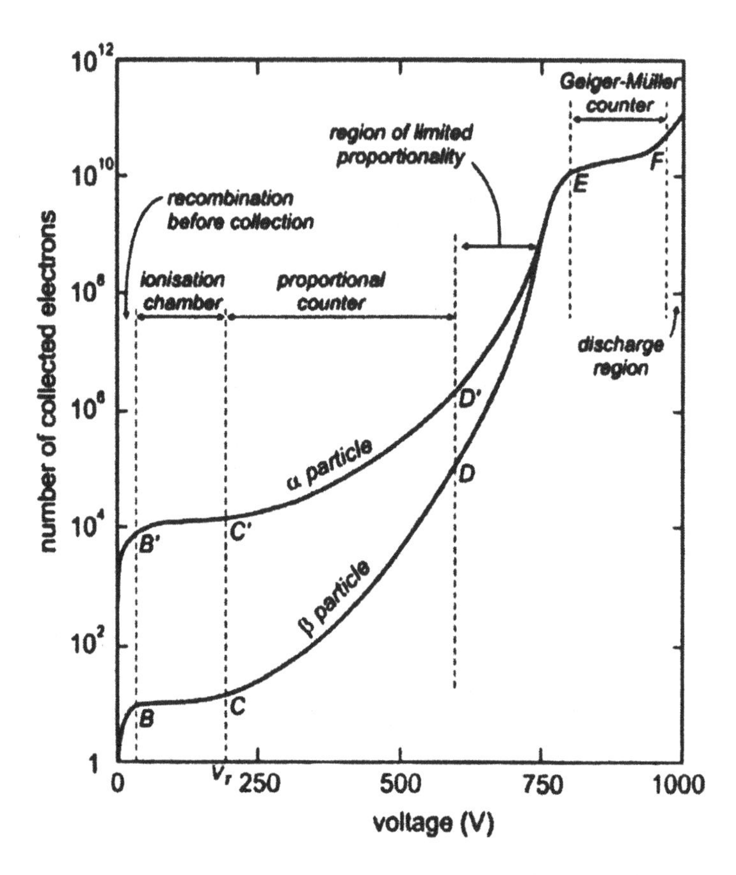

Figure 10 The gain-voltage characteristic of a proportional counter (from Ref. [18]). For a detailed discussion see text.

tween the wire and the ion cloud moving to the cathode. This is the region of 'limited proportionality'.

In region (E) the gain is saturated: the same signal is generated, independent of the original ionizing event. This is also called the 'Geiger' mode of the counter. For even higher voltages the counter discharges and the voltage breaks down.

A special role is played by the photons that are abundantly produced in the avalanche process as can be seen in Fig. 4: the fraction of energy going into excitations (thus leading to photon emission) can be of comparable magnitude to elastic interactions and ionization, depending on the value of the electric field. A small fraction of the photons will be energetic enough to ionize the gas themselves which results in further avalanche formation.

An additional effect is the 'photon feedback': if photons reaching the cathode have energies exceeding the work function of the cathode material, electrons can be extracted which will be accelerated towards the anode and generate further avalanches. For example the first excitation energy of Ar is 11.6eV and the work function of a typical cathode material like Cu is 7.7eV. Both these feedback processes can be selfsustaining and render the counter inoperational.

In order to reduce the number of photons produced in the avalanche, gases are admixed which have large photo-absorption cross sections in the visible and the UV in order to 'quench' them. Examples are CO_2 and CH_4 which provide very efficient absorption in the region (7.9-14.5)eV. Good quenchers are typically organic molecules with many vibrational and rotational degrees of freedom.

When the (usually excited) ions reach the cathode, they can also generate photons and/or electrons, producing 'ion feedback' when they are neutralized. This process is very similar to the photon feedback but will give rise to delayed pulses ('after pulses') due to the small drift velocity of the ions.

Quenchers also play a positive role in this case: usually they have a lower ionization potential than the inert carrier gases: [Ar^+ + Quencher $\rightarrow$ Ar + $Quencher^+$] such that generally only ions of the quench gas reach the cathode. Since these have lower ionization and excitation potentials, when they reach the cathode, they either neutralize or dissociate rather than stimulate a secondary emission: [Quencher + cathode $\rightarrow$ Quencher-part-a + Quencher-part-b + cathode] thus averting ion feedback. Note though that in this way the quencher gas is being used up and hence the gas has to be refreshed.

The dissociation of the quencher gas at the cathode can result in polymerization and deposition of non-conducting material on the cathode surface (see Fig. 11). Positive ions created in later avalanches attach to this non-conductive surface and will only be neutralized on a large time scale as they have to 'leak' through the insulating layer. For large irradiation fluxes the ion production can be much larger than this neutralization rate and a very high electric dipole field can develop across the thin layer. This field, if high enough, can extract electrons from the cathode through the insulator into the gas volume where they are accelerated towards the anode and generate new avalanches which will

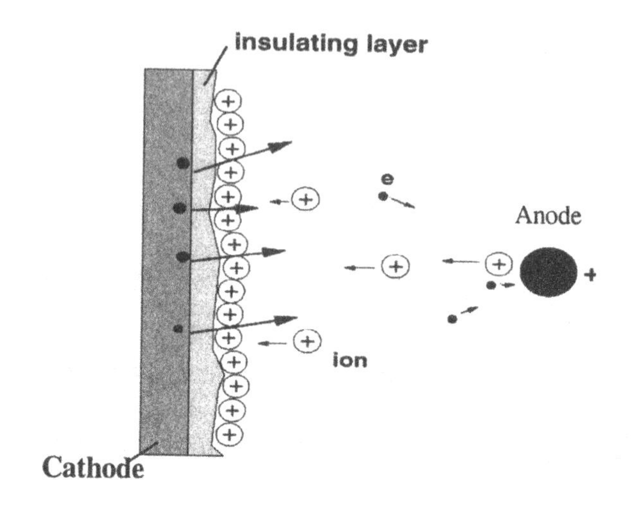

Figure 11 A schematical description of the Malter effect (see text).

in turn 'feed' the process such that a selfsustained current is observed, rendering the counter unusable. This effect is called Malter effect [22]. It is characterized by the observation of a permanent discharge of the counter that is observable even when the original source of the radiation is removed.

When the avalanche initiated by an ionizing particle comes to an end, all electrons are collected on the anode quickly and the ions are slowly moving in a cloud (forming a space charge distribution) to the cathode. In the normal case, the space charge is much smaller than the charge on the anode wire. The space charge does not affect the gain of the counter for subsequent particles and the counter stays in the proportional mode.

For large charge deposition/generation (e.g. by low energetic protons, alpha particles or at very high gains) the generated space charge is large enough to 'screen' the field of the wire locally such that the effective gain seen by the electrons coming later in the avalanche is reduced. The apparent gain of the counter at the same voltage is different for e.g. alpha particles and electrons. An example of this 'avalanche space charge effect' is shown in Fig. 12.

At high incident particle rates the ion cloud is fed continuously and can result in an effective gain reduction for subsequent particles. This is shown in Fig. 13, where the rate dependence of the peak current

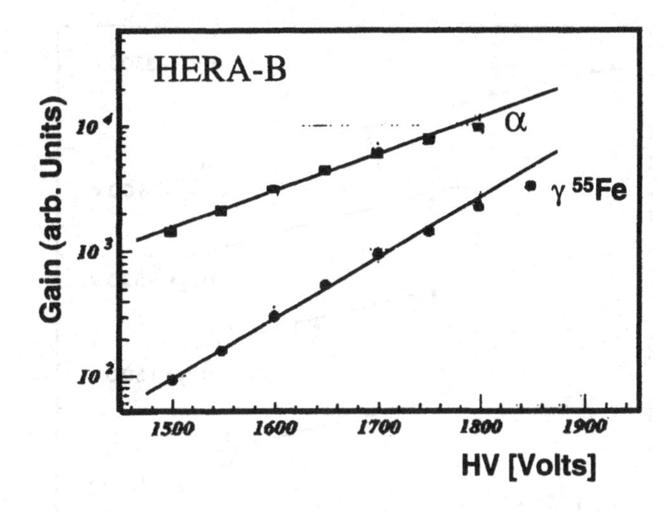

Figure 12 Gain as a function of High Voltage measured with α particles and with photons from a ^{55}Fe source, illustrating the avalanche space charge effect.

measured in a proportional counter is shown for 5.9KeV X-rays and several different anode potentials. This rate effect can become important for studies of the radiation hardness of gas counters as it sets a limit to the irradiation rates one can use.

3.4. THE CHOICE OF THE GAS MIXTURE

For good detection efficiency and good energy resolution, high primary ionization is needed. This is met in noble gases like Ar or Xe, but in order to obtain stable gain at low voltages quench gases have to be added. Typical candidates are CO_2, CH_4, iso-butane, DME (di-methyl-ether).

High drift velocities for fast timing applications are reached in gases containing CF_4. For fast signal collection and large signals high ion mobility is needed as well as low electron attachment. Good spatial resolution can be reached with gases that exhibit low diffusion properties.

To avoid anode and cathode aging problems (see Chapter 5), usually non-organic quenchers (like e.g. CO_2) are preferred. Last but not least, if possible, the gas should be non-flammable (limiting the concentration of e.g. CH_4) and affordable (limiting the use of e.g. CF_4).

As one can see, the choice is not easy and the gas mixture has to be tailored to the specific needs of the experiment.

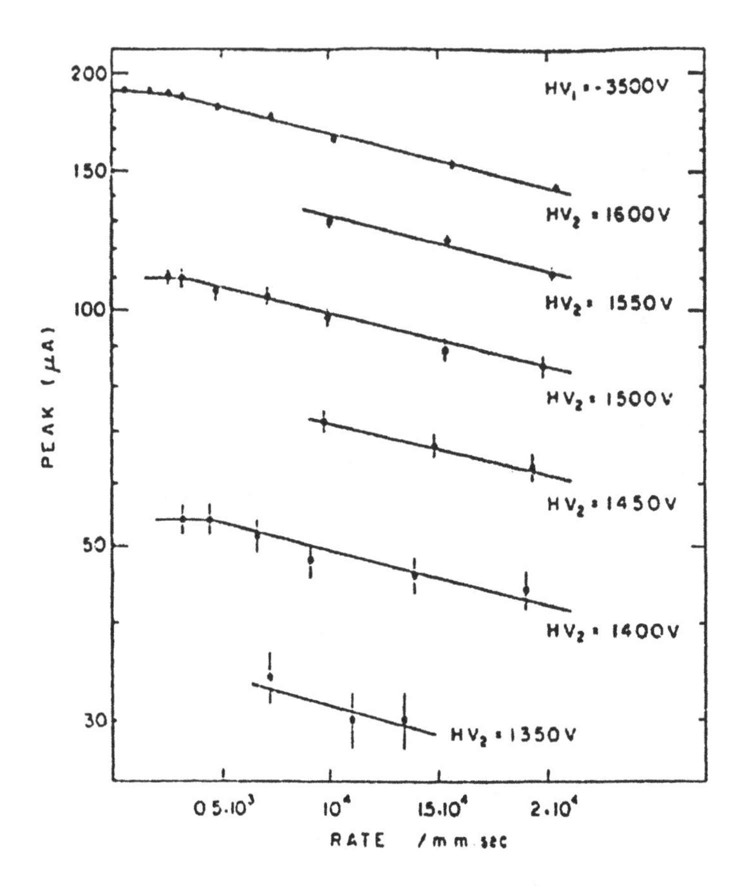

Figure 13 The rate dependence of the peak current measured in a proportional counter for 5.9KeV X-rays and several different anode potentials (from Ref. [2]).

3.5. GENERIC GASEOUS TRACKING DETECTORS

The basic design principal is to exploit the advantages of a large drift region and a short amplification region.

The simplest approach is to assemble many proportional tubes (or geometrical approximations thereof) next to each other (see Fig. 14). These types of detectors are usually used in fixed target applications and for large areas (e.g. Muon detectors). They offer the possibility for each cell to form its own individual gas volume. The electrical cross talk between cells is small. The structures are usually self-supporting and cheap. Large areas can be covered with a high degree of segmentation. Small cell sizes ($\leq$4mm diameter) can be realized allowing for high rate applications. The main disadvantage is their comparatively large material thickness.

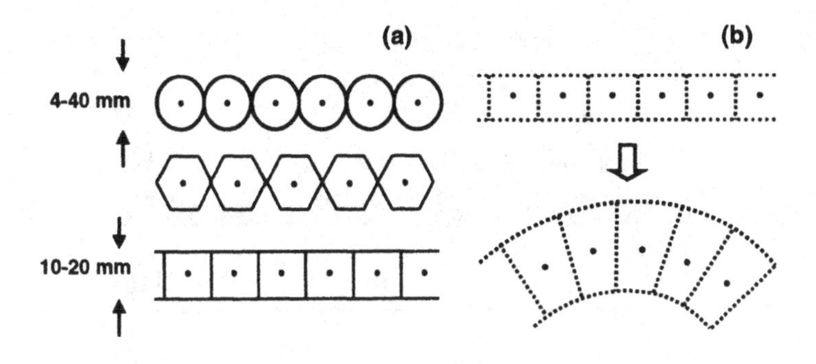

Figure 14 (a) Simple examples for straw type drift tube arrays. (b) Planar and cylindrical chambers based on arrays of field shaping wires and sense wires.

Replacing the solid cathodes by field shaping wires allows one to create a very thin chamber that lends itself easily to application in cylindrical drift chambers at colliders (Fig. 14). These detectors are well suited for low to medium rate applications. They provide good spatial resolution with very small material thickness. They are, however, not self-supporting and the spatial coverage is limited to medium sizes.

A special development is the Micro Strip Gas Chamber (Fig. 15, [23]). Particles traverse a typically few mm thick drift region filled with gas. Cathodes and anodes are thin micro strips deposited on a glass substrate. Typical cathode-anode distances are 60-100μm. Electrons drift slowly to the anode-cathode plane. The amplification occurs in the high field region close to the anode-cathode surface.

Very good spatial resolution is easily obtainable with this type of device as it is largely determined by photolithography. The detector exhibits high rate capabilities and relatively large areas (tens of m^2) can be covered with high spatial resolution. The attainable gains are relatively moderate ($O(10^3 - 10^4)$) and in this basic design the device is prone to sparking which led to modifications [24].

The devices are rather fragile (if one wants to keep the material budget reasonably small) and the working point plateau is quite limited. The detector has found application in areas where large area high precision/high granularity tracking is needed which can not easily/affordably be provided by silicon strip devices [14, 25, 26].

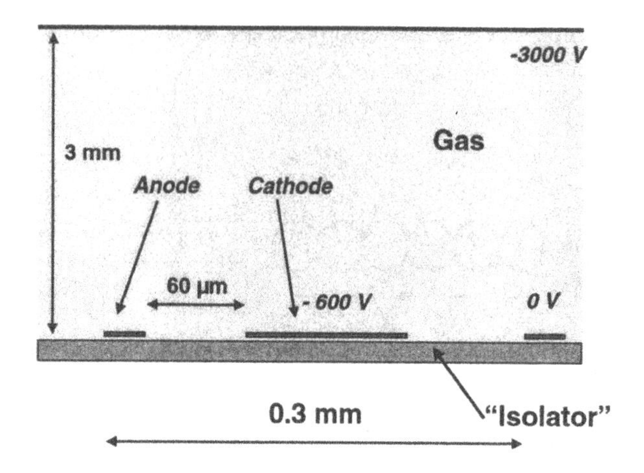

Figure 15 Cross sectional view of a Micro Strip Gas Chamber.

4. CHARGED PARTICLE TRACKING WITH SEMICONDUCTOR DETECTORS

4.1. HISTORICAL REMARKS

The story of the introduction of semiconductors as detectors in High Energy Physics research can be traced back to the year 1951, when it was first observed that an alpha particle traversing a reversed biased $p - n$ junction produces a measurable signal [27]. Until the end of the 1970's semiconductor detectors were almost exclusively used for the measurement of the energy of the traversing particle, most notably in nuclear spectroscopy.

In 1976 particles carrying the charm quantum number were discovered, having a rather short lifetime resulting in mm long decay length in typical fixed target experiments of that time. First thoughts were devoted to the question of whether or not one could use semiconductor detectors, with their compactness and high segmentation properties, for the measurement of the position of a traversing charged particle.

This development really accelerated with the introduction in 1980 of the planar process, of the semiconductor industry, to the patterning of detector grade silicon resulting in reliable production methods with high yield [28]. In 1983 the first silicon strip detectors built in this way were used very successfully for the detection of charmed particles at CERN [29].

It was soon realized that for many applications it would be advantageous, if not necessary, to miniaturize the readout electronics and in 1984/85 the first VLSI (Very Large Scale Integration) chip, the 'Microplex' was successfully tested [30]. This development paved the way for the use of silicon strip detectors in colliding beam detectors and in 1989 the Mark-II experiment successfully took data at the SLAC Linear Collider with their novel Silicon Vertex Detector [31].

In 1990 the first silicon strip detectors with patterns and readout on both sides of the wafer were commissioned for the ALEPH experiment at the LEP collider [32] and the basic ingredients for today's commissioned and planned detector systems based on silicon strip detectors had been established. Today, these type of detectors are in wide use all over the High Energy Physics scene.

4.2. BASIC SEMICONDUCTOR PHYSICS

Semiconductors provide the unique feature that the conductivity can in principal be chosen in a large region between conductors and insulators. The conductivity can depend on many factors: temperature, magnetic field, impurities and even illumination conditions.

The most prominent conductors are Germanium (Ge) and Silicon (Si) but many interesting semiconductors are made up of more than one element, e.g. GaAs.

For single atoms only discrete energy levels are available for the electrons and the ionization potential of e.g. Si is 8.1eV. When single atoms condense in crystalline form to a solid body these discrete energy levels merge to form valence and conduction 'bands' with an energy gap between them. Typical band gap values are 1.1eV for Si, 1.4eV for GaAs and 8.0eV for SiO_2. In conductors no such band gap exists.

Pure silicon is an 'intrinsic' semiconductor. At T=0°K in intrinsic semiconductors the valence band is completely filled with electrons and the conduction band is completely empty. No conduction is possible and the crystal is a perfect insulator. At elevated temperatures a few electrons will be thermally excited to the conduction band leaving free places for electrons ('holes') in the valence band and allowing thus conduction to take place. The charge carrier density at room temperature in intrinsic semiconductors is $n_i = 1.45 \times 10^{10}\text{cm}^{-3}$ which is to be compared with $\approx 5 \times 10^{22}$ Si atoms contained in 1cm^3 silicon.

The occupation probability of an electron state with energy E is given by the Fermi-Dirac function

$$F(E) = \frac{1}{1 + e^{\frac{E-E_F}{kT}}}, \tag{1.7}$$

and for an intrinsic semiconductor

$$E_F \equiv E_i = \frac{E_C + E_V}{2}, \tag{1.8}$$

where E_V and E_C denote the levels of the valence and conduction band and E_F is the Fermi energy.

Truly intrinsic semiconductors are very hard to produce due to crystal defects and impurities being essentially unavoidable in the crystal growing process. Their net effect is to provide additional energy levels in the otherwise forbidden energy gap. One can distinguish between donor and acceptor impurities.

Elements from group V of the periodic table donate an additional electron when built into the crystal lattice which can contribute to the conduction properties of the crystal (see Fig. 16 from [33]). The majority carriers in such a crystal are electrons, the minority carriers are holes and a crystal suitably doped with electron donors is called an n-type semiconductor.

At room temperature almost all electrons from donor states will move to the conduction band resulting in a shift of the Fermi level towards the conduction band.

Elements from group III of the periodic table accept an additional electron to form four covariant bonds to its partners in the crystal lattice (see Fig. 16) thus creating free electron sites (holes). The majority carriers in such a crystal are holes, the minority carriers are electrons and a crystal suitably doped with hole donors is called a p-type semiconductor. The hole density in the valence band is thus increased and the Fermi level is moved towards the valence band.

Typical donors are Phosphorus and Arsenic, a typical acceptor dopant is Boron.

4.3. THE $P-N$ DIODE JUNCTION

The $p-n$ diode junction is one of the most important electronic structures. It is obtained by joining two extrinsic semiconductors with opposite doping (see Fig. 17 from Ref. [4]).

Before joining the two crystals, the p and n regions are in electrical equilibrium. Electrons and holes respectively are distributed homogeneously. In the n region the Fermi level is shifted towards the conduction band, in the p region it is shifted towards the valence band.

After joining the two oppositely doped crystals (see Fig. 17), due to the concentration gradient electrons diffuse into the p region filling hole states there and holes diffuse into the n region, becoming available for the electrons there. As a consequence the excess nuclear charge of the

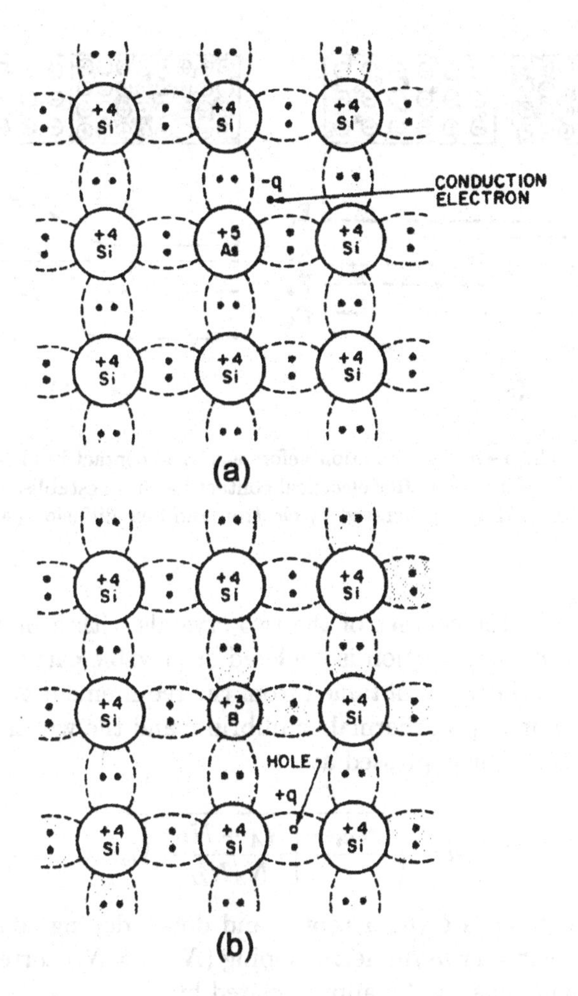

Figure 16 (a) Schematic bond picture for n type Silicon with donor (Arsenic) impurity. (b) Schematic bond picture for p type Silicon with acceptor (Boron) impurity (from Ref. [33]).

lattice sites is not neutralized any more by movable carriers and a surplus of negative charge in the p region and positive charge in the n region is built up that counteracts further diffusion. The result is a stable space

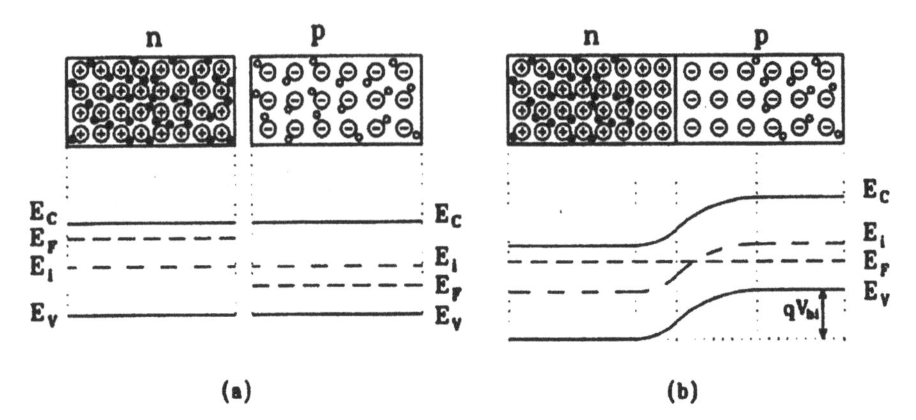

Figure 17 (a) The $p-n$ diode junction before electrical contact has been established. (b) The $p-n$ diode junction after electrical contact has been established. V_{bi} denotes the built-in voltage that is generated by electron and hole diffusion (see text). From Ref. [4].

charge region at the contact of the two crystals with a built-in voltage V_{bi}. This space charge region is depleted of movable carriers.

The built-in voltage is obtained from the requirement that the Fermi levels have to line up in thermal equilibrium and the width of the space charge region can be expressed as

$$d = \sqrt{\frac{2\epsilon_{Si}\epsilon_0}{e}\left[\frac{N_A + N_D}{N_A N_D}\right] V_{bi}}, \tag{1.9}$$

where N_A and N_D are the acceptor and donor doping concentrations, respectively. For very asymmetric doping ($N_A >> N_D$ corresponding to a p^+n junction) this can be approximated by

$$d \approx \sqrt{\frac{2\epsilon_{Si}\epsilon_0}{eN_D} V_{bi}}. \tag{1.10}$$

As an example, for $N_A = 10^{16}\text{cm}^{-3}$, $N_D = 10^{12}\text{cm}^{-3}$ the built-in voltage is $V_{bi} \approx 0.46\text{V}$ and the width of the depletion zone is $d \approx 25\mu\text{m}$. In order to get a large depletion zone, N_D should be made small and very pure high resistivity n-type silicon should be used.

When applying a voltage across the $p-n$ junction two cases have to be distinguished (see Fig. 18). In the forward biased mode, the electrons and holes are pushed into the space charge region and the width of the space charge region is reduced. Electrons and holes will recombine and a steady current will flow.

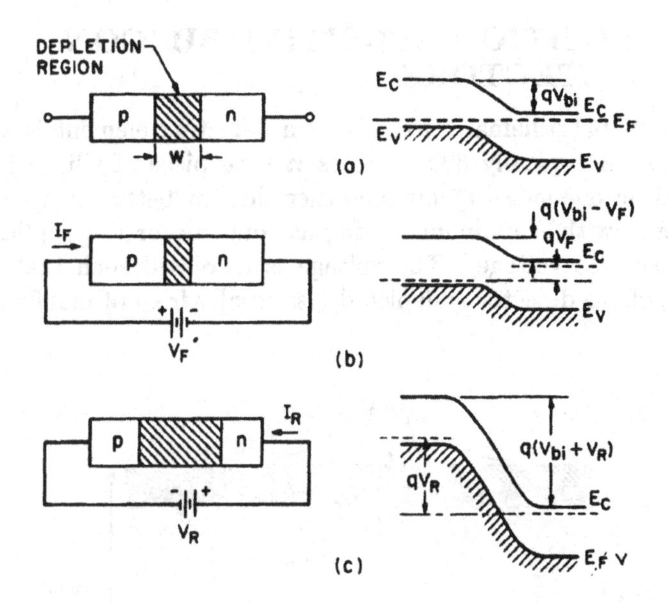

Figure 18 Schematic representations of depletion layer width and energy band diagrams of a $p - n$ junction under different biasing conditions (a) No bias voltage applied; (b) Forward-bias condition; (c) Reverse-bias condition (from Ref. [33]).

In the reversed biased mode, the electrons and holes are pulled towards the electrodes and the width of the space charge region increases correspondingly. Only a very small reverse bias current flows, produced by thermally generated charge carriers in the space charge region and the diffusion of minority carriers into the space charge region.

The width of the depletion zone for a p^+n structure under reverse bias is given by

$$d = \sqrt{\frac{2\epsilon_{Si}\epsilon_0}{eN_D}(V_{bi} - V_{Dep})} = \sqrt{2\epsilon_{Si}\epsilon_0\mu_n\rho_n(V_{bi} - V_{Dep})}. \tag{1.11}$$

where V_{Dep} is the applied voltage, μ_n is the electron mobility and ρ_n is the resistivity of the n-type silicon. For a given width of the depletion zone, higher values of the resistivity (purity) allow smaller voltages to be used. Since small voltages are advantageous for practical and stability reasons, silicon detectors are usually built from high resistivity material.

4.4. POSITION SENSITIVE SILICON DETECTORS

The principal schematic for a silicon detecting element is sketched in Fig. 19. A typically 300μm thick n type piece of silicon is highly p^+ doped on one side (n^+ on the other side for better ohmic contact) and covered with Aluminum serving as contacts for the application of the reverse bias voltage. The voltage is increased such that the full thickness of the detector is depleted (essentially free) of mobile carriers.

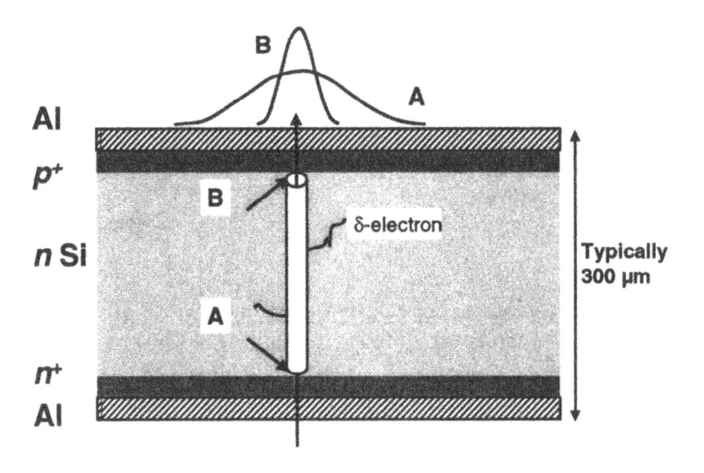

Figure 19 The principal schematic for a silicon detecting element.

At the passage of a charged particle, the majority of the electron-hole pairs are generated in a range $< 0.1\mu$m. Under the influence of the electric field the pairs are separated and the holes move to the p^+ region, the electrons move to the n^+ region. Note that in the space charge region the reverse bias current is very small and hence does not disturb the signal generated by the particle.

During the drift to the electrodes the charge cloud(s) spread due to diffusion with a typical spread of 5-10μm. The charge distribution at the electrodes is a superposition of different gaussian contributions. The charge distributions for electrons and holes differs from each other as the number of electrons and holes that traverse the low field region near the n^+ contact is different.

A position sensitive detector can now easily be made by subdividing the p^+ electrode into individual strips that are than connected to readout electronics. The resolution that can be obtained is usually much better than $d/\sqrt{12}$ where d is the strip pitch which is typically between 25 and

100μm. In special applications with analog readout of each strip, point resolutions well below 5μm have been reached.

4.5. COMPARISON OF SILICON AND GASEOUS DETECTORS

As in gaseous detectors, charged particles loose energy through elastic collisions with the electrons in the device with a subsequent generation of electron-hole pairs. The characteristic energy loss for a minimum ionizing particle in silicon is 3.8MeV/cm and the average energy needed to create an electron-hole pair in silicon is ~3.6eV resulting in the creation of ~1,000,000 charge carriers per cm of traversed material.

This has to be compared with a typical energy loss of a minimum ionizing particle in Argon of 0.0025MeV/cm. Since the energy needed to create an electron-ion pair in Argon is ~26 eV, only ~100 charge carriers are generated per cm of traversed gas.

As a consequence, in silicon very thin detectors can be realized that still produce large signals. The corresponding small range of δ electrons allows an extremely good position measurement.

Despite the high density of the material the electrons and holes can move almost freely in the depletion zone allowing for rapid charge collection, making high rate applications easy, and the mechanical rigidity of silicon makes self-supporting structures feasible.

The fact that the space charge region can be formed by suitable doping in a very selective way allows for the creation of very sophisticated field configurations resulting in new detector structures and principles [4].

There are of course also drawbacks to this technology. The fact that typical wafer sizes for detector grade silicon do not exceed 6 inch diameter require that the individual detector modules be rather small. Large area coverage can only be achieved with a very modular approach. Here gaseous detectors can have a clear advantage providing low cost tracking over large areas, provided the experimental environment allows their utilization.

5. RADIATION DAMAGE ISSUES – (A) GASEOUS DETECTORS

5.1. INTRODUCTION AND HISTORICAL REMARKS

Chamber aging can be defined as a 'deterioration of the functionality of the device as a function of the irradiation time/dose'.

Aging phenomena of gaseous detectors have been investigated as early as 1940-50 where it was found that sealed Geiger tubes with $ArCH_4$ gas seized to work because of CH_4 depletion under irradiation. The plans for the SSC/LHC colliders in the 1980's required a refinement of the understanding of chamber aging at integrated doses >100 times larger than those experienced in then running detectors. In 1986 a workshop was held at Berkeley entirely devoted to this subject [10] and many research projects were initiated in the late 1980's. To date, we witness the first large scale applications of the accumulated knowledge in this area in High Energy Physics experiments: the commissioning of the HERA-B experiment and the construction of the ATLAS, CMS and LHCb experiments for the LHC.

The aging effects that have been observed are manifold: the loss (or sometimes increase) of chamber gain, the reduction of the width of the High Voltage plateau, a loss of energy resolution, the generation of excessive or self sustained currents, the generation of dark currents, sparking etc.

A relevant quantity to calibrate the radiation hardness of a detector is the integrated charge accumulated per cm of anode wire which has been exposed to radiation

$$\frac{Q}{\Delta l} = \frac{1}{\Delta l} \int idt \qquad (1.12)$$

Chambers have been observed to 'die' at any dose between 0.01 and 5 Cb/cm and the phenomena are very much dependent on the details of the design, the materials used etc. Once a chamber has started to deteriorate, the process can often not be stopped any more.

The expertise in the field is still very limited and furthermore, the expertise required for LHC type doses exceeds what had been learned in the LEP/HERA era of experiments.

5.2. AGING MECHANISMS - CASE STUDIES

In the avalanches forming in an irradiated gaseous proportional counter a high concentration of ions and radicals is produced creating an environment very similar to that observed in plasma chemistry. Chemical reactions, polymerization, the breakup of molecules can lead to the coating of the wire and cathode surfaces with conductive or non-conductive material.

Whether these reactions occur and whether they prove detrimental to the properties of anode and/or cathode depends very much on factors like the gas composition, the gas flow, the anode and cathode material, the chamber system material other than the cathode and anode, possible contamination of the gas through the outgassing of chamber components, the chamber gain and even the type of irradiation.

As an example a coating of a 25μm anode wire with 0.5μm conductive material can result in a gain reduction in the affected region of 20-40%!

In the following, the effect of several important parameters of potential relevance for chamber aging will be illustrated.

5.2.1 The Choice of the Gas Composition. The discussion in Chapter 3 on gaseous detectors has shown that a counting gas usually consists of a noble gas (Ne, Ar, Xe) and one or more quenchers (eg. CH_4, CO_2, C_5H_{12} etc.). In Fig. 20 the dependence of the current in a gas counter under constant illumination with a source is shown as a function of the collected charge in mC/cm and for different gas mixtures (from Ref. [34]). The mixture Ar(50%)Ethane(50%) is completely stable (A) whereas already a very small admixture of 1% Ammonia leads to a rapid deterioration of the device (reduction of gain, (B)). The admixture of CF_4 and O_2 renders the device nonfunctional essentially from the start(C).

At large irradiation doses already very slight changes in the proportion of the components in an otherwise unchanged gas mixture can provoke the transition from an irradiation hard gas mixture to one that leads to a long term deterioration. This is shown in Fig. 21 from the HERA-B experiment where the normalized gain of a chamber cell is shown for three different gases as a function of accumulated charge (note that here the accumulated charge is orders of magnitude larger than in Fig. 20!).

Both in HERA-B and at the LHC fast gases are needed due to the high bunch crossing frequencies involved and the use of CF_4 has been investigated in some detail. This molecule is very inert due to the very tight bonds of the Fluor to the Carbon. Under irradiation however, the molecule breaks up and the created radicals in turn become extremely aggressive. This is in fact utilized in the semiconductor industry where CF_4 is used for plasma etching processes.

In a very systematic study [36] it was shown that the etching property depends very much on the relative amount of carrier gas. This is shown

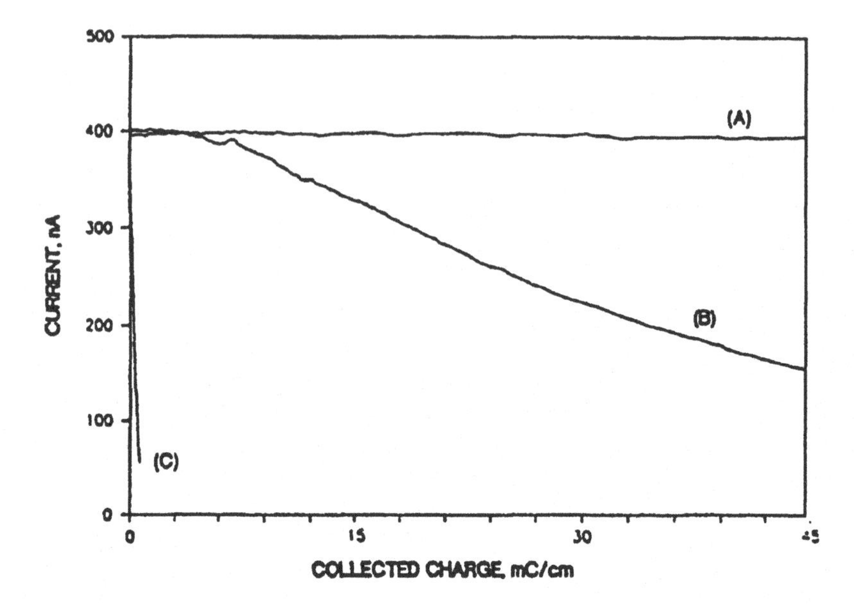

Figure 20 The dependence of the current in a gas counter under constant illumination with a radioactive source as a function of the collected charge in mC/cm and for different gas mixtures. (A) Ar(50%) Ethane(50%); (B) Ar(49.5%) Ethane(49.5%) Ammonia(1%); (C) Ar(50%) CF_4(40%) O_2(10%). From Ref. [34].

in Fig. 22, where the aging rate of a test counter is shown as a function of the percentage of CF_4 in a CF_4-isobutane gas mixture. A transition from an etching domain at low percentages to a deposition domain at large values has been observed.

It has been shown, that CF_4 can also remove deposits that stem from aging due to gas contamination and hence rejuvinate a deteriorated counter. This is shown in Fig. 23 where a previously damaged counter with an Ar-ethane gas mixture (reduced gain, before point A) is restored to full gain (point B) by adding CF_4 to the mixture.

One has to be very careful though to find a balance as CF_4 etching does not necessarily stop at the anode wire surface!

Another very important parameter in aging studies is the gas flow. As mentioned above, the gas has to be exchanged regularly, the gas composition has to be monitored with high precision (see the effects described in Fig. 21) and the gas flow has to be adjusted to the irradiation dose rate. This is specifically important for the laboratory studies of irradiation hardness of counters, since 10 years of operation at the LHC can be simulated in the laboratory in a reasonably short time only by increasing the dose rate accordingly.

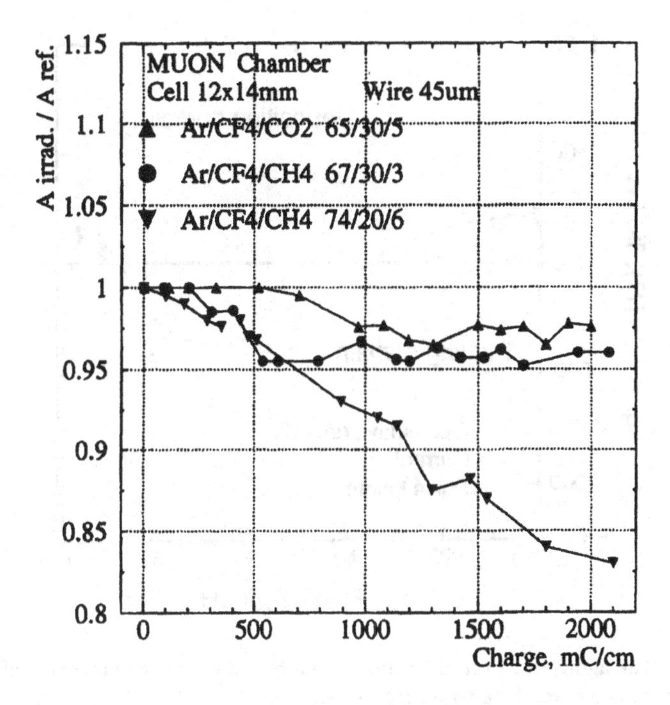

Figure 21 The normalized gain of a chamber cell of the HERA-B Muon gas chamber for three different gases as a function of accumulated charge [35].

For large gas systems, the gas has to be recirculated for cost reasons. This necessitates the construction of a gas purification system. As the removal of contaminations from the gas usually is done chemically, the purification process has to be included in the tests if one wants a realistic estimate of the radiation hardness of a given system.

5.2.2 Gas Contamination. Minor contaminations of the gas can have very large effects. Examples are the electron attachment (and subsequent signal loss) due to ppm levels of O_2 contamination or the effect of H_2O on the drift velocity of a given gas again at the ppm level.

Impurities contaminating the counting gas can come from many sources. They can already occur during the preparation of the gas bottles in industry. Impurities can diffuse through the chamber gas volume enclosure. A very probable source of impurities is the outgassing of the construction material of the chamber and of all components of the gas system (valves, hoses, seals, bubblers, purification devices etc.).

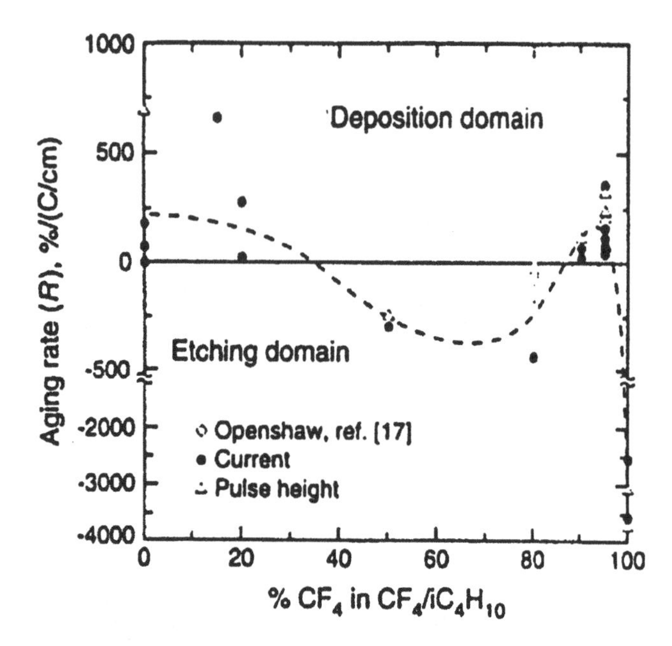

Figure 22 The aging rate of a test counter as a function of the percentage of CF_4 in a CF_4-isobutane gas mixture (from Ref. [36]).

Fig. 24 shows as an example of the effect of inserting a 10m long PVC hose into the inlet of a gas system which is otherwise made of stainless steel tubes. Shown is the normalized gain (the peak position of an ^{55}Fe source) as a function of irradiation dose in a counter with Ar-Ethane gas. The gain rapidly drops after the insertion of the PVC tube and the width of the ^{55}Fe peak (lower curve) deteriorates. Note that the counter does not recover after removal of the PVC tube but continues to deteriorate. It has been permanently damaged!

As a consequence, one needs to test all materials to be used in the chamber for natural outgassing as well as outgassing induced by irradiation if the material under question will be located in irradiated regions of the detector. This can best be done using dedicated 'outgassing setups' comprising special test chambers, ultra clean gas systems and diagnostics instruments like gas chromatographs.

In this context it is important that the irradiation tests are done with a 'final design' chamber before embarking on mass production. Obviously last minute changes to the baseline design (e.g. employing a new glue) are highly discouraged.

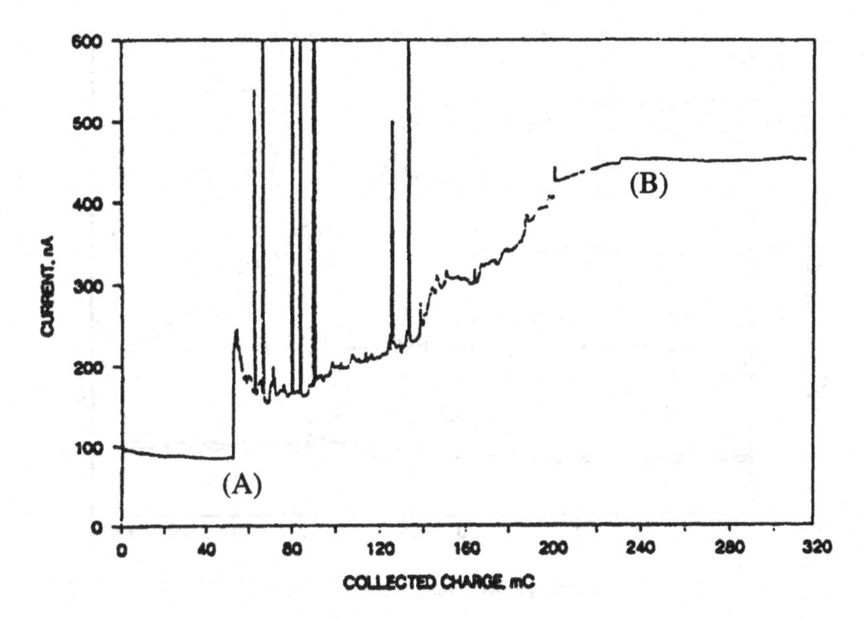

Figure 23 The effect of the admixture of CF_4 to a gas counter that had previously been damaged by bubbling the Ar-ethane mixture through silicon oil during irradiation. The CF_4 admixture is applied at the time denoted by (A) and the chamber has reached full gain at point (B) (from Ref. [34]).

5.2.3 Anode/Cathode Material. The typical anode wire for high dose applications has a Tungsten core and a thin gold coating. Early attempts to use bare Al or Ni wires have failed due to rapid deterioration of the wire. The BaBar experiment (rather low irradiation doses) is using (so far successfully) a gold coated Al wire in an attempt to minimize the material thickness of the tracking chamber.

Extensive tests done by the ATLAS experiment with a large number of wires from different manufacturers [37] have shown that gold coated tungsten wires can survive doses exceeding 3C/cm even when using CF_4 based gases, provided care is taken that the gold surface of the wires is intact after assembly.

The choice of the cathode material is usually rather strongly dictated by the requirements of the experiment's design. As an example in the HERA-B experiment (see Chapter 6) a cathode material was needed, that is moldable/foldable, flexible but self supporting, thin in radiation lengths, conductive and last but not least, affordable. The choice fell on a carbon loaded polycarbonate foil.

Tests with prototype chambers in the laboratory showed no problem but when the modules were installed in the HERA environment, they

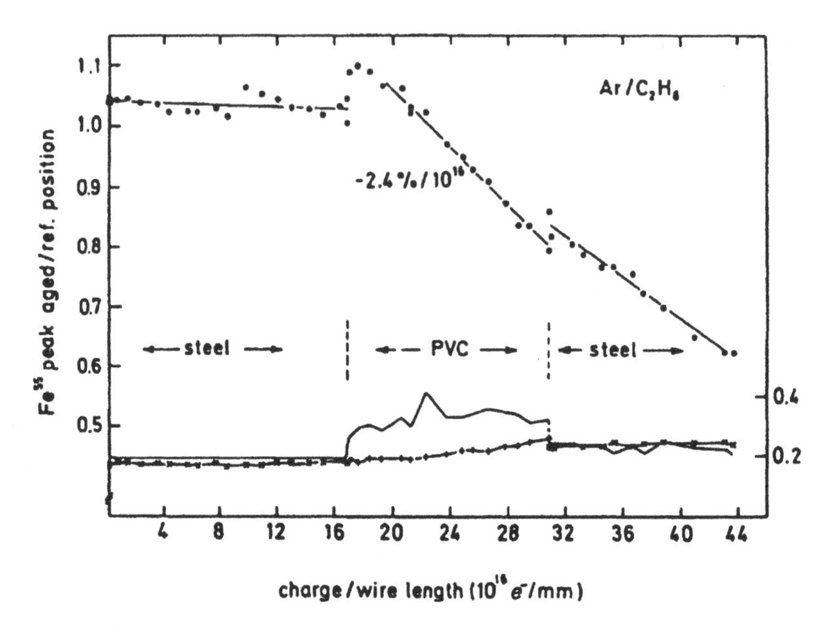

Figure 24 The effect of inserting a 10m long PVC hose into a gas inlet that otherwise consists entirely of stainless steel tubes. For details see the text. From Ref. [3].

rapidly showed signs typical of the Malter effect (see Chapter 3). Fig. 25 shows the current measured in a group of wires as a function of time. The value of the High Voltage is also shown. The module is irradiated with a spray of particles created in the interaction of the HERA proton beam with a wire target. The target rate (and hence the illumination rate of the chamber with ionizing particles) was constant and is shown at the bottom of the figure.

At the beginning, the chamber current follows the target rate and stays constant. After $\sim$ 1/2 hour, the current in the chamber rises without any change in the illumination rate. When the illumination is removed, the current is selfsustaining until the High Voltage is turned off. The effect is reproducible when the High Voltage is turned on again. The counter has been rendered unusable.

The problem could eventually be traced to a combination of several points: the surface properties of the carbon loaded polycarbonate, the presence of heavily ionizing particles and the properties of the specific gas mixture used. A thin coating of the cathode foils with a layer of copper followed with a layer of gold and a modification of the gas mixture solved the problem.

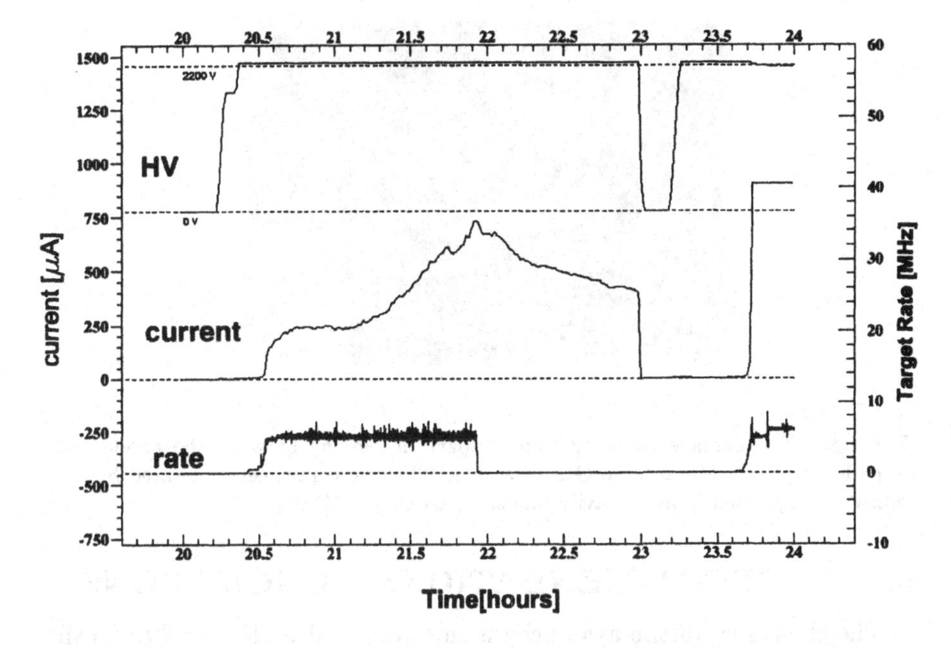

Figure 25 Observation of Malter type effect in the HERA-B experiment. Top: High voltage status as a function of time. Bottom: target rate (proportional to chamber illumination rate). Middle: Current in a group of 256 adjacent wires.

5.2.4 Gain and Irradiation Type.

Usually aging tests have been performed with easily available laboratory sources like Xray tubes and radioactive sources. Unfortunately some types of radiation damage can strongly depend on the particle type used (see e.g. the observation of the Malter effect in the HERA-B type environment). A famous example is the damage of Micro Strip Gas Chambers by highly ionizing particles [38]. Typically, these devices were tested for radiation damage intensively using Xray sources in the laboratory and no damage was observed for considerable integrated doses. In a test beam to study spatial resolutions at the PSI, however, it was found that the detectors rapidly deteriorated through sparks damaging the fragile anodes.

This effect could be reproduced in the laboratory using a radioactive gas laced with alpha particles. An example showing the resulting damage of such a spark is shown in Fig. 26. The spark is initiated by streamers originating in the large ionization by the heavy alpha particles.

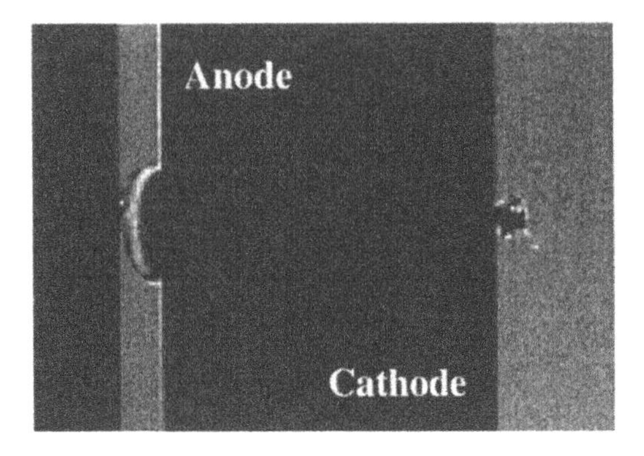

Figure 26 The damage resulting from a spark developing between the anode and cathode strips of a Micro Strip Gas Chamber. The spark has been generated by the ionization deposited from a heavily ionizing particle (Ref.[38]).

5.3. RECOMMENDATIONS/CONCLUSIONS

The chemistry in the avalanche is not very well understood and can be highly dependent on the actual boundary conditions as there are: contamination with impurities, the gas flow, the choice of gas, the area of illumination, the actual geometry of the field shaping electrodes, the irradiation load, the particle species etc. and one is left with the pragmatic approach to learn as much as one can from what others have done and to test, test, test, test, test...

Nevertheless, there are some important lessons that are useful to keep in mind

- Test more than one prototype chamber (statistics).
- 'Simulate' the final environment as well as possible (bracket the extremes, get information on the expected particle mix, test with different sources (hadrons, electrons/photons)).
- Carefully select and test all (!) materials that are in contact with the gas.
- Be extremely careful that nobody 'sneaks in' last minute changes in the production.
- Provide lots of monitoring (gain tubes, gas flow monitors, gas chromatographs for mixture and impurity control).
- Provide interlocks (note: one wrong gas bottle can screw up the entire system forever!)

- Don't forget system aspects. Small test chambers are usually not enough as system aspects often force you to modify the design.

- Think carefully about the implications of the necessity to speed up the aging tests (space charge effects, outgassing time constants, gas flow vs. irradiation load, load dependence of chemistry in the avalanche).

6. RADIATION DAMAGE ISSUES – (B) SILICON DETECTORS

When an ionizing particle traverses a solid body, it can interact both with the electron cloud and with the nuclei of the lattice atoms. The interaction with the electron cloud is a transient effect and is used for the detection of radiation in silicon (i.e. the generation of electron-hole pairs). The interaction with the nuclei of the lattice atoms can instead lead to long lasting changes of the material properties.

Fig. 27 illustrates the various point defects that can occur: the displacement of lattice atoms from their original sites can result in interstitials as well as vacancies. Nuclear interactions can result in neutron capture and nuclear transmutations.

When the recoil energy of the Silicon lattice atom is high enough ($>>$ 2KeV), both point defects and clusters of defects can be generated. A cluster is a dense agglomeration of point defects that appear at the end of a recoil silicon tracks where the atom loses its last 5-10KeV of energy and the elastic scattering cross section increases by several orders of magnitude. The typical size of a cluster is $\sim$100 lattice displacements.

Most primary effects are unstable, e.g. interstitials and vacancies are mobile and will anneal at room temperature. Some effects are, however, stable at room temperature and can change the electrical properties of the semiconductor. Examples are so called 'A-centers' and 'E-centers'.

A-centers are a combination of a vacancy with an Oxygen interstitial (the Oxygen being present as impurity from the crystal growing process). The Oxygen, being previously inactive in its interstitial position, becomes electrically active when a vacancy is created next to it resulting in the formation of acceptor states that act as trapping centers for electrons. The formation of A-centers hence results in an effective doping depletion.

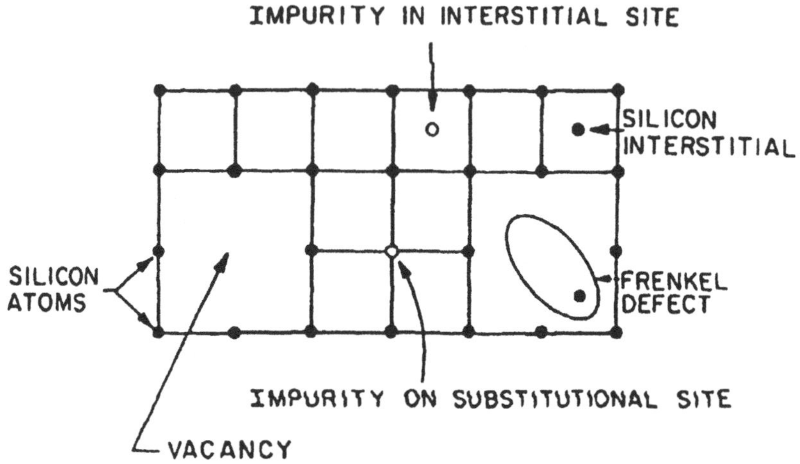

Figure 27 The various point defects that can occur when radiation interacts with the Si lattice. The displacement of lattice atoms from their original sites can result in interstitials, vacancies and defect centers.

E-centers are generated when a vacancy is combined with a Phosphorus atom sitting at a lattice site. The Phosphorus atom does not fulfill ist original role as donor any more and the result is an effective donor removal. The space charge changes from positive to neutral.

Vacancy-Boron complexes are similar but here the Boron becomes inactive as an acceptor resulting in an effective acceptor removal.

The probability for cluster formation depends on the radiation type. This is illustrated in Table 4 for electrons, protons, neutrons and Si ions. Electrons hence will produce point defects but almost no defect clusters.

The main consequences of defect complexes are

- They can change the charge density in the space charge region (e.g. E-centers) resulting in the necessity to increase or decrease the detector bias voltage in order to keep the entire detector bulk sensitive for maximum signal collection.

- They act as recombination-generation centers (they can capture and emit electrons and holes) resulting in an increase of the reverse-bias current in the space charge region of a detector.

- They act as trapping centers (electrons/holes are captured and then emitted with some time delay) resulting in a trapping of part of the signal charge in the space charge region that may be released too late for efficient detection.

Table 4 The parameters affecting cluster formation for different types of radiation. T_{max} is the maximum kinematically possible recoil energy, T_{av} is the mean recoil energy and E_{min} is the minimum radiation energy needed for the creation of a point defect and a defect cluster (from Ref. [4]).

Radiation	Electrons	Protons	Neutrons	Si^+
Interaction	Coulomb Scattering	Coulomb and nuclear Scattering	Elastic nuclear scattering	Coulomb scattering
T_{max} [eV]	155	133,700	133,900	1,000,000
T_{av} [eV]	46	210	50,000	265
E_{min} [eV]				
point defect	260,000	190	190	25
defect cluster	4,600,000	15,000	15,000	2,000

Fig. 28 shows the effective doping concentration of a n-type silicon detector as a function of radiation fluence. Note that the depletion voltage of the detector is related to the effective doping concentration N_{eff} through

$$V_{Dep} = \frac{e}{2} \frac{d^2}{\epsilon_{Si}\epsilon_0} |N_{eff}|, \tag{1.13}$$

where $N_{eff} = N_D^+ - N_A^-$ is the net difference between ionized shallow donor levels and the charged deep acceptor levels. The numbers in Fig. 28 are corrected for the effect of self-annealing during irradiation (see below) and scaled to the non-ionizing energy loss equivalent to 1MeV neutrons.

As a function of irradiation the effective doping decreases with increasing dose as the creation of e.g. vacancy-Phosphorus (E-) centers results in the removal of donors. At a flux of approx. $10^{12}\mathrm{n/cm^2}$ the material becomes 'intrinsic', the effective doping concentration (see Eq. 1.13) vanishes.

Above this value, the doping becomes effectively p-type (e.g. through the continuing creation of A-centers which act as acceptors, the defect centers assuming negative charge states in the space charge region).

One hence observes a 'type inversion' of the n-type silicon to p-type silicon and in a silicon strip detector the diode junction moves to the opposite side of the wafer. The curve can be parameterized as

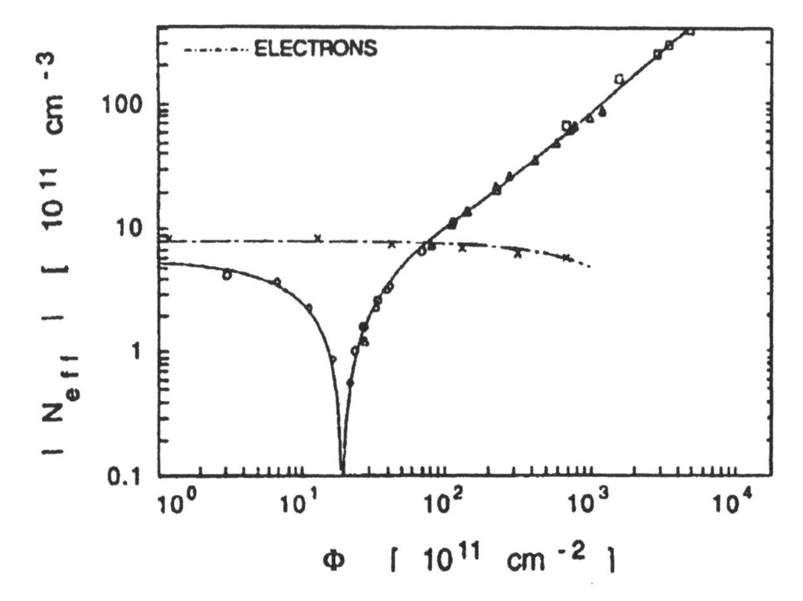

Figure 28 The effective doping concentration N_{eff} of a n-type silicon detector as a function of radiation fluence (from Ref. [4]).

$$N_{eff}(\Phi) = N_{D,0}\ e^{-c_D\Phi} - N_{A,0}\ e^{-c_A\Phi} + b_D\Phi - b_A\Phi, \tag{1.14}$$

where $N_{D,0}$, $N_{A,0}$ are the donor and acceptor concentrations before the irradiation, the constants c_D and c_A describe the removal of donors and acceptors, respectively, and b_D, b_A account for the creation of acceptor and donor states.

Fig. 29 shows the reverse-bias current of a n-type silicon detector as a function of irradiation fluence. The reverse-bias current generated in a volume V can, below type inversion, be parameterized as

$$\frac{\Delta I_{Vol}}{V} = \alpha \cdot \Phi. \tag{1.15}$$

Above type inversion a stronger rise is observed. A typical value for the damage constant alpha is $\alpha \approx 8 \times 10^{-17}$A/cm. Note that the damage constant depends rather strongly on temperature: a factor of 2 increase is observed for every 7 degrees of temperature increase and cooling of the detectors during operation is beneficial.

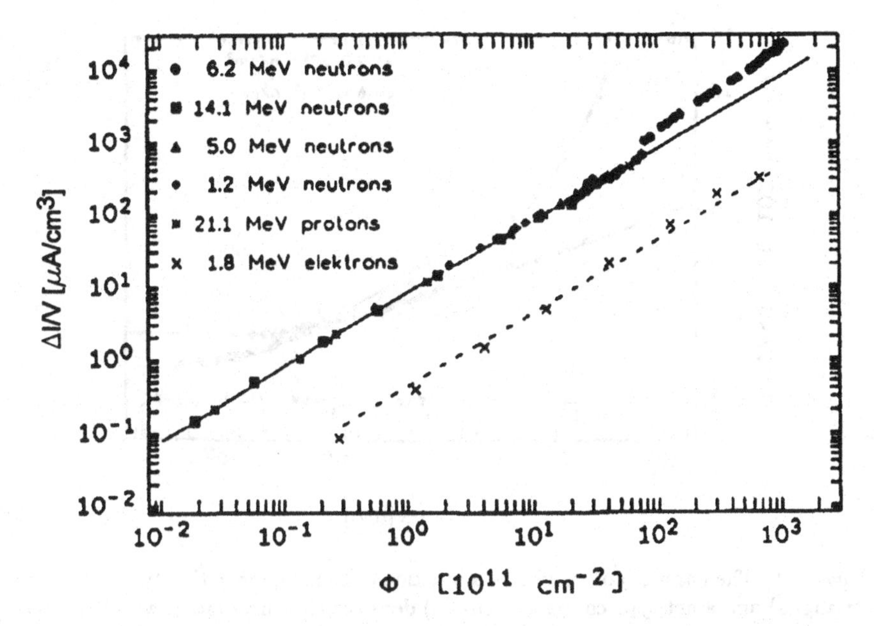

Figure 29 The reverse-bias current of a n-type silicon detector as a function of irradiation fluence (from Ref. [4]).

The damage constant α is essentially not dependent on the wafer resistivity, on impurities or on the type of radiation that generates the damage. It is entirely determined by the formation probability of defect clusters.

It has been observed, that the damage (increase in leakage current, effective doping changes, charge trapping) diminishes with the time that has passed after the irradiation. This effect is called annealing. The rate of annealing depends strongly on the storage temperature of the detector after irradiation. Fig. 30 shows the damage constant α as a function of time [4] for both a type converted and a not type converted detector after irradiation with neutrons.

It is thought that the annealing is due to the partial disappearance of radiation generated crystal effects, e.g. through the filling of a vacancy by an interstitial as well as the transformation of defect complexes into other, more stable ones with changed properties.

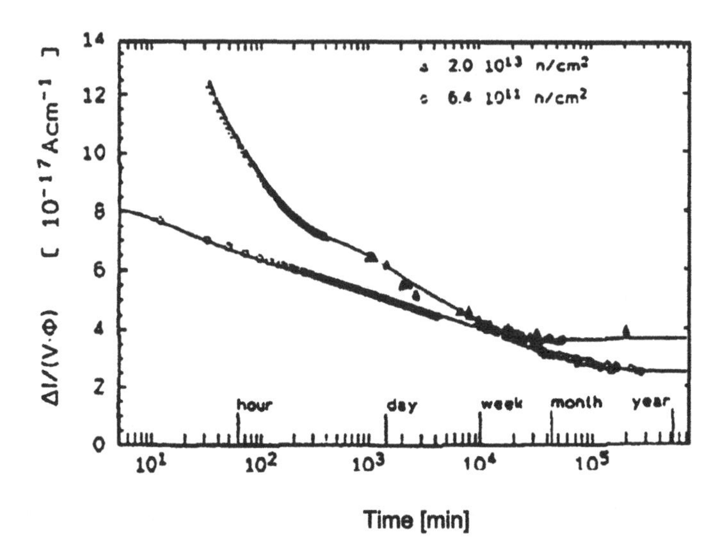

Figure 30 The damage constant α as a function of time [4] for both a type converted (triangles) and a not type converted (circles) detector after irradiation with neutrons.

An initial decrease of the doping change during the first weeks of room temperature annealing is unfortunately followed by a new increase even without irradiation (see Fig. 31). This effect, termed 'reverse-annealing' can be delayed by running and annealing at reduced temperatures. As an example, reverse annealing effects that occur over weeks at 25°C are slowed down to years at -10°C.This effect is thought to be due to the long term transformation of inactive defects into active defects and the interplay of several annealing and effective doping processes with different time constants.

What are now the consequences for the silicon strip detector design at HERA-B and the LHC?

The effect of charge trapping is small, even at irradiation doses expected at the LHC, and can be coped with as long as a safety margin in the sensitivity of the readout electronics exists.

The effective doping changes resulting in large depletion voltages require the design of guard ring structures that allow stable operation of silicon strip detectors at voltages up to 500V and more. The effective type inversion has led people to investigate whether it is not advantageous to start with n^+n detectors rather than p^+n detectors or even to live with the higher voltages needed when starting with p^+p detectors before irradiation.

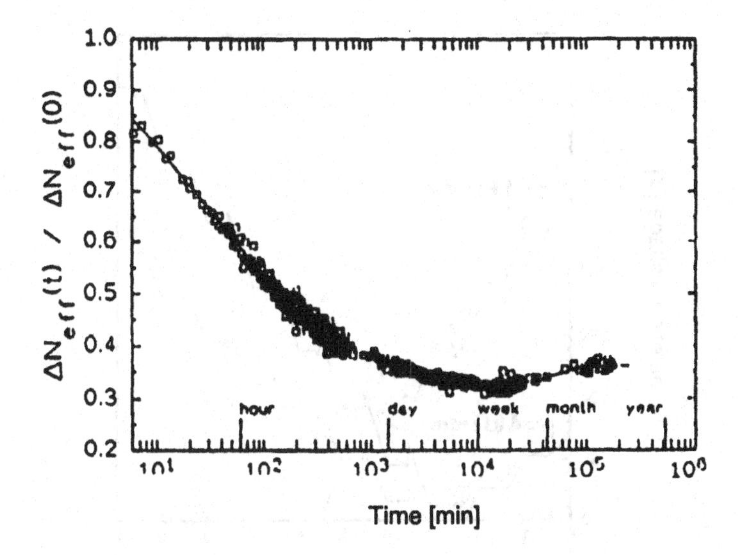

Figure 31 The change of the effective doping concentration of a previously irradiated *n* type silicon detector after the irradiation showing the effect of annealing and the onset of reverse annealing (from Ref. [4]).

The choice of the resistivity of the bulk silicon allows one to balance the time of occurance of the type inversion with the maximum voltage needed for full depletion. This is shown in Fig. 32 where the depletion voltage is shown as a function of irradiation time for two silicon detectors with different resistivity. For the low resistivity material the initial depletion voltage is larger than for the higher resistivity wafer but the maximum depletion voltage ever needed can stay lower even after 10 years of running [39].

The large leakage currents after high irradiation dose lead to the requirement that the readout electronics be AC coupled to the silicon strip detectors. Nevertheless, the power consumption and dissipation due to these currents can become a factor: at a depletion voltage of 500V a current of 1μA corresponds to 0.5mW per strip and there are thousands of strips concentrated in a very small volume!

The temperature dependence of the leakage currents and the reverse annealing effects force one to run at decreased temperatures (typically -10°C at the LHC). This poses serious complications for the operation of the detectors, as access, maintenance and repair scenarios have to be carefully planned out.

Over the past years, much work has been devoted to make silicon detectors less prone to irradiation damage. One promising development is the work done by the ROSE collaboration [40]. They have investigated

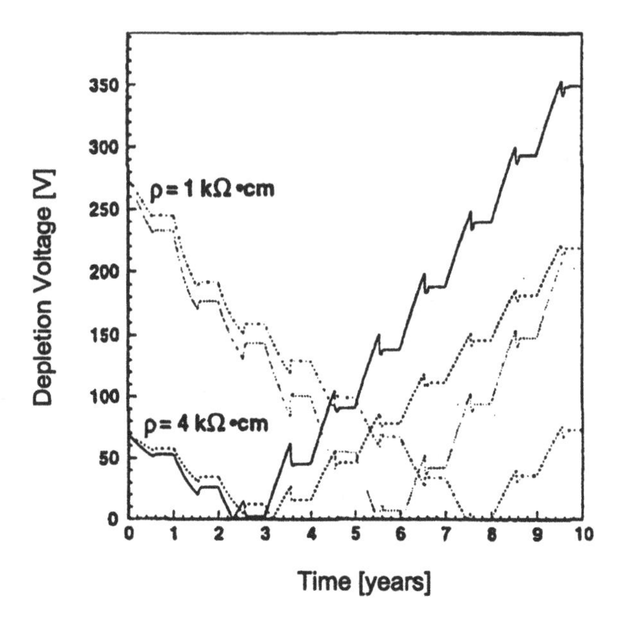

Figure 32 The depletion voltage as a function of irradiation time for two silicon detectors with different resistivity ρ [39].

in detail the various defects that occur under irradiation and attempt to improve the radiation hardness of silicon through 'defect engineering'.

Fig. 33 shows that oxygenated wafers show lower depletion voltages at the same irradiation dose than normal silicon wafers. The oxygenation can be done both at the ingot wafer growing stage and with the finished wafer by 'driving' the Oxygen into the wafer at elevated temperatures. Fig. 34 shows that also the reverse annealing can be slowed down and even seems to saturate at high fluences. Hence detectors could stay at room temperature for a longer time during access and repair periods thus providing an additional safety margin.

As was said above, the leakage current damage constant α (Eq. 1.15) has been found to be material independent, so in this area defect engineering will not lead to improvements in the detector performance.

7. NEW TRACKING SYSTEMS - SELECTED EXAMPLE

In the following, two new tracking systems will be briefly sketched: the ATLAS Semiconductor Tracker exemplifying the progress made in the field of silicon strip detector technology and the HERA-B Outer Tracker as an example of gaseous tracking detectors at high rate facili-

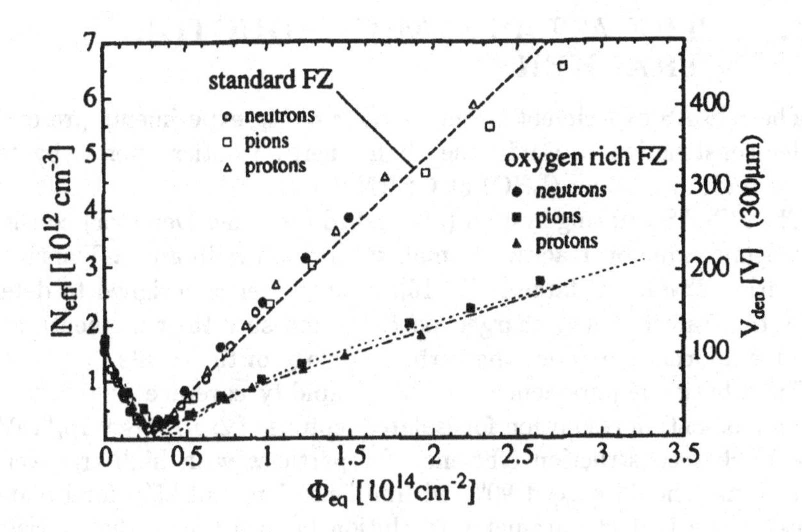

Figure 33 The effective doping concentration N_{eff} as a function of irradiation fluence for standard and for oxygenated silicon wafers [40].

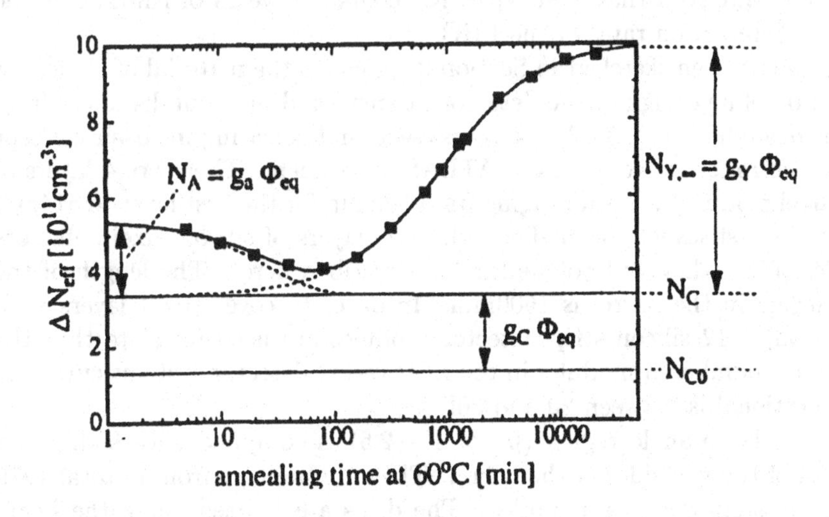

Figure 34 The effect of reverse annealing for oxygenated wafers [40].

ties. Apologies to all the experiments I have not have had the time to describe in equal detail.

7.1. THE ATLAS SEMICONDUCTOR TRACKER

The ATLAS experiment is one of the two big experiments presently under construction to study the High Energy frontier opened by the Large Hadron Collider (LHC) at CERN.

The ATLAS tracking system (also called the Inner Detector) consists of a Semiconductor Tracker at small to medium radii and a Transition Radiation Tracker at large radii [16]. The tracker is designed to determine the trajectories of charged particles, measure their momenta and provide information about the vertex topology of the events.

The physics requirements call for a rapidity coverage of $|\eta| < 2.5$ and a momentum resolution for isolated leptons of $\Delta p_t/p_t \sim 0.1 p_t[\text{TeV}]$. The track reconstruction efficiency for particles with high transverse momentum should exceed 90% within dense jets and 95% for isolated tracks. The impact parameter resolution (a measure of the precision with which primary and secondary vertices can be measured) should follow $\sigma_{r-\phi} < 20\mu$m and $\sigma_z < 100\mu$m. In order not to spoil the energy measurement with the electromagnetic calorimeter, the material budget of the tracker should be kept as low as possible. Finally, the tracker has to be built such that it survives more than 10 years of running at the large interaction rates of the LHC.

As has been sketched in Section 2, silicon is the material of choice for the baseline design of the detector at small and medium distances from the beam line. Fig. 35 shows a cross section illustrating the basic concept for the Inner Detector of the ATLAS experiment. There are 4 layers of barrel modules at radii ranging from 300mm for the first layer to 514mm for the last layer. At smaller radii, two layers of silicon pixel detectors are added (they will not be further described here). The length of the 4 layers in the barrel is 1600mm. In order to cover the 4 layers with silicon, 2112 silicon strip detector modules are needed. Note that the typical number of modules in the LEP vertex detector systems currently operational is between 20 and 30!

The large angle region ($|\eta| \approx 1 - 2.5$) is equipped with 9 disks on each of the two sides of the barrel. They are built up from in total 1976 silicon strip detector modules. The disks are located along the beam line starting at z=835mm up to z=2788mm. The total length of the ATLAS silicon tracker hence exceeds 5.5m! Also in the forward region the silicon strips are complemented with pixel detectors at smaller radii.

Fig. 36 shows a sketch of the barrel module support structure. It consists of a carbon fiber cage to minimize material. The modules are

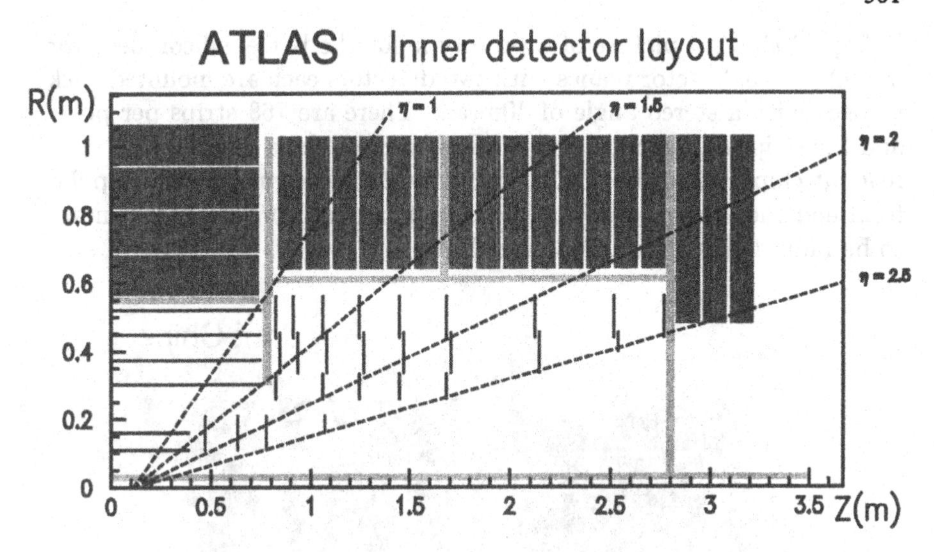

Figure 35 **A cross section illustrating the basic concept for the Inner Detector of the ATLAS experiment. The $r-z$ view of one quandrant of the detector is shown (Ref. [16]).**

tilted and staggered with respect to each other in order to achieve overlap important for solid angle coverage and alignment purposes.

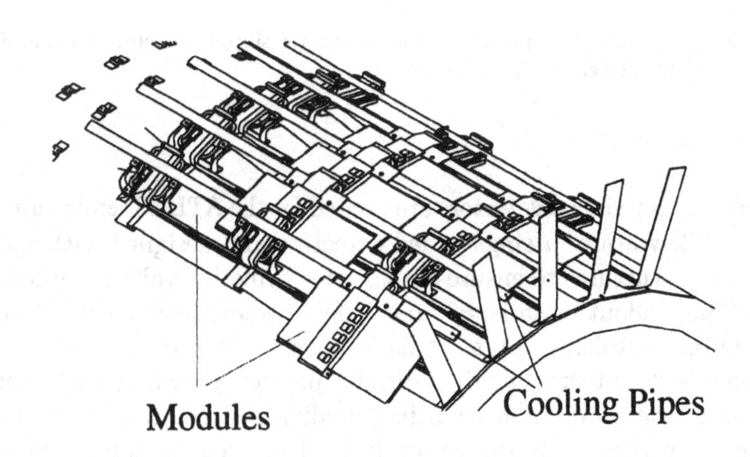

Figure 36 **A sketch of the barrel module support structure for the ATLAS Silicon Tracker.**

Fig. 37 shows a picture of a prototype for the barrel silicon detector module. Two detector planes with two detectors each are mounted back to back with a stereo angle of 40mrad. There are 768 strips per plane and the strips are readout via six VLSI readout chips on each side. The readout chip (called the ABCD-2T) is custom designed with a bipolar front end and binary readout. It is manufactured using a process known to be radiation hard to the desired levels.

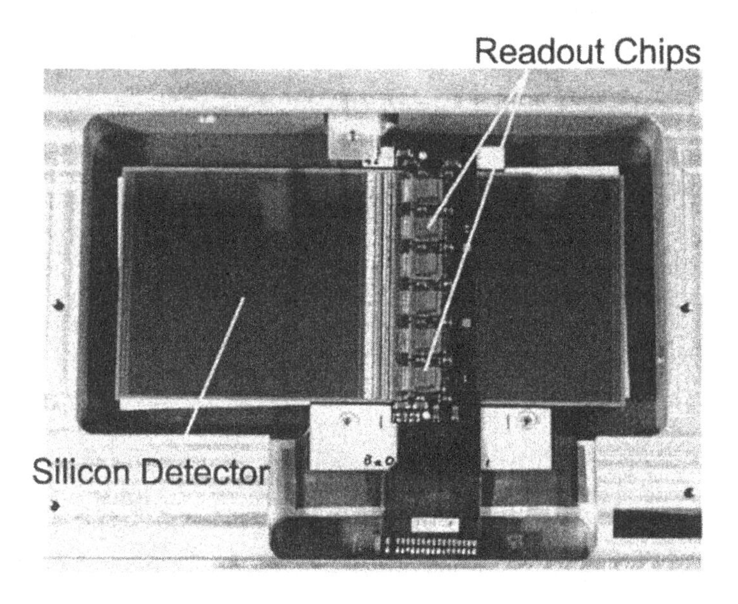

Figure 37 A picture of a prototype for the barrel silicon detector module of the ATLAS Silicon Tracker.

In total ≈20,000 silicon detectors are needed for the ATLAS semiconductor tracker. They are p^+n single sided detectors and designed with special guard ring structures to ensure stability for depletion voltages exceeding 500V. The readout pitch is ≈80μm. The detectors have been measured to be radiation tolerant up to radiation doses of 3×10^{14}p/cm^2.

Such a large detector requires specific and new measures for the organization of the production regarding quality control, material flow etc. among the various laboratories involved. This organization is presently being set up. In total, eleven centers will build the roughly 4,000 silicon strip detector modules. The mounting of the modules to the mechanical mounting structures will then be performed in three regional centers in Europe.

The ATLAS Semiconductor Tracker represents a major departure from the well known track paved by the design and construction of many silicon strip vertex detectors for collider experiments in the last decade. Silicon is for the first time used also in the main tracking system. The extrapolation lever arm is large and new problems in the areas of radiation hardness, large system aspects (cooling, mechanical stability, material flow, quality control, mass production management etc.) have to be tackled.

7.2. THE HERA-B OUTER TRACKER

As described in Section 2, the HERA-B experimental environment is very similar to that expected at the LHC. For 4-5 interactions of the 920 GeV HERA protons with the wire target per bunch crossing and with a bunch crossing time of 96ns, the particle flux can be approximated to a good degree to be $2 \times 10^7/R^2$ [sec^{-1}] where R is the radial distance from the beam line. The radiation load in the silicon vertex detectors closest to the beam approaches 5Mrad and in the hottest regions of the gaseous tracking chambers the accumulated charge per year of running is O(0.5)C/cm.

The charged particle tracking is performed in a multi-layered tracking system covering the angular range between 10 and 160mrad (in the nonbending view) and ranging over a length of 11m between the target and the electromagnetic calorimeter (see Fig. 2). The largest chambers in front of the calorimeter have to cover an area of $\sim$(4.5x6)m^2.

The tracking system is built of two different technologies: the Inner Tracker is built out of Microstrip Gas Chambers, the Outer Tracker consists of Honeycomb Drift Chambers. Due to the very large areas to be covered, silicon has not been an option. In the following the Outer Tracking system will be described in more detail [41].

The Outer Tracker of HERA-B consists of 13 'superlayers', each superlayer in itself consisting of 3 (respectively 6) single layers oriented in three different views (0, +80 mrad, -80 mrad with respect to the vertical coordinate).

Owing to the rapidly varying particle occupancy as a function of the distance to the beam line, each single layer is electrically segmented (see Fig. 38) such that the occupancy is kept approximately at the same level for each segment and that a maximum occupancy of $<$30% is not exceeded. In addition, the segmentation is done in two interleaving halves in order to accommodate the proton and electron beam pipes of the HERA ring. For channel minimization reasons, two drift cell sizes (5mm and 10mm diameter) are chosen.

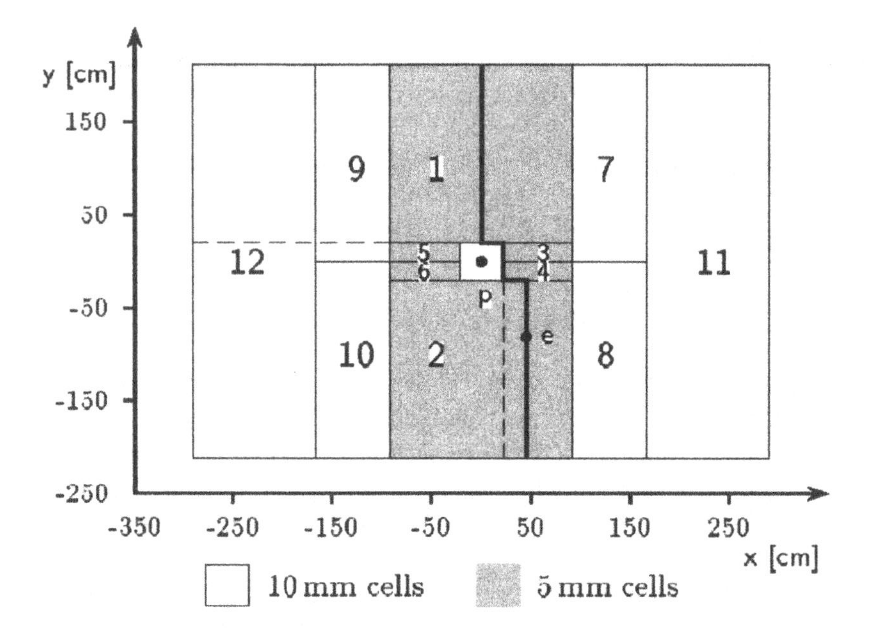

Figure 38 The electrical segmentation of a layer of the HERA-B Outer Tracker. p and e denote the hole for the proton and electron beam pipe, respectively. The shaded area denotes 5 mm diameter cells. In total 4 different active wire lengths are required for each of the 13 superlayers in order to distribute the charged particle occupancy approximately evenly [41].

As a consequence of these design requirements, in total 978 modules of length varying between 0.8 and 4.5m and with a typical width of ≈30cm with varying electrical segmentation are needed. The electrical segmentation is done by dividing the cell length into a region with thin wire (providing gain) and a thick wire (with negligible gain, used for signal transport). This necessitates means for wire interruption along the cell. The 5 and 10mm cells are up to 4.5m long and for electrostatic stability the wire has to be supported at short (50-65cm) intervals along its length within the cell.

This goal has been reached utilizing planar drift chambers built in the honeycomb design using a special carbon loaded polycarbonate foil. The design principle is shown in Fig. 39. Single folded layers of polycarbonate foil are mounted in specially formed templates, wires are strung and soldered to special G10 strips spaced by 50-65cm. Thin or thick wires are used according to the segmentation scheme in Fig. 38. When a layer is completed, another folded polycarbonate foil is mounted and the Honeycomb structure of one (mono)layer is completed.

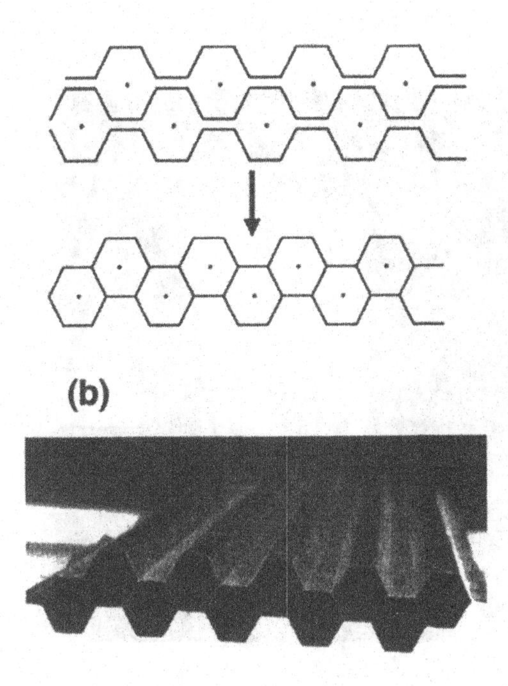

Figure 39 The design principle of the honeycomb tracker modules for the HERA-B Outer Tracker (for details see text). (a) Schematic of the assembly of one monolayer. (b) A picture of one monolayer made of carbon loaded polycarbonate foil.

The production of the ≈1,000 modules has been performed at five different labs situated in Russia, China and Germany. The mass production has posed major logistical problems (quality control, material flow, module inspection, timing), but the production of the entire set of modules could finally be achieved in a little over 9 months.

The mechanical mounting structure of a given superlayer is shown schematically in Fig. 40. The individual honeycomb modules are mounted in a separate gas box which is built of honeycomb material cover plates and thin carbon fiber caps in the overlap region of the two layer halves near the proton and electron beam pipes. The gas box in turn is mounted onto a C-shaped 'Outer Frame' which provides for the stability of the superlayer and carries the cables routed from the electronics to the TDC crates mounted to its side.

Fig. 41 shows the assembly of one half of the largest chamber in the laboratory. Also shown is a detail of the readout end illustrating the

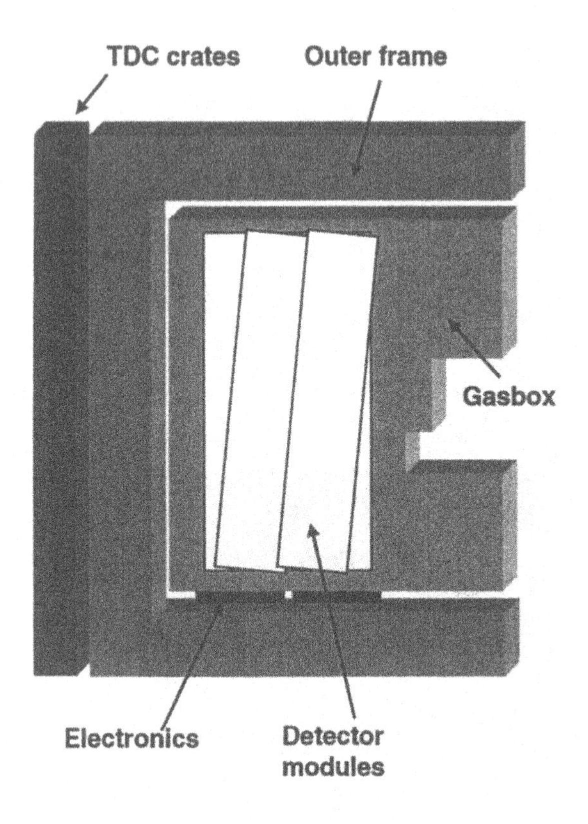

Figure 40 A schematical view of the mechanical mounting structure for one superlayer halve of the HERA-B Outer Tracker.

three different stereo angles. The chambers are read out via custom designed 8 channel chips with amplifier, shaper and discriminator. The signal cables are then routed through the cable frame to the TDC electronics.

The Outer Tracker of the HERA-B experiment has been completed in December 1999 and has taken data since February 2000. It is the first large detector designed and constructed for an LHC-like environment, which has taken data. It works at a 10 MHz readout rate and occupancies exceeding 30%. So far it has accumulated ~0.1 C/cm in the hottest regions without visible degradation.

Figure 41 The assembly of one half of the largest chamber of the HERA-B Outer Tracker in the laboratory. Also shown is a detail of the readout end with the three different stereo angles (marked by the circle in the top picture).

8. SUMMARY

The development in charged particle tracking has accelerated over the last decade, largely prompted by the demands of the new experiments coming into operation and being planned at the hadronic B factories and the Large Hadron Collider.

New developments for gaseous tracking detectors have focused on high rate performance and tolerance of the high radiation doses expected. Silicon has matured well and will be used both for vertexing and tracking due to its very high precision and high radiation tolerance.

The new systems represent an extrapolation from those used in the 90's that extends over more than two orders of magnitude. The system aspects and managerial problems are formidable.

The HERA-B experiment is the first to test the ideas incorporated into the designs for the LHC experiments. The first run this year indicates that the ideas are working out. Both silicon and gaseous tracking detectors work at unprecedented rates and irradiation loads, so far without measurable deterioration in performance.

We are looking forward to the next two decades of developments in charged particle tracking.

Acknowledgments

I would like to thank all of those who have provided me with material for these lectures: Francesco Forti and Natalie Roe for the BaBar silicon vertex detector, Torsten Akesson and Hans-Günther Moser for the ATLAS silicon tracker, Gigi Rolandi for the CMS Tracker, Hiroyasu Tajima from the BELLE silicon detector and Herbert Kapitza and Maxim Titov from the HERA-B Outer Tracker and Muon detector systems, respectively.

My special thanks go to Anatoli Romaniouk and Valeri Saveliev for sharing their expertise and to Bernhard Schmidt for many illuminating discussions on gaseous detectors and sharing his lecture notes for the Maria Laach School of Physics with me.

Last but not least I would like to thank Tom Ferbel, Harrison Prosper and Misha Danilov for inviting me to this school. It has been a pleasure to interact with the young colleagues and appreciate their inquisitive minds in such a stunning environment. The St.Croix school of 2000 will remain a beautiful memory.

References

[1] TECHNIQUES AND CONCEPTS OF HIGH-ENERGY PHYSICS. PROCEEDINGS, 1-9TH NATO ADVANCED STUDY INSTITUTE, St.Croix, USA. By T. Ferbel, (ed.) (Rochester U.). 1981-1997. (NATO ASI series B, Physics: 351).

[2] F. Sauli, *Principles of Operation of Multiwire Proportional and Drift Chambers*, Lectures given in the Academic Training Programme of CERN 1975-1976, CERN-77-09, 1977.

[3] PARTICLE DETECTION WITH DRIFT CHAMBERS. By W. Blum (Munich, Max Planck Inst.), L. Rolandi (CERN). 1993. Berlin, Germany: Springer (1993) 348 p.

[4] SEMICONDUCTOR RADIATION DETECTORS: DEVICE PHYSICS. By G. Lutz (Munich, Max Planck, MPE and Munich, Max Planck Inst.). 1999. 353pp. Berlin, Germany: Springer (1999) 353 p.

[5] WIRE CHAMBER. PROCEEDINGS, VIENNA, AUSTRIA
Nucl. Inst. and Meth **A419** (1998) 189-769,
Nucl. Inst. and Meth **A367** (1995) 1-455,
Nucl. Inst. and Meth **A323** (1992) 1-551,
Nucl. Inst. and Meth **A283** (1989) 371-815,
Nucl. Inst. and Meth **A252** (1986) Nos.2-3,
Nucl. Inst. and Meth **217** (1983) Nos.1-2,
Nucl. Inst. and Meth **176** (1980) Nos.1-2.

[6] *2nd London Conference on Position Sensitive Detectors*, London, U.K., 1990; Nucl. Inst. and Meth. **A310** (1991) 1-571.

[7] Proceedings of the *7th* International Workshop on Vertex Detectors, 1998, Nucl. Inst. and Meth. **A435** (1999) 1.

[8] Proceedings of the Symposia on Semiconductor Detectors,
Nucl. Inst. and Meth. **226** (1984) 1-218,
Nucl. Inst. and Meth. **A253** (1984) 309-586,
Nucl. Inst. and Meth. **A288** (1984) 1-292,
Nucl. Inst. and Meth. **A326** (1993) 1-389,
Nucl. Inst. and Meth. **A377** (1996) 177-564,
Nucl. Inst. and Meth. **A439** (2000) 199-665.

[9] A. S. Schwarz, Phys.Rep. **238** Nos.1-2, 1994.

[10] Workshop on Radiation Damage to Wire Chambers, Berkeley, California, Jan 16-17, 1986. PROCEEDINGS. Lawrence Berkeley Lab, 1986. 344p. (LBL-21170-mc,C86/01/16).

[11] See e.g. T. Iijima and E. Prebys (BELLE Collaboration),"Commissioning and first results from BELLE," in Nucl. Inst. and Meth. **A446** (2000) 75 and references therein.

[12] See e.g. P. F. Harrison, "An Introduction to the BABAR experiment," in Nucl. Inst. and Meth. **A368** (1995) 81 and references therein.

[13] ALICE - A Large Ion Collider Experiment, Technical Proposal, CERN/LHCC 95-71, December 1995.

[14] See e.g. HERA-B: AN EXPERIMENT TO STUDY CP VIOLATION AT THE HERA PROTON RING USING AN INTERNAL TARGET. By HERA-B Collaboration (W. Schmidt-Parzefall for the collaboration). 1995 in Nucl.Inst. and Meth **A368** (1995) 124-132 and references therein.

[15] D. Denegri, *Standard Model Physics at the LHC (pp Collisions)*, Proceedings of the *LHC Workshop*, Aachen, 1990, CERN-90-10, 1990 p.56.

[16] ATLAS Inner Detector Technical Design Report, CERN/LHCC/97-17, April 1997.

[17] CMS - The Tracker Project, Technical Design Report, CERN/LHCC 98-6, April 1998.

[18] F.D. van den Berg, *Gas-filled micro-patterned radiation detectors*, Delft University Press, 2000, ISBN 90-407-1997-7.

[19] DETECTORS FOR PARTICLE RADIATION. 2nd ed. By Konrad Kleinknecht. Cambridge Univ. Press, 1998. 246p.

[20] See Ref. [18], p.36.

[21] See Ref. [18], p.26.

[22] L. Malter, Phys.Rev.**50** (1936) 48.

[23] A. Oed, Nucl. Inst. and Meth. **A263** (1988) 351-359.

[24] F. Sauli, IEEE Trans.Nucl.Sci. Vol.**44** No.3 (1997) 646.

[25] See http://wwwcompass.cern.ch/

[26] See http://lhcb.web.cern.ch/lhcb/

[27] K.G. McKay, Phys. Rev. **84** (1951) 829.

[28] J. Kemmer, Nucl. Inst. and Meth. **169** (1980) 499.

[29] R. Bailey et al. (ACCMOR Collaboration), Nucl. Inst. and Meth. **226** (1984) 56-58.

[30] J.T. Walker et al., Nucl. Inst. and Meth. **226** (1984).

[31] C. Adolphsen *et al.*, (Mark II Collaboration), Nucl. Inst. and Meth. **A313** (1992) 63.

[32] B. Mours et al., Nucl. Inst. and Meth. **A379** (1996) 101-115.

[33] S.M. Sze, *Semiconductor Devices*, Physics and Technology, John Wiley and Sons, 1985.

[34] J.A. Kadyk, Nucl. Inst. and Meth. **A300** (1991) 436-479.

[35] Maxim Titov, ITEP Moscow, HERA-B Collaboration, private communication.

[36] J. Wise, J. Kadyk and D. Hess, J.Appl.Phys. **74** (9), p.5327, November 1993.

[37] A. Romaniouk, CERN and MEPhI, ATLAS collaboration, private communication.

[38] B. Schmidt, Nucl. Inst. and Meth. **A419** (1998) 230-238 and private communication.

[39] See Ref. [17], p.96.

[40] See http://rd48.web.cern.ch/RD48/

[41] See e.g. M. Capeans, Nucl. Inst. and Meth. **A446** (2000) 317 and references therein.

ISSUES IN CALORIMETRY

Fabiola Gianotti
CERN, EP Division
1211 Genève 23, Switzerland
fabiola.gianotti@cern.ch

Abstract

These lectures present an overview of calorimetry in high energy physics. The physical principles of electromagnetic and hadronic calorimeters, as well as the main techniques used to build these detectors, are discussed. Examples of calorimeters operating in present experiments or in construction for future experiments are given, with the purpose of illustrating the various techniques, the most recent developments, and the different choices motivated by the different physics goals.

1. INTRODUCTION

Calorimeters are used in high energy physics experiments mainly to measure the energy of electrons, photons and hadrons. They are blocks of instrumented material where these particles are fully absorbed and their energy is transformed into a measurable quantity. The interaction of the incident particle with the detector (through electromagnetic or strong processes) produces a shower of secondary particles with progressively degraded energy. The energy deposited by the charged particles of the shower in the active part of the calorimeter, which can be collected and measured in the form of charge or light, is proportional to the energy of the incident particle.

Calorimeters can be broadly divided into electromagnetic calorimeters, used to measure mainly electrons and photons through their electromagnetic interactions (e.g. Bremsstrahlung, pair production), and hadronic calorimeters, used to measure mainly hadrons through their strong and electromagnetic interactions. They can be further classified according to the construction technique into sampling calorimeters and homogeneous calorimeters. Sampling calorimeters consist of alternating layers of an absorber, a dense material used to degrade the energy of the

H.B. Prosper and M. Danilov (eds.), Techniques and Concepts of High-Energy Physics, 313–367.

incident particle, and an active medium, which provides the detectable signal. Homogeneous calorimeters, on the other hand, are built of only one type of material which performs both tasks, energy degradation and signal generation.

Calorimeters play an increasingly important rôle in high energy physics for various reasons:

- In contrast to magnetic spectrometers, where the momentum resolution deteriorates linearly with the particle momentum, the calorimeter energy resolution improves with energy as $1/\sqrt{E}$, where E is the energy of the incident particle. Therefore calorimeters are very well suited to high energy physics experiments.

- In contrast to magnetic spectrometers, calorimeters are sensitive to all types of particles, charged and neutral (e.g. neutrons). They can even provide an indirect detection of neutrinos and their energy through a measurement of the event missing energy.

- They are versatile detectors. Although they were originally conceived as devices for energy measurements, operation in high energy physics experiments has introduced increasingly demanding requirements. Therefore calorimeters can be used today also to determine the shower position and direction, to identify different particles (for instance to distinguish electrons and photons from pions and muons on the basis of their different interactions with the detector), and to measure the particle arrival time. Finally, calorimeters are commonly used for trigger purposes, since they can provide fast signals, easy to process and to interpret.

- They are space (and therefore cost) effective. Since the shower length increases only logarithmically with energy, the detector thickness must increase only logarithmically with the energy of the studied particles. In contrast, to obtain a fixed momentum resolution, the bending power BL^2 of a magnetic spectrometer (where B is the magnetic field and L the size) must increase linearly with the particle momentum p.

For these reasons calorimeters are well suited detectors for the demanding environment of modern high energy physics experiments.

This paper is organised as follows. Sections 2 and 3 describe the physics principles of the detection of electromagnetic showers, and the electromagnetic energy measurement and resolution. Sections 4 and 5 discuss similar issues for hadronic showers. Performance requirements of modern calorimeters are listed in Section 6. Section 7 describes the main construction techniques and compares their strengths and weaknesses. Examples of calorimeters operating in present high energy physics experiments or being built for future experiments are also given. Calorimeter calibration methods are discussed in Section 8 and issues related to integration in big experiments in Section 9. Finally Section 10 is devoted to the conclusions.

2. PHYSICS OF ELECTROMAGNETIC SHOWERS

Despite the apparently complicated phenomenology of shower development in a material, electrons and photons interact with matter via only a few well-understood QED processes, and the main shower features can be parametrised with simple empirical formulae.

The average energy lost by electrons in a dense material and the photon interaction cross-section are shown in Fig. 1 as a function of energy.

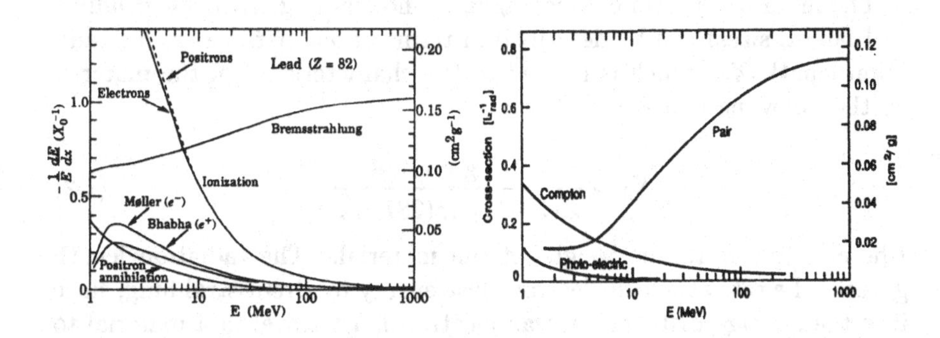

Figure 1 Left: fractional energy lost in Pb by electrons and positrons as a function of energy (from Ref. [1]). Right: photon interaction cross-section in Pb as a function of energy.

Two main regimes can be identified. For energies larger than ~ 10 MeV, the main source of electron energy loss is Bremsstrahlung (i.e. photon radiation off a nucleus). In this same energy range, photon interactions lead mainly to electron-positron pair production. For energies above 1 GeV both these processes become essentially energy independent. For E <10 MeV, on the other hand, electrons lose their energy mainly through collisions with the atoms and molecules of the material, giving rise to ionisation or thermal excitation, and photons through Compton scattering and the photoelectric effect. The energy at which the electron ionisation losses and Bremsstrahlung losses become equal is called the critical energy (ϵ). This energy depends on the material type and is approximately given by

$$\epsilon \sim \frac{800 \text{ MeV}}{Z}, \tag{1.1}$$

where Z is the atomic number of the absorbing material. It can be seen from Fig. 1 that $\epsilon \sim 10$ MeV in lead.

Therefore, electrons and photons of sufficiently high energy ($\geq$1 GeV) impinging on a block of material produce secondary photons by Bremsstrahlung or secondary electrons by pair production, which in turn produce other particles by the same mechanisms, thus giving rise to a cascade (shower) of particles with progressively degraded energies. The number of particles in the shower increases until the electron energies fall below the critical energy. At this point energy is mainly dissipated by ionisation and atomic and molecular excitation, and not in the generation of other particles, and the shower stops.

The main features of electromagnetic showers (e.g. their longitudinal and lateral sizes) can be described in terms of one parameter, the radiation length X_0, which is related to the characteristics of the material by the following relation

$$X_0 \sim \frac{716 \text{ g cm}^{-2} A}{Z(Z+1)\ \ln(287/\sqrt{Z})}, \tag{1.2}$$

where A is the atomic weight of the material. The radiation length governs the rate at which electrons lose energy by Bremsstrahlung, since it is the average distance that an electron must travel in a material to reduce its energy to 1/e of its original energy E_0

$$< E(x) > = E_0\ e^{-\frac{x}{X_0}}. \tag{1.3}$$

Similarly, a photon beam of original intensity I_0 incident on a block of material is absorbed mainly through pair production. After traveling a distance $x = \frac{9}{7}X_0$, its intensity is reduced to 1/e of its original intensity

$$< I(x) > = I_0\ e^{-\frac{7}{9}\frac{x}{X_0}} \tag{1.4}$$

. These relations show that the physical scale over which the showers develop is similar for incident electrons and photons. Therefore electromagnetic showers can be described in an universal way by using simple functions of X_0.

For instance, the longitudinal shower profile is to first order independent of the type of material if expressed in terms of X_0. The shower maximum, i.e. the depth at which the largest number of secondary particles is produced, occurs at

$$t_{max}(X_0) \sim \ln \frac{E_0}{\epsilon} + t_0, \tag{1.5}$$

where t_{max} is measured in radiation lengths, E_0 is the incident particle energy and t_0=-0.5 (+0.5) for electrons (photons). This formula demonstrates the anticipated logarithmic dependence of the shower length, and therefore of the detector thickness needed to absorb a shower, on the incident particle energy. Shower profiles for different incident particle energies are shown in Fig. 2 (left plot). The detector thickness needed to contain 95% of the shower energy is given by

$$t_{95\%}(X_0) \sim t_{max} + 0.08Z + 9.6, \tag{1.6}$$

which indicates that calorimeter thicknesses of $\sim 25\ X_0$ are sufficient to contain showers in the energy range up to a few hundred GeV. Therefore even at the LHC energies (~TeV) electromagnetic calorimeters are very compact devices: the ATLAS lead-liquid argon electromagnetic calorimeter [2] and the CMS crystal electromagnetic calorimeter [3] have active thicknesses of ~ 50 cm and ~ 23 cm respectively.

The lateral size of an electromagnetic shower is mainly due to the fact that particles undergo multiple Coulomb scattering in the absorbing material and therefore can travel away from the shower axis. Since the deflection is larger the smaller the particle energy, the shower becomes broader towards the end of its development. A measurement of this transverse size, integrated over the full shower depth, is given by the Molière radius (R_M)

$$R_M(\mathrm{g/cm^2}) \sim 21\ \mathrm{MeV} \frac{X_0}{\epsilon(\mathrm{MeV})}, \tag{1.7}$$

which is inversely proportional to the material density and describes the average lateral deflection of electrons at the critical energy after traversing one X_0. Since in most calorimeters R_M is of order 1 cm, and since about 90% of the shower energy is contained in a cylinder of radius R_M, electromagnetic showers are quite narrow (see the right plot in Fig. 2). It can also be seen that the shower lateral size is essentially energy independent. Obviously the cell size of a segmented calorimeter must be comparable to or smaller than one R_M if the calorimeter is to be used for precision measurements of the shower position.

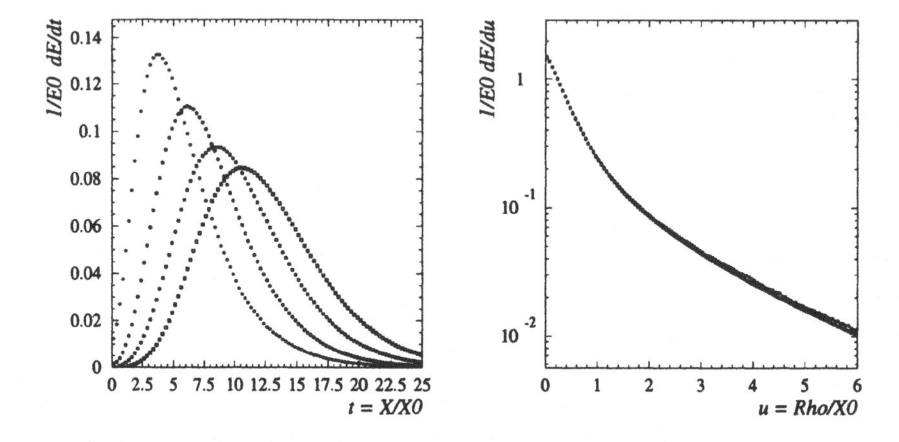

Figure 2 Left: simulated shower longitudinal profiles in $PbWO_4$, as a function of the material thickness in radiation lengths, for incident electrons of energy (from left to right) 1 GeV, 10 GeV, 100 GeV, 1 TeV. Right: simulated lateral shower profiles, as a function of the distance from the shower axis in Molière radii, for 1 GeV (closed circles) and 1 TeV (open circles) electrons. From Ref. [4].

3. ENERGY RESOLUTION OF ELECTROMAGNETIC CALORIMETERS

The energy measurement with a calorimeter is based on the fact that the energy released in the detector material, mainly through ionisation, by the charged particles in the shower is proportional to the energy of the incident particle (E_0).

The total track length of the shower (T_0), defined as the sum of all ionisation tracks due to all charged particles in the cascade, can be approximated by

$$T_0 \sim X_0 \frac{E_0}{\epsilon}, \tag{1.8}$$

since X_0 is related to the mean free path of a charged particle and E_0/ϵ to the number of particles in the shower. The above formula shows that a measurement of the signal produced by the charged tracks in an electromagnetic shower provides a measurement of the original particle energy E_0. This measurement can be performed for instance by detecting the light produced in a scintillating material, or by collecting the electric signal produced in a gas or in a liquid.

The intrinsic energy resolution of an ideal calorimeter, that is a calorimeter with infinite size and no losses due to instrumental effects (e.g. inefficiencies in the signal collection, cracks), is due to fluctuations of the track length T_0. Since T_0 is proportional to the number of track segments in the shower, and the process is random in nature, the intrinsic energy resolution is given, from purely statistical arguments, by

$$\sigma(E) \sim \sqrt{T_0}, \tag{1.9}$$

from which the well known behaviour of the fractional energy resolution

$$\frac{\sigma(E)}{E} \sim \frac{1}{\sqrt{T_0}} \sim \frac{1}{\sqrt{E_0}}, \tag{1.10}$$

can be derived. As discussed in more detail in Section 3.1, in homogeneous calorimeters the intrinsic fluctuations can be smaller than those given by this relation.

The actual energy resolution of a realistic calorimeter is worsened by other contributions and can be written in a general way as

$$\frac{\sigma}{E} = \frac{a}{\sqrt{E}} \oplus \frac{b}{E} \oplus c \tag{1.11}$$

where the first term on the right-hand side is called "stochastic term" (or "sampling term"), and includes the intrinsic fluctuations mentioned above, the second term is the "noise term", and the third term is the "constant term". Owing to their different energy dependence, the relative importance of the various terms varies with the energy of the incident particle. Therefore the choice of the calorimeter technique and the detector design optimisation can be very different for experiments operating in different energy ranges, since the energy resolution is dominated by different contributions. These terms are discussed one by one below.

3.1. STOCHASTIC TERM

This term is due to the fluctuations related to the physical development of the shower. One contribution to this term are the intrinsic fluctuations given in Eq. 1.10.

In homogeneous calorimeters intrinsic fluctuations are very small because the energy deposited in the active volume of the detector by an incident monochromatic beam of particles does not fluctuate event by event. Therefore the energy resolution can be better than the statistical expectation given in Eq. 1.10 by a factor, called the "Fano factor", which is discussed in Section 7.1. Typical stochastic terms of homogeneous electromagnetic calorimeters are at the level of a few percent in units of $1/\sqrt{E(\mathrm{GeV})}$.

On the other hand, in sampling calorimeters the energy deposited in the active medium fluctuates event by event, because the active layers are interleaved with the absorber layers. These fluctuations, which are called "sampling fluctuations" and are the most important limitation to the energy resolution of these detectors, are due to variations in the number of charged particles N_{ch} which cross the active layers. This number is given by

$$N_{ch} \sim \frac{E_0}{t}, \tag{1.12}$$

where t is the thickness of the absorber layers in radiation lengths. Assuming statistically independent crossings of the active layers, which is satisfied if the absorber layers are not too thin, a "sampling" contribution to the energy resolution comes from the fluctuations on N_{ch}, that is [5]

$$\frac{\sigma}{E} \sim \frac{1}{\sqrt{N_{ch}}} \sim \sqrt{\frac{t}{E_0(\mathrm{GeV})}}. \tag{1.13}$$

For a given calorimeter thickness, the smaller is t the larger is the number of times the shower is sampled by the active layers (i.e. the sampling frequency), the larger is the number of detected particles, and the better is the energy resolution. Therefore, in principle the energy resolution of a sampling calorimeter can be improved by reducing the thickness of the absorber layers. However, in order to achieve resolutions comparable to those typical of homogeneous calorimeters, absorber thicknesses of a few percent of a radiation length are needed, which is not feasible in practice.

In sampling calorimeters using gas as the active medium two additional contributions (Landau fluctuations and path length fluctuations) limit the energy resolution. A discussion of these effects can be found for instance in Ref. [6].

The typical energy resolution of sampling electromagnetic calorimeters is in the range 6-20%$/\sqrt{E(\mathrm{GeV})}$.

Another important parameter of sampling calorimeters is the sampling fraction (F_{mip}), defined as the fraction of energy deposited by a minimum ionising particle in the active part of the detector:

$$F_{mip} = \frac{E_{mip}(vis)}{(E_{mip}(vis) + E_{mip}(inv.))} \tag{1.14}$$

where $E_{mip}(vis)$ and $E_{mip}(inv.)$ are the energies deposited by the minimum ionizing particle in the active and passive layers of the calorimeter respectively. The equivalent quantity for electrons (F_e) is usually smaller than F_{mip} because electromagnetic showers contain a large number of low-energy photons at the end of the cascade development which tend to be absorbed by the passive material by the photoelectric effect (whose cross-section goes like $\sigma \sim Z^5$).

3.2. NOISE TERM

This contribution to the energy resolution comes from the electronic noise of the readout chain and depends on the detector technique and on the features of the readout circuit (detector capacitance, cables, etc.). Calorimeters where the signal is collected in the form of light, such as scintillator-based sampling or homogeneous calorimeters, are characterised by small levels of noise, because the first step of the electronic chain, a photosensitive device like a phototube, provides a high-gain multiplication of the original signal with essentially no noise. On the other hand, the noise is larger in detectors where the signal is collected in the form of charge because the first element of the readout chain is a preamplifier. Techniques like signal shaping and optimal filtering can be used to minimise the electronic noise in these detectors.

The noise contribution to the energy resolution increases with decreasing energy of the incident particles (see Eq. 1.11) and at energies of a few GeV becomes dominant. Therefore, the equivalent noise energy is usually required to be much smaller than 100 MeV per channel.

In sampling calorimeters the noise term can be decreased by increasing the sampling fraction, because the larger the sampling fraction the larger is the signal from the active medium and therefore the larger is the signal-to-noise ratio.

3.3. CONSTANT TERM

This term includes all contributions which do not depend on the particle energy. Any instrumental effect which produces a variation of the calorimeter response as a function of the shower position in the detector gives rise to response non-uniformities. The latter contribute an additional smearing to the energy reconstructed over large calorimeter areas, which shows up in a constant term. Non-uniformities can originate from the detector intrinsic geometry (for instance if the absorber and active layers have non-regular shapes), from imperfections in the detector mechanical structure and readout system, from temperature gradients, from the detector ageing, from radiation damages, etc. These non-uniformities can be cured (to a large extent) if they exhibit a periodical pattern, as it is the case if they are related to the detector geometry, or if they originate from the readout chain (this is the task of the calibration system discussed in Section 8). On the other hand other effects, such as mechanical imperfections, are randomly distributed and therefore more difficult to correct.

With the increasing energy of present and future accelerators, the constant term becomes more and more the dominant contribution to the energy resolution of electromagnetic calorimeters. Tight construction tolerances are therefore imposed for the mechanics and readout systems of modern calorimeters, for instance LHC calorimeters.

Typically the constant term of an electromagnetic calorimeter should be kept at the level of a percent or smaller. This is particularly true for homogeneous calorimeters, where the stochastic term is small.

3.4. ADDITIONAL CONTRIBUTIONS

Additional contributions to the energy resolution come from the constraints to which a calorimeter is subject when integrated into an experiment (see Section 9). Examples are:

- Longitudinal leakage. Space and cost constraints limit the thickness of a calorimeter operating in a high energy physics experiment. Therefore energetic showers can lose part of their energy beyond the end of the active calorimeter volume. These leakage losses fluctuate event by event, thus deteriorating the energy resolution. Although general parametrisations are difficult to derive in this case, often the stochastic term of crystal calorimeters is described as $1/E^{1/4}$ (instead of $1/E^{1/2}$) to take into account the additional smearing caused by leakage. The leakage can be partially compensated for by weighting the energy deposited by the showers in the last compartment of a longitudinally segmented calorimeter.

- Upstream energy losses. Inside an experiment calorimeters are usually preceded by other detectors, such as tracking devices. In addition, they are supported by mechanical structures and equipped with cables and various types of connections which can bring inactive material in front of the active volume. Although experiments are carefully designed so as to minimise the thickness of dead layers in front of the calorimeter, usually electrons and photons coming from the interaction region have to traverse a non-negligible amount of material before reaching the active part of the calorimeter. The energy lost in this material fluctuates event by event, and these fluctuations deteriorate the energy resolution. Possible techniques to recover part of these losses (e.g. the use of dedicated devices like presamplers and massless gaps) are discussed in Section 9.

- Non-hermetic coverage. Cracks and dead regions are often present inside the calorimeter volume, because big detectors are usually built of mechanically independent modules and divided into barrel and forward parts. The quality of the energy measurement is degraded for showers developing in these inactive areas. This leads to the deterioration of the energy resolution and to the formation of low-energy tails in the reconstructed energy spectra of electrons and photons.

Finally, other sources can potentially degrade the energy resolution of a calorimeter, for instance the presence of coherent noise or cross-talk between the electronic channels. These will not be discussed here.

Figure 3 shows the energy resolution measured with a prototype of the NA48 liquid krypton electromagnetic calorimeter (see Section 7.1.4). A fit to the experimental points with the form given in Eq. 1.11 gives a small stochastic term, as expected for an homogeneous calorimeter.

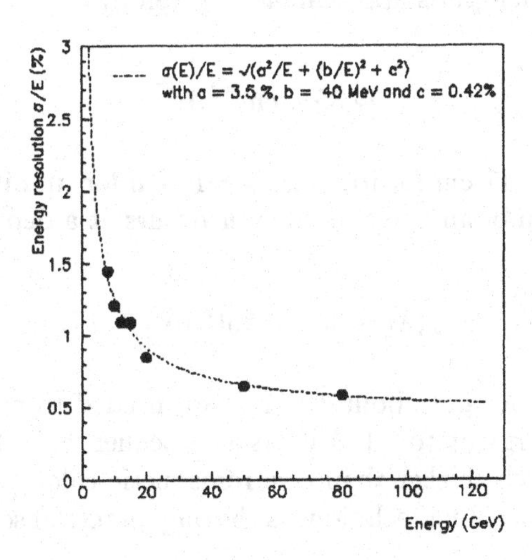

Figure 3 Fractional electron energy resolution as a function of energy measured with a prototype of the NA48 liquid krypton homogeneous electromagnetic calorimeter [7]. The full line is a fit to the experimental points with the form and the parameters given on the top of the plot.

4. PHYSICS OF HADRONIC SHOWERS

Hadrons incident on a block of material produce a cascade of secondary particles (neutral and charged pions, neutrons, protons, etc.) through electromagnetic and strong interactions. Hadronic showers are more complex than electromagnetic showers because of the nature of strong interactions and because they are made of two components. The electromagnetic component, mainly due to π^0's which decay into photon pairs, carries ~30% of the shower energy at 10 GeV and increases with increasing energy of the incident particle. This component develops fast and over a short range in space. The hadronic component, on the other hand, which is produced by a large variety of complex processes (fission, nuclear excitation, etc.), is characterised by a slower and longer-range development.

Owing to the poorer knowledge of strong nuclear interactions, the phenomenology of hadronic showers is less well understood than that of electromagnetic showers, as indicated also by the proliferation of Monte Carlo simulation packages with different physics content and often rather different predictions.

The development of an hadronic shower can be parametrised in terms of the interaction length λ, which is defined as the mean free path between two inelastic nuclear collisions, and which plays a similar rôle for hadronic showers as the radiation length for electromagnetic showers. The interaction length is approximately given by

$$\lambda \sim 35 \text{ g cm}^{-2} A^{1/3}. \tag{1.15}$$

For example, λ =17 cm for iron and λ =10 cm for uranium.

The maximum of an hadronic shower occurs at a depth

$$t_{max}(\lambda) \sim 0.2 \ \ln E_0(\text{GeV}) + 0.7, \tag{1.16}$$

that is after 1-2 λ, and about 10-11 λ are needed to contain hadronic showers with energy up to ~1 TeV. As a consequence, hadronic calorimeters must be much thicker than electromagnetic calorimeters (typically 1-2 m) and the sampling technique is the only practical solution for these detectors.

Concerning the lateral shower size, about 95% of the shower is contained in a cylinder of radius 1 λ. Therefore, in contrast to electromagnetic calorimeter, a fine lateral granularity is not needed in hadronic calorimeters and cell sizes of several cm are fully adequate.

5. ENERGY RESOLUTION OF HADRONIC CALORIMETERS

In addition to the sources already mentioned for electromagnetic showers (sampling fluctuations, constant term and noise), there are other (dominant) contributions to the energy resolution for hadronic showers. These are discussed one by one below.

5.1. MUONS AND NEUTRINOS

They are produced for instance from pion decays inside the shower and their energy can not be easily measured by the calorimeter. If a magnetic spectrometer is available in the experiment, the muon momentum can be determined by the spectrometer and added to the shower energy measured by the calorimeter. These escaping particles account for $\sim 1\%$ (on average) of the incident hadron energy at 40 GeV.

5.2. STRONG INTERACTIONS

The large variety of strong interactions occurring in an hadronic shower gives rise to several products. Ionising particles, such as protons, α-particles and nuclear fragments, are detected. On the other hand neutrons are often invisible, because they can travel for a long time ($\sim \mu$s) and a long distance before being moderated and captured, and therefore can escape out of the time window and space region used to measure the shower. Finally, part of the initial particle energy (a few percent) goes into nuclear break-up and is therefore invisible. As a consequence, only a fraction of the incident energy is detected, with the undetected energy amounting up to 40% of the hadronic component of the shower and being subject to large fluctuations. Therefore, even if one ignored the presence of two components (electromagnetic and hadronic) and non-compensation effects (see below), the energy resolution for hadronic showers could not be better than $\sim 30\%/\sqrt{E(\mathrm{GeV})}$. It is possible to minimise these energy losses by, for instance, choosing an active medium rich in protons. In this case a large part of the otherwise poorly detected neutrons transfer their energy through neutron-proton scattering to protons (which are detected through ionisation). As an example, in plastic scintillators, which contain hydrogen, almost 100% of the neutrons transfer their energy to protons, whereas in liquid argon such a fraction is only 10%.

5.3. SATURATION EFFECTS

Nuclear fragments produce a large amount of ionisation along their path and can therefore saturate the response of the active medium through molecule break-up (e.g. in scintillators), recombination (e.g. in liquid argon), etc.. For instance, the light L emitted per unit path by an ionising particle in a scintillator is given by

$$\frac{dL}{dx} = S\frac{dE}{dx}, \tag{1.17}$$

where S is the scintillation efficiency. The density of damaged molecules is proportional to the particle energy loss dE/dx. Assuming that a fraction k of these damaged molecules give no signal (quenching), the previous formula becomes

$$\frac{dL}{dx} = \frac{SdE/dx}{1 + kBdE/dx}, \tag{1.18}$$

where B is a proportionality factor. The above equation, called Birk's law, describes the response suppression of a medium as a function of the particle energy loss. Therefore, the presence of highly-ionising nuclear fragments among the debris of a hadronic shower produces additional energy losses.

5.4. NON COMPENSATION

The response to an incident hadron (R_h) can be written as:

$$R_h = \epsilon_e E_e + \epsilon_h E_h, \tag{1.19}$$

where E_e is the electromagnetic energy of the shower (i.e. the part of the shower energy which is carried by electrons, positrons and photons), E_h is the hadronic energy of the shower (i.e. the part of the shower energy which is not carried by electrons, positrons and photons), and ϵ_e and ϵ_h are the fractions of the electromagnetic and hadronic energies (respectively) which are detected. For the reasons mentioned above, the response of a calorimeter to the electromagnetic component of a shower is larger than the response to the hadronic component, therefore in general $\epsilon_e > \epsilon_h$. Hence the e/h ratio, defined as

$$\frac{e}{h} = \frac{\epsilon_e}{\epsilon_h}, \tag{1.20}$$

is larger than unity in most calorimeters. Such calorimeters are called "non-compensated", meaning that the poor response to the hadronic part of the shower is not compensated for. On the other hand a calorimeter with $e/h \sim 1$ is called "compensated".

Since the fractions of electromagnetic and hadronic energies in an hadronic shower fluctuate event by event, and the response to these two components is not the same in a calorimeter with $e/h \neq 1$, the energy resolution receives additional contributions from these (non-gaussian) fluctuations. And the further distant is e/h from one, the worse the energy resolution. Therefore, calorimeters aiming at good energy resolutions for hadronic showers should be built to have e/h as close as possible to one (see Section 5.5).

In addition, the electromagnetic fraction of a shower (mainly due to π^0) increases with energy as [8]

$$F(\pi^0) \sim 0.11 \ \ln E_0, \tag{1.21}$$

and is typically 30% (60%) in hadronic showers initiated by $\pi^{\pm}$ of 10 GeV (150 GeV). This is due to the fact that charged pions in the shower can produce $\pi^0 \to \gamma\gamma$ whereas the opposite is not true. Therefore, if $e/h > 1$ the calorimeter response is non-linear, because the ratio between the measured energy and the incident energy increases with energy.

It should be noted that the e/h ratio is an intrinsic property of a calorimeter, since it is essentially energy independent. It can not be measured directly because the calorimeter response to a purely hadronic shower can not be evaluated (all realistic showers have an electromagnetic component). Another frequently used parameter is the e/π ratio, defined as the ratio of the average calorimeter responses to an electron and to a charged pion of the same incident energy. In contrast to e/h, the e/π ratio is a directly measurable quantity and depends on the incident particle energy, since the electromagnetic component of a pion shower increases with energy. As a consequence, the e/π ratio approaches unity at very high energies. An example is shown in Fig. 4.

The e/h ratio of a calorimeter can be obtained from a measurement of e/π by using the following relation

$$\frac{e}{\pi} = \frac{e/h}{1 + (e/h - 1)F(\pi^0)}. \tag{1.22}$$

For instance, by fitting the experimental points in Fig. 4 with Eq. 1.22 one obtains $e/h = 1.362 \pm 0.006$.

Finally, one can also define in a similar way a e/jet ratio. Since jets contain a large electromagnetic component, in general the following inequalities hold:

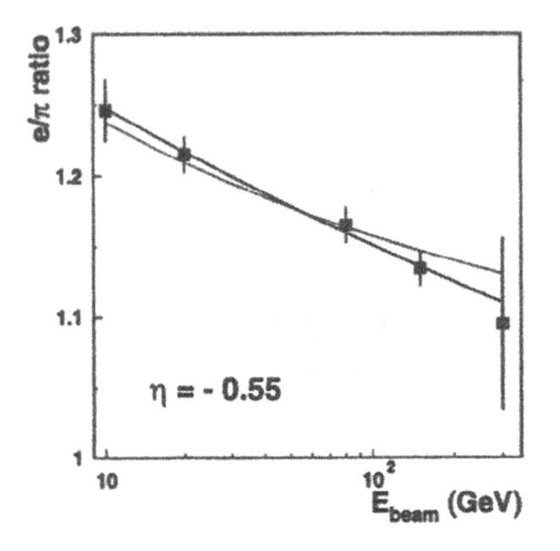

Figure 4 The e/π ratio as a function of the incident particle energy, as measured with a module of the ATLAS Fe-scintillator Tilecal calorimeter [9]. The full line is a fit with Eq. 1.22.

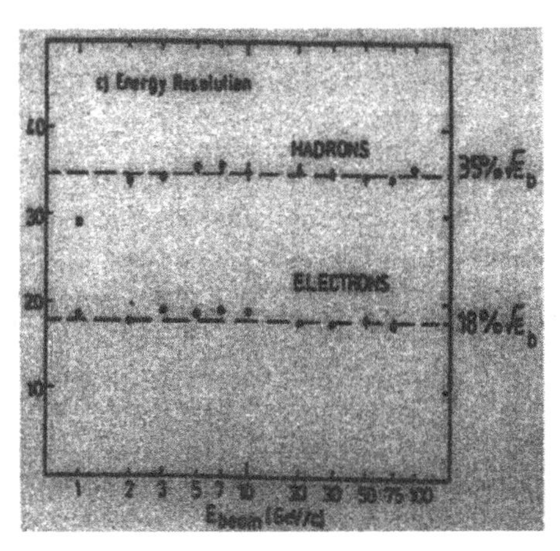

Figure 5 Sampling term of the energy resolution (in percent) as a function of energy as measured with test beam electrons and hadrons incident on a prototype of the ZEUS uranium-scintillator forward calorimeter.

$$\frac{e}{h} > \frac{e}{\pi} > \frac{e}{jet} > 1. \tag{1.23}$$

As a consequence, a calorimeter with $e/h \sim 1.5$ has a bad energy resolution for (hypothetical) "purely hadronic" showers and a good energy resolution for jets.

5.5. COMPENSATION TECHNIQUES

To build a compensated calorimeter, or to increase the degree of compensation of an uncompensated calorimeter, one has to obtain an e/h ratio as close to unity as possible. This could be achieved by either increasing the response to the hadronic component of the shower, or by suppressing the response to the electromagnetic component, or by acting on both.

To increase the hadronic response one can choose an active medium rich in hydrogen, as already mentioned, because in this way part of the otherwise poorly detected energy carried by neutrons is transferred to protons. One of the best absorbers for compensation is ^{238}U (used for instance in the D0 [10] and ZEUS [11] calorimeters), which is fissionable by neutrons of a few MeV and therefore yields many neutrons through chain reactions. If ^{238}U is coupled to an hydrogenated active medium, this large number of neutrons produces a large number of (detectable) protons and the e/h ratio can be even brought below one. Increasing the shaping time of the calorimeter response also favours compensation, since this allows the late signals coming from slow neutrons to be collected. This method is however not suited to detectors operating at high-rate machines like the LHC.

The electromagnetic response can be suppressed by using a thick high-Z absorber, which prevents low-energy photons and electrons from reaching the active layers. In addition, if a calorimeter has a good lateral and longitudinal segmentation, it is possible to give a small weight, at the level of the off-line analysis, to the energy deposited in the cells close to the shower core in the first longitudinal compartments. These cells contain most of the electromagnetic part of the shower.

Figure 5 shows the energy resolution measured with a prototype of the Uranium-scintillator forward calorimeter of the ZEUS experiment. Thanks to the high degree of compensation ($e/h \sim 1$) the hadron energy resolution is excellent ($\sim 35\%/\sqrt{E(\mathrm{GeV})}$). On the other hand, the electron energy resolution is relatively modest ($\sim 18\%/\sqrt{E(\mathrm{GeV})}$). This is due to the fact that the calorimeter is made of 3.3 mm thick U plates interleaved with 2.6 mm scintillator plates. These thicknesses were optimised to achieve compensation, at the prize of a good sampling frequency.

Hadronic calorimeters are only built with the sampling technique, one of the reasons being that the choice of two materials offers more flexibility and allows absorber and active medium to be coupled in an optimal way (e.g. Uranium and scintillator).

6. CALORIMETER PERFORMANCE REQUIREMENTS

Operation in modern experiments sets a large number of stringent requirements, both from the technical (e.g. fast response, low noise, dynamic range) and physics performance (e.g. good energy, space, time resolutions, particle identification) points of view. Constraints come also from the integration in the rest of the experiment (e.g. the presence of a magnetic field) and from the environment conditions (e.g. radiation levels). Therefore the choice of the optimal technique and of the best calorimeter layout is a multi-dimensional problem which takes into account all the above issues as well as cost. The requirements, and therefore the detector choice, can obviously be very different for the different applications. Since the LHC environment will be unprecedently demanding in terms of physics and technical performance, the main requirements of the ATLAS and CMS calorimeters [3, 12] are listed below as examples.

- Fast response. At the LHC design luminosity of 10^{34} cm^{-2}s^{-1}, at each bunch crossing, i.e. every 25 ns, on average 25 events are expected to be produced. These are mostly soft interactions (called "minimum-bias events"), characterised by low-p_T particles in the final state. When, occasionally, an interesting high-p_T physics event occurs, for instance the production of a W boson or a Higgs boson, this event will be overlapped with (on average) 25 minimum-bias events produced in the same bunch crossing and, if the detector response is not fast enough, possibly also with other soft interactions from the preceding and following bunch crossings. This overlap of events is called "pile-up". At each crossing, the pile-up of $\sim$ 25 minimum-bias events produces about 2000 particles in the region $|\eta| < 2.5$ with average transverse momentum $\sim$ 500 MeV, giving a total average transverse energy of $\sim$ 1 TeV. Although this average transverse energy can be subtracted, the event-by-event fluctuations can not and produce a smearing of the calorimeter response (called pile-up noise) which contributes an additional term of the form $\sim$ (pile-up r.m.s.)/E to the energy resolution. To reduce the magnitude of this contribution, i.e. the pile-up r.m.s., a fast calorimeter response, at the level of 50 ns or faster, is needed, in order to integrate over a minimum number of bunch crossings. This requires also high performance readout electronics. With this response time, the typical pile-up r.m.s. over a region containing an electromagnetic shower is $E_T \sim 250$ MeV, which gives a contribution of 2.5% to the energy resolution of p_T=10 GeV electrons. In addition, a fine calorimeter granularity is also important to minimise the probability that pile-up particles hit the same cell as an interesting object (e.g. an electron from a possible $H \rightarrow 4e$ decay).

- Radiation hardness. Because of the huge flux of particles produced by the high-rate pp collisions, the LHC experimental environment will be characterised by high levels of radiation. Over ten years of operation, the calorimeter regions at $|\eta| \sim 5$, where the radiation is the largest because of the high energy density of the particles hitting the forward parts of the detector, will be exposed to a flux of up to $\sim 10^{17}$ neutrons/cm^2 and to a dose of up to 10^7 Gy (1 Gy = 1 Joule/Kg is a unit of absorbed energy). In addition to radiation hard detectors and electronics, this requires quality control and radiation tests of every single piece of material which will be installed in the detector.

- Angular coverage. LHC calorimeters must be as hermetic as possible, and cover the full azimuthal angle and the rapidity region $|\eta| < 5$ (i.e. down to 1° from the beam axis). This is required mainly for a reliable measurement of the event transverse energy, which is in turn needed to detect neutrinos (or other hypothetical weakly interacting neutral particles). These particles can only be detected indirectly by observing a significant amount of missing transverse energy in the final state.

  At e^+e^- colliders the presence of a neutrino in an event and the neutrino energy are simply deduced by comparing the total energy measured in the final state with the centre-of-mass energy of the colliding beams and by applying energy conservation criteria. At hadron colliders, on the other hand, the energy of the interacting quarks and gluons is not known (only the proton energy is known), and therefore the missing energy in the final state cannot be determined. However, since the transverse momentum (perpendicular to the beam axis) of the incident quarks and gluons is negligible, the total transverse momentum in the initial state is close to zero, and so has to be the total transverse energy in the final state. This constraint is used to detect the presence of neutrinos or other weakly interacting neutral particles. In order to minimise energy losses in poorly-instrumented regions of the apparatus or along the beam axis, which could fake a neutrino signal, the calorimeters have to cover as much of the solid angle as possible. For example, the simulation of events containing a supersymmetric Higgs with $m_H = 150$ GeV decaying into τ-pairs in an ideal calorimeter with infinite resolution showed that the resolution on the missing transverse energy would be ~2 GeV if the calorimeter coverage extends over $|\eta| < 5$ and ~8 GeV if the coverage extends over $|\eta| < 3$. This deterioration is obviously due to particles escaping detection because they are produced at small angles from the beam line.

- Excellent electromagnetic energy resolution. This is needed for instance to extract a possible $H \to \gamma\gamma$ signal from the irreducible background of $\gamma\gamma$ events produced by known processes. Since the irreducible background cross-section is a factor of ~40 larger than the signal cross-section for a Higgs mass of ~120 GeV, a very good mass resolution, at the level of 1%, is needed in order to observe a narrow resonance above the large $\gamma\gamma$ continuum background. A mass resolution of 1% requires in particular a constant term of the energy resolution smaller than 1%.

- Angle measurements. The energy resolution is not the only contribution to the width of the reconstructed $\gamma\gamma$ invariant mass distribution from H$\rightarrow\gamma\gamma$ decays. To reconstruct the two-photon invariant mass it is necessary to know the directions of both photons and therefore the position of the primary vertex. At the LHC the vertex position will be well localised in the transverse plane (spread $\sim 15\ \mu$m) but not along the beam axis (spread $\sim$5.6 cm). If the electromagnetic calorimeter has longitudinal segmentation as in ATLAS, or sits behind a preshower detector as in the CMS end-cap region, then the photon direction in θ (which is the angle with the beam line) can be measured with the calorimeter or from a combination of the calorimeter and preshower informations. The angular resolution required to achieve a $\gamma\gamma$ mass resolution of 1% is $\sigma_\theta \sim 50$ mrad$/\sqrt{E(\text{GeV})}$.

  It should be noted that in the initial phase at low luminosity (10^{33} cm^{-2}s^{-1}) the charged tracks belonging to the underlying event associated with the H$\rightarrow\gamma\gamma$ production should allow a determination of the vertex position with high accuracy and good efficiency. This is the method which will be used by the CMS experiment if both photons are in the barrel region. However, at 10^{34} cm^{-2}s^{-1}, due to the pile-up of many tracks from the same bunch crossing, often the vertex reconstruction will be wrong, leading to tails in the $\gamma\gamma$ mass distribution.

- Large dynamic range. Electrons should be measured over an unprecedented energy range going from a few GeV up to $\sim$3 TeV. This is needed on the one hand to detect the soft electrons produced in the decays of B-hadrons, and on the other hand to look for heavy particles decaying into electrons (e.g. additional gauge bosons W' and Z') up to masses of $\sim$ 6 TeV.

  The readout system of the electromagnetic calorimeter should be sensitive to signals as low as $\sim$ 50 MeV (which is the typical electronic noise per channel) and as high as to 3 TeV (which is the maximum energy deposited per calorimeter cell by electrons produced in the decays of Z' and W' with masses $\sim$ 6 TeV). This corresponds to a dynamic range of 16 bits, which is realised in practice by using multi-gain electronic chains. A smaller dynamic range would increase the electronic noise due to a significant contribution of the quantization noise.

- Jet energy resolution and linearity. The jet energy resolution should be at the level of $\sim 50\%/\sqrt{E(\mathrm{GeV})} \oplus 3\%$ at the LHC for various physics goals [13]. These include a precise measurement of the top mass, the search for the Higgs boson in the ttH channel with $H \to b\bar{b}$, which requires a good $b\bar{b}$ mass resolution to observe a resonant peak over the large background, searches for Supersymmetric particles and for new heavy resonances (e.g. a Z') decaying into two jets. In particular in the last two cases a small constant term is important since this is the dominant contribution to the energy resolution at high energy.

  The linearity of the reconstructed jet energy should be better than 2% up to ~4 TeV, which sets constraints on the level of non-compensation of the calorimeter response to hadrons. This requirement comes from the fact that an uncorrected calorimeter non-linearity could produce an enhancement of the (steeply falling) QCD jet cross-section at high energy, which is the signature expected from quark compositeness. For instance, an uncorrected non-linearity of 5% (2%) could fake a compositeness scale $\Lambda \sim$20 TeV ($\Lambda \sim$30 TeV). The calibration of the jet energy scale is discussed in Section 8.

- Particle identification. An efficient rejection of jets faking electrons and photons is needed for several physics studies at the LHC (e.g. Higgs searches). Usually a jet consists of many particles, and therefore it can be easily recognised from a single electron or photon for instance from the broader shower size in the calorimeters or from the presence of several tracks in the inner detector. There are however cases in which a jet of particles can fake for instance a single photon. This happens when a quark fragments into a very hard π^0 plus a few other very soft particles. The two photons from the subsequent π^0 decay are very close in space because the parent π^0 is usually produced with a large boost. For instance, for a π^0 of $p_T \sim$ 50 GeV the distance between the two decay photons is smaller than 1 cm at 150 cm from the interaction point, which is approximately the radius at which electromagnetic calorimeters are located. Therefore the two photons appear as a single photon in the electromagnetic calorimeter, unless the latter has a fine enough granularity to be able to detect two distinct showers. Although the probability for a jet to fragment into a single isolated π^0 is small, the cross-section for di-jet production is for instance $\sim 10^8$ times larger than the $H \to \gamma\gamma$ cross-section, which makes such a background potentially dangerous. This sets requirements on the calorimeter granularity, which in particular should be adequate to provide γ/π^0 separation capabilities.

7. MAIN CALORIMETER TECHNIQUES

In this Section, the main techniques used to build homogeneous and sampling calorimeters are briefly summarised, and advantages and drawbacks of the various solutions are discussed. Examples of calorimeters operating in present high energy physics experiments, or in construction for future machines (e.g. LHC), are given.

7.1. HOMOGENEOUS CALORIMETERS

As already mentioned, the main advantage of these detectors is their excellent energy resolution, which is due to the fact that essentially the whole energy of an incident particle is deposited in the active medium, in contrast to sampling calorimeters. On the other hand, homogeneous calorimeters are less easy to segment laterally and longitudinally, which is a drawback when position measurements and particle identification are needed, and compensation is more difficult to achieve than in sampling calorimeters. For this last reason, and because of the prohibitive thickness needed to contain hadron showers in an homogeneous detector, homogeneous calorimeters are widely used in high energy physics experiments as electromagnetic calorimeters, and much less as hadronic calorimeters. They are also employed in neutrino and astroparticle physics experiments, where large volumes are needed to detect rare events and therefore only homogeneous detectors made of inexpensive materials (like water or air) are affordable.

Homogeneous calorimeters can be broadly divided into four classes:

- Semiconductor calorimeters. In this case the ionisation tracks produce hole-electron pairs in the semiconductor valence and conduction bands, which give rise to an electric signal. These detectors give the best energy resolution of all calorimeters. Examples are Silicon and Germanium crystals, used in many nuclear physics applications.

- Cerenkov calorimeters. The medium is a material with high refraction index where relativistic $e^{\pm}$ tracks in the shower produce Cerenkov photons. The signal is therefore collected in the form of light. An example are lead-glass calorimeters.

- Scintillator calorimeters. The medium is a material where ionisation tracks are converted into light (fluorescence). Examples are BGO and CsI crystals.

- Noble liquid calorimeters. The medium is a noble gas (Ar, Kr, Xe) operated at cryogenic temperature. Both ionisation and scintillation signals can in principle be collected, however large-scale calorimeters for high energy physics applications are based so far on the charge measurement.

For detectors where the signal is collected in the form of light (Cerenkov, scintillators), photons from the active volume are converted into electrons (usually called photoelectrons) through a photosensitive device like a photomultiplier. Therefore the main contribution to the energy resolution can come from statistical fluctuations in the number of photoelectrons. This contribution goes like $\sim 1/\sqrt{N_{pe}}$, where N_{pe} is the number of photoelectrons, and is important if N_{pe} is small. The number of photoelectrons can be small if the number of photons produced in the active medium is small, as it is the case in Cerenkov calorimeters, or if there are losses in the light collection. Furthermore, the efficiency of the device converting photons into electrons (the photocathode in the case of a photomultiplier) is usually in the range 20%-70%. In addition, if the amplification of the electric signal is relatively small, the contribution of the electronic noise to the energy resolution becomes important. This is not a problem with photomultipliers, which have gains of order 10^6, but could be a problem with photodiodes, which have gains in the range 1-10. Unfortunately, when operating in high magnetic field standard photomultipliers cannot be used, since their gain and linearity are affected by the field. Maximisation of the light yield is one of the most critical issue in the design, construction and operation of homogeneous calorimeters.

Another critical aspect is the minimisation and accurate control of all possible sources of response non-unformities, which otherwise could give rise to a large constant term of the energy resolution, thereby spoiling the excellent intrinsic resolution of these detectors.

7.1.1 Semiconductor calorimeters. These calorimeters are not much used in high energy physics experiments for various reasons. They are expensive and therefore not suited to large systems. They have an excellent intrinsic resolution which is best suited to low-energy particles, whereas at high energy other effects, like leakage and response non-uniformities, can spoil the resolution. On the other hand semiconductor calorimeters are extensively used as photon detectors for nuclear physics applications, in particular for spectroscopy measurements. They are briefly mentioned here because of their excellent energy resolution, better than the intrinsic limit discussed in Section 3.

The energy W needed to create an electron-hole pair is about 3.8 eV in Si and 2.9 eV in Ge (at a temperature of 77 ^{0}K). Since the whole incident particle energy is absorbed and goes into electron-hole pair creation, an incident monochromatic photon beam of fixed energy E_0 produces an almost fixed number of electron-hole pairs given by $N_{eh} = E_0/W$. Therefore the intrinsic energy resolution is

$$\frac{\sigma}{E} \sim \frac{\sqrt{F}}{\sqrt{N}} \tag{1.24}$$

where F, the Fano factor [14], is smaller than one (e.g. $F \simeq 0.13$ in Ge). For instance, for photons of energy 1 MeV incident on a Ge crystal the number of produced electron-hole pairs is $N_{eh} \sim 3.3 \times 10^5$. An energy resolution going like $1/\sqrt{N_{eh}}$ (see 1.10) would give σ(E) ~1.7 keV, whereas one obtains σ(E) ~630 eV by including the Fano factor. The measured value [15] of σ(E) ~550 eV is in good agreement with this latter prediction. The much superior energy resolution of semiconductor detectors compared to other calorimeters is illustrated in Fig. 6, which shows the spectral lines of a Ag γ source as measured with a NaI(Tl) scintillator (one of the scintillator calorimeters with the best resolution) and with a Ge crystal.

7.1.2 Cerenkov calorimeters. Detectable Cerenkov light is produced whenever a particle traverses a medium with a speed $v > c/n$, where c/n is the speed of light in that medium and n is the refractive index of the medium. Dielectric materials with $n >> 1$ are good candidates for Cerenkov detectors.

Cerenkov detectors are usually employed for particle identification purposes, since the particle velocity, and therefore the possibility of exceeding or not the Cerenkov "threshold" c/n, depends on the particle mass for a given momentum p. However these detectors can also be used as calorimeters by collecting the light produced by relativistic $e^{\pm}$ tracks in the showers. The most common material in high energy physics applications is lead glass (PbO).

The advantage of Cerenkov calorimeters is that they are cheap (e.g. $<$ 1 USD/cc for PbO) and easy to handle. Drawbacks are the modest radiation resistance (with serious light output deterioration for doses of ~100 Gy), the poor compensation (slow nuclear fragments usually travel below the Cerenkov threshold), and the fact that these detectors are, among homogeneous calorimeters, those with the worst energy resolu-

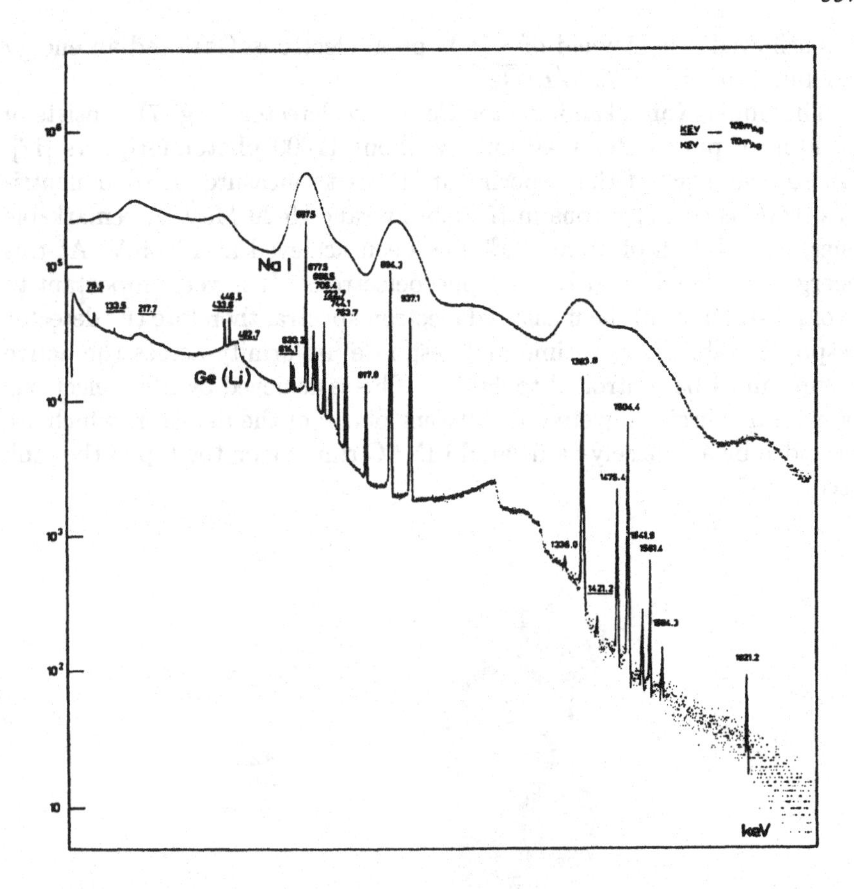

Figure 6 Spectral lines of a Ag γ source as measured with a Germanium crystal and a NaI(Tl) scintillator. The peaks are labeled in keV.

tion. The reason is that the light yield is small ($\sim 10^4$ times smaller than in scintillator), because only tracks in the shower with $v > c/n$ produce a detectable signal. In addition, the maximum photon intensity is obtained for short wavelengths (typically $\lambda <$ 300-350 nm), whereas most photocathodes are sensitive to the region 300-600 nm. For instance, in lead glass about 1000 photoelectrons are produced per deposited GeV, which alone gives an energy resolution of $\sim 3.2\%/\sqrt{E(\text{GeV})}$. Optimisation efforts have been made to maximise the light yield in large-scale detectors operating at colliders. The OPAL end-cap lead glass calorimeter [16] has achieved a light yield of ~1800 photoelectrons/GeV and an energy resolution of order $5\%/\sqrt{E(\mathrm{GeV})}$.

The Super-Kamiokande water Cerenkov detector (Fig. 7) consists of 50 kton of pure water read out by about 12000 photomultipliers [17]. One of the goals of the experiment [18] is to measure ^{8}B solar neutrinos by detecting electrons in the energy range 5-20 MeV. A remarkable energy resolution of about 20% has been achieved at 10 MeV. At this energy the signal yield is $\sim$ 60 photoelectrons. It is very important to avoid distortions in the measured electron spectra, therefore the detector response stability with time and response uniformity across the active volume must be controlled to $\pm 0.5\%$. This is achieved by using electrons of several energies injected at different places of the detector, which are provided by a precisely-calibrated LINAC running on the top of the tank (see Fig. 7).

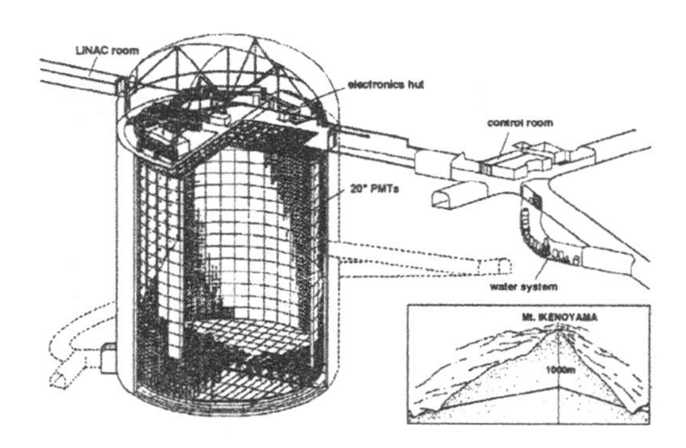

Figure 7 View of the Super-Kamiokande water Cerenkov detector.

7.1.3 Scintillation calorimeters. Scintillators can be divided into organic and inorganic. They are characterised by two different physical mechanisms for light emission, therefore advantages and drawbacks are different in the two cases: inorganic scintillators offer large light yield and good linearity but usually have a slow response, organic scintillators are fast but suffer from a poorer light yield.

In organic scintillators light emission is a molecular process. The incident charged particles bring the electrons of the medium to the highest molecular levels. The excited electrons fall back to the first excited level, usually through radiationless internal conversion or thermalisation, and then decay to the ground level. It is the transition from the first level to the ground level which produces radiation emission. Thanks to this mechanism, the emission spectrum and the absorption spectrum are well separated in wavelength, and therefore organic scintillators are usually transparent to their own emission and exhibit a long light attenuation length. To further improve this separation, often a wavelength shifting material is added, which absorbs the light emitted by the scintillator and emits it again at a longer wavelength. Usually organic scintillators are binary or ternary systems, consisting of an organic solvent (e.g. a mineral oil) with a small fraction (typically $\leq 1\%$) of a scintillating solute. The molecules of the solvent are excited by the incident charged particles and transfer the excitation to the solute (for instance through dipole-dipole interactions) which scintillates. A wavelength shifter can also be added as a third component. The process of excitation, molecular transfer and light emission is very fast, of order a few ns. However, the light output is relatively small because the solute concentration is small. The use of organic scintillators for homogeneous calorimeters is very limited, whereas they are commonly chosen as the active medium of sampling calorimeters (see Section 7.2.1).

In inorganic scintillators light emission is related to the crystal structure of the material. Ionisation tracks produce electron-hole pairs in the conduction and valence bands of the medium, and photons are emitted when electrons return to the valence band. The frequency of the emitted radiation and the response time depend on the gap between the valence and the conduction bands and on the details of the electron migration in the lattice structure. They vary a lot from material to material. In most cases, however, in order to increase the light yield (e.g. by matching the radiation wavelength to the photocathode spectral sensitivity) and in order to obtain faster emission, crystals are doped with tiny amounts of impurities. The task of these dopants, of which the most commonly used is Tallium (Tl), is to create additional activation sites in the gap between the valence and the conduction band, which in turn can be excited and can decay, and thus to increase the probability of light emission, change the light wavelength and the material decay time.

The intrinsic energy resolution is better than in Cerenkov calorimeters because the light yield is several orders of magnitude larger thanks to the lower energy needed to produce an electron-hole pair compared to the Cerenkov threshold. Nevertheless, minimisation of inefficiencies in the light collection, which could arise from reflections, photon absorption, bad matching between optical elements, is a crucial issue also in scintillator calorimeters. The energy resolution is given by Eq. 1.24, as for semiconductor detectors.

One of the drawbacks of the crystal technique is that it is not intrinsically uniform. Indeed, it is not easy to grow the thousands of crystals needed for a big calorimeter system in an identical way, nor to ensure the same light collection efficiency in all of them. This could give rise to response variations from cell to cell which, if not minimised and controlled with adequate calibration systems, could translate into a large constant term of the energy resolution.

Table 1 summarises the main properties of the most commonly used crystals for high energy physics applications.

Table 1 Main properties of crystals commonly used in homogeneous calorimeters.

	NaI(Tl)	*CsI(Tl)*	*CsI*	*BGO*	$PbWO_4$
Density (g/cm^3)	3.67	4.51	4.51	7.13	8.28
X_0 (cm)	2.59	1.85	1.85	1.12	0.89
R_M (cm)	4.8	3.8	3.5	2.3	2.2
Decay time (ns)	230	680	6	60	5
slow component			35	300	15
Emission peak (nm)	410	560	420	480	440
slow component			310		
Light yield γ/MeV	4×10^4	5×10^4	4×10^4	8×10^3	1.5×10^2
Photoelectron yield relative to NaI	1	0.45	0.056	0.09	0.013
Rad. hardness (Gy)	1	10	10^3	1	10^5

NaI(Tl) has been widely employed in the past because of its low cost and large light yield. However, it is hygroscopic and has a relatively long radiation length, therefore it is not well suited to modern experiments where denser crystals like BGO and $PbWO_4$, allowing more compact detectors, are preferred. CsI is also very popular (BaBar, Belle, KTeV). It has the second largest light yield of all crystals after NaI, a shorter radiation length, and is easier to handle. Pure CsI has a fast component (6 ns), which is well adapted to high-rate experiments and therefore is used for instance by the KTeV [19] experiment. If doped with Tallium it becomes much slower, but the light yield increases significantly. For this reason CsI(Tl) has been chosen for the BaBar experiment (see below) which does not need a fast response but needs large light yields to detect low energy signals. Finally $PbWO_4$, which is very dense, fast and radiation hard, is the best suited crystal to the LHC environment and has been adopted by the CMS experiment.

Examples of homogeneous scintillator calorimeters are discussed in more detail below.

The BaBar CsI(Tl) electromagnetic calorimeter

The choice of the electromagnetic calorimeter for the BaBar experiment [20] at the SLAC PEP-II B-factory was dictated by the goal of reconstructing low-energy (down to $\sim$ 10 MeV) photons and π^0's from B-meson decays with high efficiency, in order to be sensitive to rare decays, and very good energy and position resolutions, in order to achieve a good signal-to-background ratio for these decays.

A CsI(Tl) calorimeter offers excellent energy resolution and large light yield. The signal yield of the BaBar calorimeter is $\sim$ 7000 photoelectrons/MeV, which allows small noise levels ($\sim$ 230 keV per crystal) and high detection efficiency at very low energies ($\sim$ 95% for 20 MeV photons). The long decay time of CsI(Tl), 680 ns, is not a limitation given the relatively low interaction rate ($\sim$100 Hz).

The calorimeter consists of 6580 crystals covering the polar angle $-0.78 < \cos\theta < 0.96$ (PEP II is an asymmetric machine), as schematically shown in Fig. 8. The (tapered) crystals have a constant thickness of $\sim 17\ X_0$ for particles coming from the interaction region, a trapezoidal face of transverse size $5 \times 5\ \mathrm{cm}^2$, and no longitudinal segmentation. They do not point to the interaction centre, so that photon losses in gaps between crystals are minimised. Each channel is read out by two (for redundancy and efficiency reasons) Si photodiodes (the calorimeter is inside a magnetic field of 1.5 T), followed by preamplifiers, shapers and an ADC. A cell-to-cell calibration uniformity of $\sim$ 0.25% should be achieved by using an electronic calibration system, a Xe pulser, a radioactive source, and physics events at the collider (Bhabha, $e^+e^- \to \gamma\gamma$, etc.).

A preliminary energy resolution of $\sim 2\%/E^{1/4} \oplus 1.8\%$ has been achieved [21], with contributions from photoelectron statistics, longitudinal leakage, lateral shower fluctuations outside the cluster size of 5×5 crystals, energy losses in the upstream dead material ($< 0.5\ X_0$), light collection non-uniformities, cell-to-cell calibration spread, electronic and beam background noise.

Figure 9 shows the reconstructed two-photon invariant mass for hadronic B-meson events [21]. A nice π^0 peak is visible. The π^0 mass resolution measured in the data (6.9 MeV) is in excellent agreement with the Monte Carlo expectation (6.8 MeV).

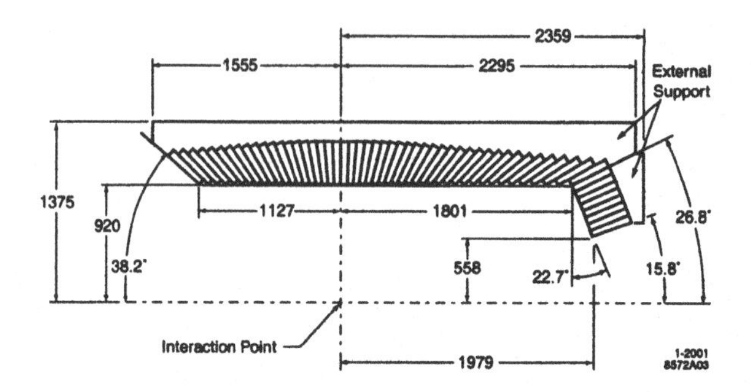

Figure 8 View of the BaBar electromagnetic crystal calorimeter. All dimensions are in mm.

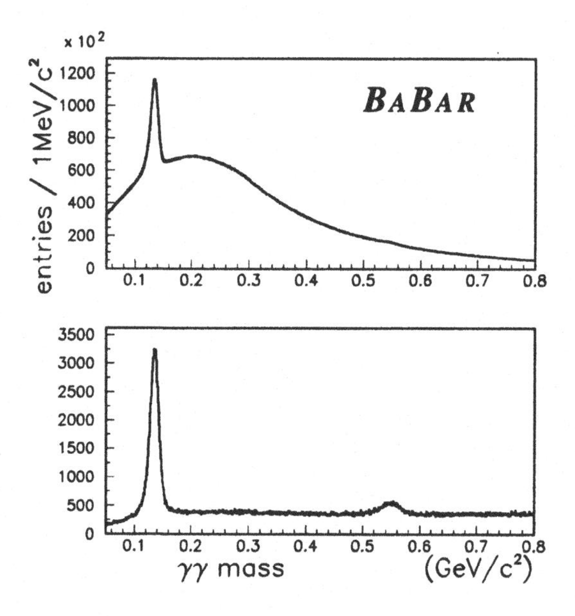

Figure 9 The invariant mass of $\gamma\gamma$ pairs from hadronic B-meson events with E_γ >30 MeV and $E_{\gamma\gamma}$ >300 MeV (top plot), and E_γ >100 MeV and $E_{\gamma\gamma}$ >1 GeV (bottom plot). Peaks due to π^0 and η production are visible.

The L3 BGO electromagnetic calorimeter

Bismuth germanate ($Bi_4Ge_3O_{12}$, or BGO) has been used for the electromagnetic calorimeter of the L3 experiment at LEP [22]. This calorimeter, which consists of about 10000 crystals of transverse size 2×2 cm^2, is the crystal detector with the largest number of channels operated so far.

The energy resolution obtained with test beam data is $1.5\%/\sqrt{E(\mathrm{GeV})} \oplus 0.4\%$, whereas the resolution measured at LEP over the full calorimeter coverage by using Bhabha electrons ($E \sim 45$ GeV) is 1.2%, thus indicating a constant term of order 1% (see Fig. 10). This larger constant

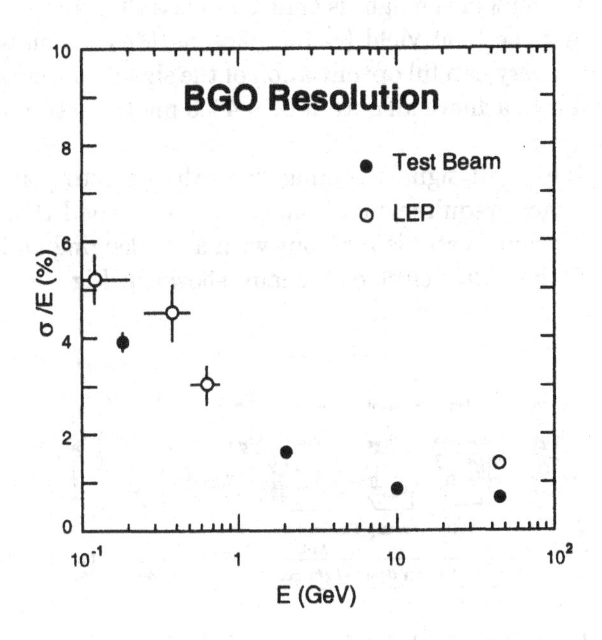

Figure 10 Energy resolution of the L3 calorimeter as a function of energy, as obtained with test beam electrons (closed circles) and with physics data (e.g. Bhabha electrons) at LEP [23].

term than expected from test beam measurements has been attributed to temperature effects, cell-to-cell intercalibration errors, electrons impinging near the crystal boundaries, etc.. It demonstrates that the control of the response uniformity in crystal calorimeters is a crucial and difficult issue, especially in big systems.

The CMS $PBWO_4$ electromagnetic calorimeter

The CMS electromagnetic calorimeter [3] consists of 83000 lead tungstate ($PBWO_4$) crystals covering the rapidity region $|\eta| < 3$. The crystals have a transverse size of $\sim 2 \times 2$ cm^2 and no longitudinal segmentation. This technique has been chosen because of the excellent energy resolution, which is important for instance in the search for a possible $H \rightarrow \gamma\gamma$ signal.

As shown in Table 1, lead tungstate exhibits some features which make it particularly suited to the LHC environment. It has a very short radiation length, which allows a calorimeter active thickness of $\sim 26\ X_0$ to be fit in a radial space as short as 23 cm, a small Molière radius, which ensures small lateral shower spread, high radiation resistance, and fast response since $\sim$ 80% of the light is emitted in less than 15 ns. The main drawback is that the light yield ($\sim$ 150 photons/MeV) is quite modest, which requires a very careful optimisation of the signal collection system. The CMS goal is to achieve an output of $\sim$ 4000 photoelectrons per GeV.

Owing to the small signals coming from the detector, and in general to the stringent requirements from operation at the LHC, the CMS electromagnetic calorimeter is read out with a challenging and sophisticated almost-fully-digital electronic chain, shown in Fig. 11. The light

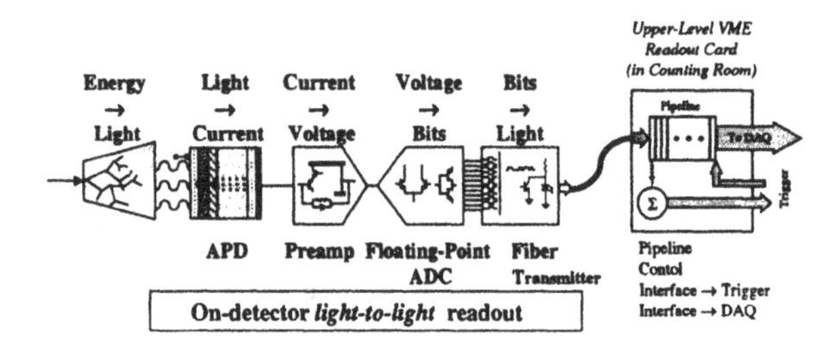

Figure 11 The readout chain of the CMS electromagnetic calorimeter.

signal from the crystal is transformed into an electric signal using (in the barrel part) Avalanche Photo Diodes (APD). These devices are p-n junctions where photoelectrons undergo avalanche multiplication. They are therefore characterised by a high gain ($\sim$50), which is needed because of the small $PbWO_4$ light yield. They are also radiation hard and able to work in the 4 T field in which the CMS calorimeter is immersed. The main drawback is their sensitivity to the temperature (-2% gain variation per degree) and to the applied bias voltage (-2% gain variation per volt), which requires temperature regulation and voltage control to better than 0.1 ^{0}C degree and 40 mV respectively. The APD is followed by a 4-gain preamplifier-shaper system (shaping time 40 ns), and by a 40 MHz 12-bit ADC. Digital optical links transform the signal back into light and transfer it from the detector to the counting room at a rate of 800 Mbit/s. In the counting room the information is stored in digital pipelines while waiting for the trigger decision (which takes $\sim$ 2.5 μs). It is interesting to notice that the part of the readout chain sitting on the detector, i.e. all components up to the ADC (see Fig. 11), plus the temperature regulation system take a radial space of ~25 cm, which is comparable to the crystal thickness.

The energy resolution obtained from the beam tests of a matrix of crystals is shown in Fig. 12. The electron data can be fit with a stochastic term of $3.3\%/\sqrt{E(\text{GeV})}$, a local constant term of 0.27% and a noise term of $0.19/E(\text{GeV})$.

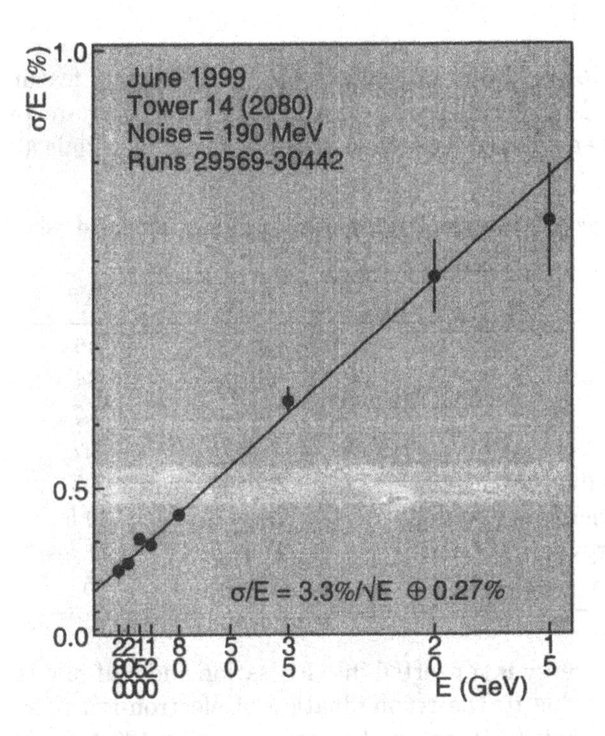

Figure 12 The fractional energy resolution as a function of energy measured with electrons incident on a matrix of pre-series crystals of the CMS calorimeter. The line is a fit with the function given in the plot.

One of the main challenges of the CMS calorimeter is to achieve an overall constant term over the full detector coverage at the level of 0.5%, as needed at the LHC (see Section 6). This requires a very good control of all possible sources of non-uniformities. Two effects are particularly important with $PbWO_4$ crystals. Radiation damage, which affects the transparency of the crystal (not the light yield), is expected to produce a response drop of ~2% over the duration of an LHC fill (~15 hours). This drop can give rise to non-uniformities because it is not constant along the crystal depth (the front part of the crystal being more damaged) and the shower longitudinal profile fluctuates. The response is expected to recover in between two fills. In addition, as it is the case for the APD's, also the crystal response has a temperature dependence (2% light drop per degree at room temperature), which requires temperature regulation and monitoring to better than 0.1 ^{0}C. These effects should be controlled mainly by means of a laser system, by calibrating each crystal with test beams before installation in the final detector, and by using control physics samples (e.g. $Z \to ee$ events) when running at the LHC (see Section 8).

7.1.4 Noble liquid calorimeters. The main features of noble liquids used for calorimetry applications (Ar, Kr, Xe) are presented in Table 2. When a charged particle traverses these materials about half of

Table 2 Main properties of liquid argon, krypton and xenon.

	Ar	*Kr*	*Xe*
Z	18	36	58
A	40	84	131
X_0 (cm)	14	4.7	2.8
R_M (cm)	7.2	4.7	4.2
Density (g/cm^3)	1.4	2.5	3.0
Ionisation energy (eV/pair)	23.3	20.5	15.6
Critical energy ϵ (MeV)	41.7	21.5	14.5
Drift velocity (mm/μs)	10	5	3

the released energy is converted into ionisation and half into scintillation. This latter is due to the recombination of electron-ion pairs and gives rise to fast signals ($\sim$10 ns) in the spectral region 120-170 nm. The best energy resolution in homogeneous liquid calorimeters would be obviously obtained by collecting both, charge and light. This is illustrated in Fig. 13, which shows the (anti)-correlation between the ionisation and scintillation signals as measured in liquid argon. However, no large-scale calorimeters based on both readout principles have been realised, due to the technical and geometrical difficulties of extracting charge and light at the same time.

Excellent energy resolution can be nevertheless achieved by collecting the ionisation signal alone, as demonstrated by the following simple calculation. Assuming that the full particle energy is absorbed by the liquid, the total signal is given by $N = N_{ion} + N_{scint}$, where N_{ion} is the number of electron-ion pairs and N_{scint} the number of photons. The

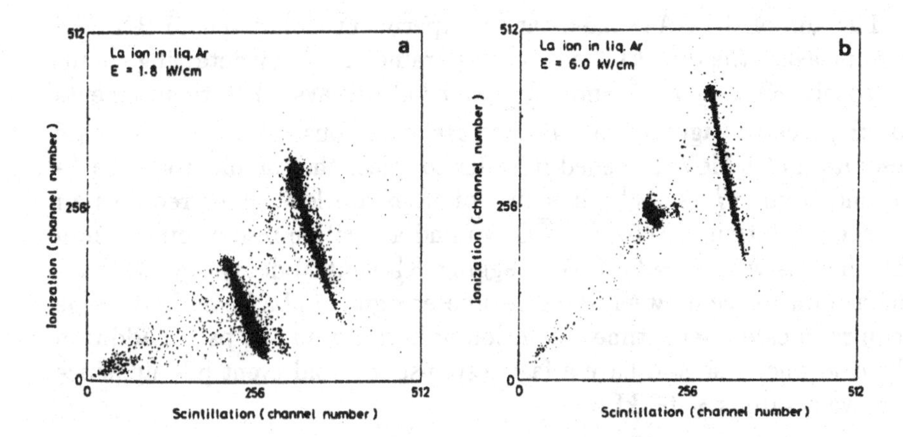

Figure 13 The anti-correlation between the ionisation signal and the scintillation signal produced in liquid argon by a beam of La ions, for two high voltages. From Ref. [24].

fluctuation on N_{ion}, i.e. the part of the signal which is read out, will be (from binomial statistics)

$$\sigma(N_{ion}) = \sqrt{N \frac{N_{ion}}{N} \frac{N_{scint}}{N}}. \tag{1.25}$$

If for instance 80% of the released energy goes into ionisation, i.e. $N_{ion}/N = 0.8$, and 20% into scintillation, then the previous equation gives $\sigma(N_{ion}) \sim 0.4\sqrt{N}$. Thus the energy resolution is a factor of 2.5 better than that expected from a pure $1/\sqrt{N}$ behaviour. This Fano factor is again due to the fact that N does not fluctuate event by event, as already mentioned.

Liquid argon is the most commonly employed noble liquid for sampling calorimeters (see Section 7.2.4) because of its low cost and high purity. On the other hand krypton is usually preferred for homogeneous calorimeters, mainly because of its much shorter radiation length which allows more compact detectors. Xenon would be an even better choice, however it is very rare in nature and therefore it is expensive and procurement for large-scale detectors is a serious issue.

Noble liquid calorimeters offer good radiation resistance and good response uniformity by construction, since the liquid distributes in a uniform way throughout the detector. The disadvantage of this technique is that it requires massive cryogenics and purification equipments.

The NA48 liquid krypton electromagnetic calorimeter

The aim of the NA48 fixed-target experiment [25] at the CERN SPS is to measure the direct CP-violation parameter ϵ'/ϵ by detecting simultaneously $K^0_{S,L} \to \pi^0\pi^0$ and $K^0_{S,L} \to \pi^+\pi^-$ decays. The requirements for the electromagnetic calorimeter are numerous. A $\pi^0 \to \gamma\gamma$ mass resolution of 1 MeV is needed in order to reject the combinatorial background from $K^0_L \to 3\pi^0$ when two photons are lost. This requires an energy resolution of $\sim 5\%/\sqrt{E(\mathrm{GeV})}$ and a position resolution of 1 mm for photons with $E \sim 25$ GeV. Tagging K^0_S neutral decays by demanding a coincidence between the calorimeter signal and a beam hodoscope requires a calorimeter time resolution of better than 500 ps. In addition the calorimeter should have a fast response to avoid event pile-up, since the event rate is ~100 kHz.

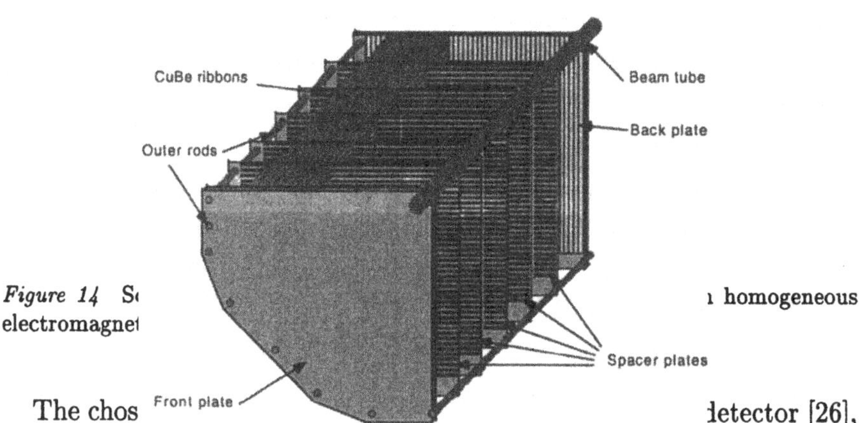

Figure 14 S[illegible] [illegible] homogeneous electromagne[illegible]

The chos[illegible] detector [26], shown in Fig. 14. It has a length of ~1.2 m ($\sim 25\ X_0$), and is segmented in cells of transverse size 2×2 cm^2 with no longitudinal segmentation. The total number of channels is ~13500. The readout electrodes run parallel to the beam direction and have a zig-zag shape, which ensures that the collected signal is independent of the shower distance from the electrodes. This is needed to achieve a good response uniformity and a constant term of $\sim 0.5\%$. The calorimeter is read out with Si preamplifiers located on the detector faces, and therefore operating at the liquid krypton temperature of 120 ^{0}K, followed outside the cryostat by shapers (shaping time 80 ns), 40 MHz ADCs and pipelines.

Examples of the achieved performance with data [27] are shown in Fig. 15. The reconstructed π^0 mass has a resolution of $\sim$ 1.1 MeV and the calorimeter time resolution for K^0 events is $\sim$ 230 ps. This is in agreement with the requirements listed above.

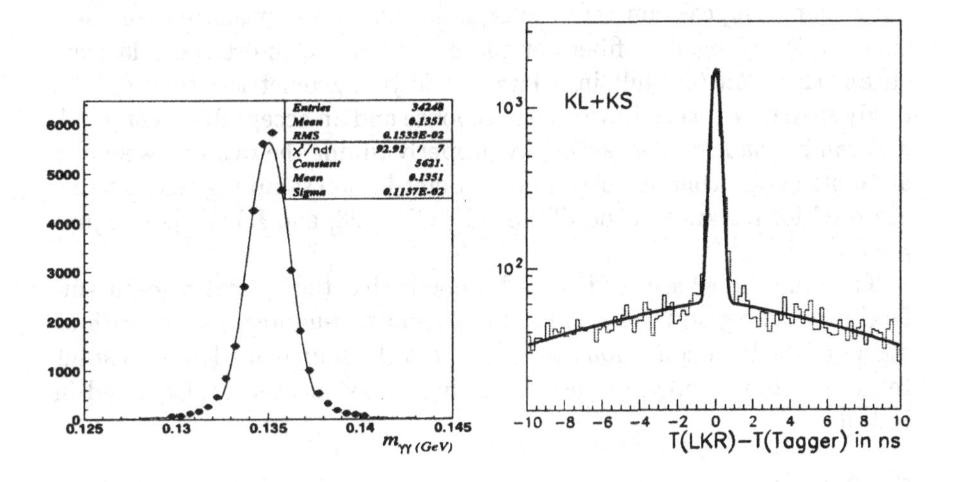

Figure 15 Left: reconstructed π^0 mass in the NA48 calorimeter. Right: the difference between the time measured by the NA48 calorimeter and the time given by a K_S^0 beam tagger. The peak due to K_S^0 events is shown on top of the background due K_L^0 events.

7.2. SAMPLING CALORIMETERS

As discussed in Section 3.1, due to the sampling fluctuations produced by the absorber layers interleaved with the active layers, the energy resolution of sampling calorimeters is in general worse than that of homogeneous calorimeters, and typically in the range $6 - 20\%/\sqrt{E(\mathrm{GeV})}$ for electromagnetic calorimeters. On the other hand, sampling calorimeters are relatively easy to segment longitudinally and laterally, and therefore they usually offer better space resolution and particle identification than homogeneous detectors. In addition, they are most commonly used to measure hadronic showers, since they provide enough interaction lengths with a reasonable detector thickness (< 2 m), they are easier to compensate, and the hadronic energy resolution is limited by the nature of strong interactions rather than by sampling fluctuations.

Sampling calorimeters can be classified, according to the type of active medium, into scintillation calorimeters, gas calorimeters, solid state calorimeters, and liquid calorimeters. In the first case the signal is collected in the form of light, in the last three cases in the form of electric charge. The most commonly used absorber materials are lead, iron and copper.

7.2.1 Scintillation sampling calorimeters. A large number of sampling calorimeters use organic (plastic) scintillators (see Section 7.1.3) arranged in fibers or plates. These detectors are relatively cheap, they can be built in a large variety of geometries, they can be easily segmented, they have a fast response and an acceptable light yield, and can be made compensating by properly tuning the ratio between the amounts of absorber and scintillator. Scintillation sampling calorimeters are used for instance in the ZEUS [11], CDF [28] and KLOE [29] experiments.

The main drawbacks of this technique is that the optical readout suffers from ageing and radiation damage, and non-uniformities at various stages of the light collection chain are often the source of a large constant term. A more detailed discussion of these calorimeters can be found in Ref. [6].

7.2.2 Gas sampling calorimeters. Gas calorimeters have been widely employed until very recently (e.g. for LEP experiments), mainly because of their low cost and segmentation flexibility. However they are not well suited to present and future machines because of their modest energy resolution ($> 10\%/\sqrt{E(\mathrm{GeV})}$), which receives contributions also from Landau and path length fluctuations, and to the low density of the active medium. This latter implies a small sampling fraction ($< 1\%$) and therefore requires operation in proportional mode to obtain enough signal and an acceptable signal-to-noise ratio. Operating a gas calorimeter in proportional mode, that is with wire planes in the active layers and a large voltage on the wires to produce avalanche multiplication of the electron signal, yields signal gains of $10^3 - 10^5$ but entails poor stability and uniformity of the detector response. This is because the gain is very sensitive to several factors, such as the exact diameter and position of the wires, the gas pressure, temperature and purity, the exact high voltage setting, etc.. Therefore achieving a response stability and uniformity at the level of a few permill, as it is needed for instance at the LHC, is difficult with this technique. These detectors will not be discussed here in detail.

7.2.3 Solid-state sampling calorimeters. The active medium is Silicon in most cases. The main advantage of these detectors is that the density of the active layers is a factor of about 1000 larger than in gas calorimeters, which allows the construction of more compact devices and a higher signal-to-noise ratio. This latter is also due to the fact that only 3.6 eV are needed to produce an electron-hole pair in Si, compared to $\sim$ 30 eV in gas. Therefore solid state calorimeters can be operated with unity gain, thus avoiding the drawbacks of charge multiplication (see above). The main disadvantages of this technique are the high cost, which prevents its use in large-scale detectors, and the poor radiation resistance. Small and compact Si sampling calorimeters, often employing a very dense absorber like tungsten, have been widely used as luminosity monitors at LEP.

7.2.4 Liquid sampling calorimeters. These detectors are discussed here in some details because they offer good application perspectives also for future experiments.

Warm liquid (e.g. tetramethylpentane or TMP) calorimeters work at room temperature, and therefore do not bring the burden of cryogenics. However they are characterised by a poor radiation resistance and they suffer from purity problems.

Cryogenic liquid sampling calorimeters have been and are widely employed in high energy physics experiments (e.g. R807/ISR, Mark II, Cello, NA31, SLD, Helios, D0, H1), mainly with argon as the active medium. This well established technique offers several advantages. The liquid density (see Table 2) gives enough charge to operate the detector with unity gain, which ensures a better response uniformity than in calorimeters with electron amplification. Liquid sampling calorimeters are relatively uniform and easy to calibrate because the active medium is homogeneously distributed inside the volume and the signal collection is not subject to the cell-to-cell variations which characterise detectors with optical readout [1]. They provides good energy resolution ($\sim 10\%/\sqrt{E(\mathrm{GeV})}$ and a stable response with time. They are radiation hard. The drawbacks are the cryogenic equipment, which complicates the operation and introduces additional dead material in front of the

[1] Since the liquid response exhibits a temperature dependence, and typically drops by 2% for a T increase of one degree, care must be taken to ensure a temperature distribution inside the cryostat uniform to a fraction of a degree.

calorimeter (cryostat) and between different detector parts (e.g. between barrel and end-cap); the need to achieve and maintain high purity conditions, which in turn requires a purification system; and the fact that classical liquid calorimeters have a slow response. This last disadvantage, which would render these detectors not suitable for operation at high-rate machines, has been recently overcome by the introduction of a novel geometry, the Accordion geometry, which has been chosen for the ATLAS electromagnetic calorimeter.

The ATLAS lead-liquid argon Accordion electromagnetic calorimeter

A novel design [30] providing a fast detector response has been developed for the ATLAS lead-liquid argon electromagnetic calorimeter. In standard liquid argon sampling calorimeters, the alternating absorber and active layers are disposed perpendicular to the direction of the incident particle, as illustrated in Fig. 16 a). The ionisation signal produced

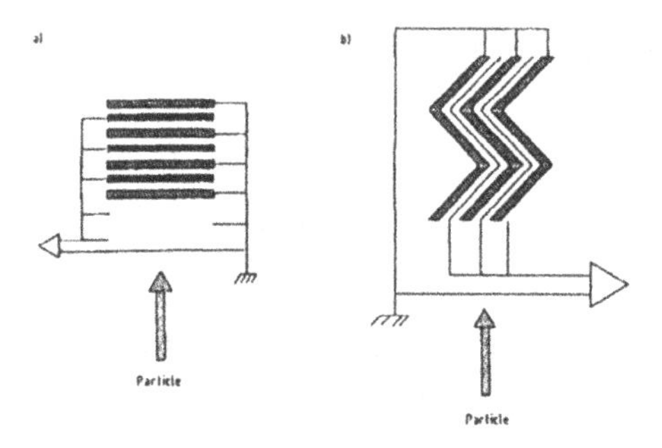

Figure 16 Schematic view of a traditional sampling calorimeter geometry (a) and of the Accordion calorimeter geometry (b).

by the shower in the liquid argon gaps is collected by electrodes located in the middle of the gaps. These electrodes carry the high voltage, whereas the absorbers are at ground. For a typical liquid argon gap of 2 mm on either sides of the collection electrode and an high voltage of ~2 kV across the gap, the electron drift time to the electrode is ~400 ns. This time, which is also the time needed to collect the total ionisation charge, is obviously too slow for operation at the LHC where detector responses of 50 ns or smaller are needed (see Section 6). A possible solution is to integrate the ionisation current over a time (t_p) of only 40-50 ns, thus collecting only a fraction of the total charge. This solution has the drawback that the signal-to-noise ratio is degraded, and can only work if the signal transfer time from the electrodes to the readout chain is much smaller than t_p, i.e. if cables and connections (which introduce capacitance and inductance and therefore give rise to a long time constant of the circuit) are minimised. With the standard electrode geometry shown in Fig. 16 a) long cables are needed to gang together successive longitudinal layers to form calorimeter towers and to transfer the signal from these towers to the electronic chain which is in general located at the end of one calorimeter module. As a consequence, the charge transfer time from the electrodes to the first element of the readout chain (usually a preamplifier) is of several tens ns, i.e. comparable to t_p, and a very tiny signal is collected if $t_p \simeq 40 - 50$ ns. In addition, these cables introduce dead spaces between calorimeter towers at the expense of the detector hermeticity.

These problems can be solved by placing the absorber and gap layers perpendicular to the particle direction. In this way the signal from the collection electrodes can be extracted directly from the front and back faces of the calorimeter and sent to the readout chain with a minimum amount of cables and connections. Dead spaces inside the detector active volume are also minimised with this geometry. However, in order to avoid that the incident particles escape through the liquid argon gaps without crossing the absorber, the electrodes are bent into an accordion shape, as illustrated in Fig. 16 b). This is the technique developed for the ATLAS electromagnetic calorimeter [2].

The structure of this calorimeter is shown in Fig. 17. The lead layers have a thickness of 1.1-2.2 mm, depending on the rapidity region, and are separated by 4 mm liquid argon gaps.

The calorimeter covers the rapidity region $|\eta| < 3.2$ and is divided longitudinally in three compartments (Fig. 18). The first compartment is segmented into fine strips of pitch 4 mm in the η direction, which provide good γ/π^0 separation capabilities, the second compartment has square towers of size 4×4 cm^2, and the third compartment has a factor of two coarser granularity in η than the second compartment. The approximately 200000 channels are read out with a 3-gain electronic chain located outside the cryostat and consisting of a preamplifier, a shaper (peaking time ~40 ns), a 40 MHz analog pipeline and a 12-bit ADC. Unlike in the CMS calorimeter, digitisation is performed in the last stage of the chain, after the first-level trigger has accepted an event.

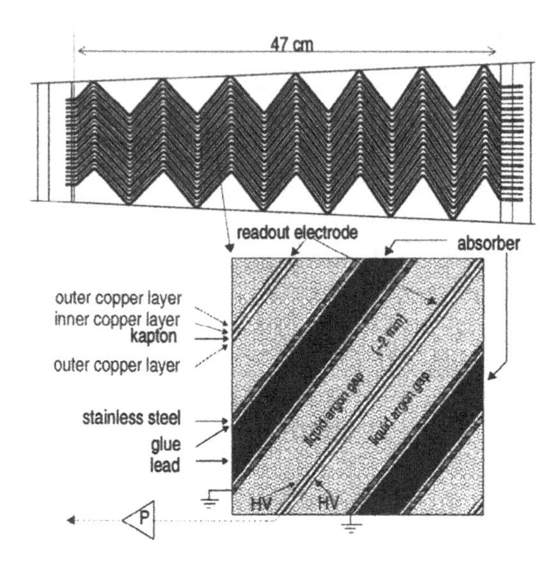

Figure 17 Schematic view of the electrode structure of the ATLAS electromagnetic calorimeter. In the top picture particles enter the calorimeter from the left.

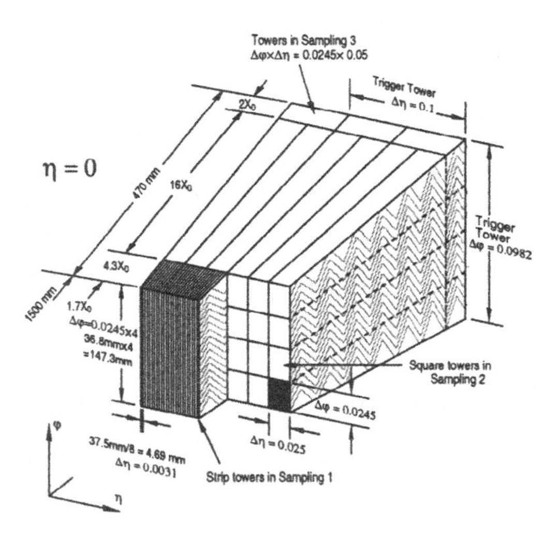

Figure 18 Schematic view of the segmentation of the ATLAS electromagnetic calorimeter.

Figure 19 shows two examples of expected performance. The energy resolution measured with the so-called "module zero" of the calorimeter, that is a module identical to the final calorimeter modules, is about

$10\%/\sqrt{E(\mathrm{GeV})}$, with a local constant term of $\sim$0.3% and a noise term of $\sim 0.25/E$(GeV) (left plot). By using the longitudinal and lateral segmentation of the calorimeter it will be possible to measure the position of the primary vertex as explained in Section 6, with an expected resolution of about 1 cm (right plot).

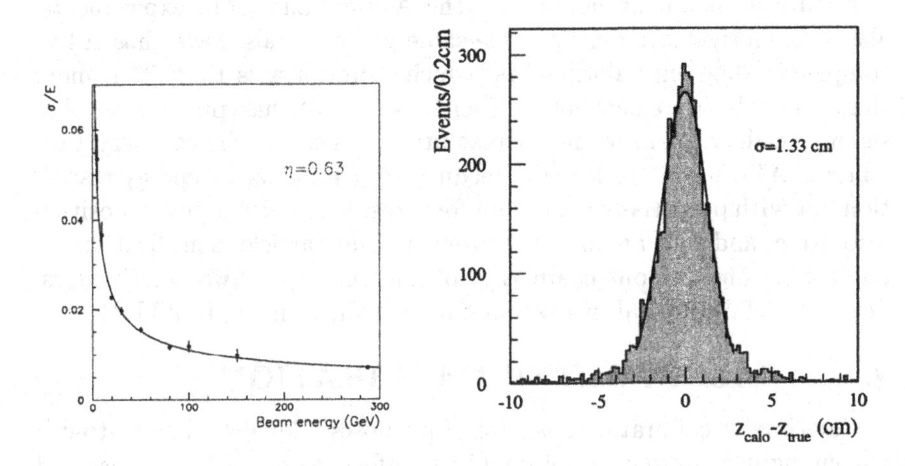

Figure 19 Left: fractional electron energy resolution as a function of energy as obtained from a beam test of the module zero of the ATLAS barrel electromagnetic calorimeter (the electronic noise has been subtracted). Right: difference between the generated primary vertex and the vertex reconstructed by the electromagnetic calorimeter, as obtained from GEANT-simulated $H \rightarrow \gamma\gamma$ events with m_H =100 GeV.

As already mentioned, one aspect which requires a careful control when operating a liquid calorimeter is purity. This is because electronegative molecules dissolved in the liquid, such as oxygen molecules or unsaturated carbon composites, can capture the ionisation electrons thus reducing the collected signal. Therefore it is important to avoid any oxygen leaks inside the cryostat, and to make sure that no piece of material which goes in the detector emits impurities by outgassing, in particular if exposed to high radiation doses. Argon is the best liquid from this point of view. It has the lowest boiling temperature (87° K compared to 120° K for krypton), and therefore outgassing is reduced. Commercial liquid argon is very pure (impurities concentration below $\sim$ 0.5 ppm) whereas krypton needs purification. It is relatively cheap, and therefore can be easily replaced in the event of a major pollution, whereas krypton and xenon are more expensive and rare. Experience with the H1 and D0 calorimeters, which have been equipped with sophisticated systems of probes and purity monitors, has shown a very good control of the liquid quality. For instance, the response of the D0 calorimeter has dropped by only 0.5% over ten years of operation. In addition, purity is less of an issue if the calorimeter is operated with a fast shaping time, because electrons are allowed to drift over very short distances ($\sim$ 200 μm for t_p $\sim$40 ns) before being sampled. For instance, in the ATLAS calorimeter the sensitivity to impurities is reduced by a factor of ten compared to the H1 and D0 calorimeters which have integration times of $\sim$ 450 ns.

Finally it should be noted that the ATLAS and CMS experiments, although motivated by exactly the same physics goals, have chosen two completely different calorimeters, which demonstrates that often more than one solution exists for a given case. CMS has put the emphasis on excellent intrinsic energy resolution, hence the choice of crystals, whereas ATLAS has preferred a technique with moderate energy resolution but with potentially a more uniform response, with a better position resolution, and with angular measurement and particle identification capabilities. The readout chain is also different in the two calorimeters, i.e. almost fully digital in CMS and almost fully analog in ATLAS.

8. CALORIMETER CALIBRATION

Calorimeter calibration has several purposes. Equalise the cell-to-cell output signals in order to obtain an as uniform as possible response and therefore a small constant term of the energy resolution; set the absolute energy scale for e.g. electrons, photons, single hadrons, jets; monitor the detector response variations with time. No single calibration system is able to achieve all these goals, therefore usually several methods are combined.

It should be noted that with the increasing energy of present and future machines, response uniformity, and therefore calibration, become more and more important issues since at high energy the calorimeter energy resolution is dominated by the constant term. Furthermore, the increasing size and complexity of the experiments render the calibration and monitoring tasks very challenging, given that calorimeters are often equipped with a large number of channels (up to several hundred thousands).

Three main tools are usually employed to calibrate a calorimeter:

- Hardware calibration. It is mainly used to equalise and monitor the response of the detector and of the associated electronics. The electronics calibration system injects a known pulse at the input of the readout chain. The aim is to equalise the channel-to-channel response of the calorimeter electronics and to monitor variations with time (due for instance to temperature effects). Channel-to-channel dispersions as small as 0.2% can be achieved. This system does not allow a calibration of the detector response, since the pulse is injected at the level of the readout chain. Therefore most calorimeters are equipped with other devices (e.g. lasers, radioactive sources), which inject a well known light or charge signal in the active elements of the detector. Their aim is to equalize the detector response, this time at the detector cell level, and monitor variations of this response with time. In the ZEUS uranium-scintillator calorimeter a natural calibration source is provided by the radioactivity of the absorber.

- Test beam calibration. Some calorimeter modules are usually exposed to test beams before being installed in the final detector. One of the main aims of this step is to set a preliminary absolute energy scale for electrons and pions, given that the incident beam energy is well known.

- *In situ* calibration with physics. A further calibration step is needed after installation in the experiment. This is because the experiment environment, e.g. the presence of material in front of the calorimeter due to tracking devices, is not the same as at the test beam and is not seen by the hardware calibration. Furthermore, the calorimeter response to physical objects like jets and missing transverse energy can not be measured at the test beam where only single particles are accessible. Therefore, *in situ* calibration is needed in order to cancel out residual non-uniformities, understand the impact of the upstream material and of the environment in general, follow the detector response variations with time, and set the final absolute energy scale in the experiment conditions. This is achieved by using well-known control physics samples, such as $Z \to ee$ decays.

Setting the absolute energy scale using physics samples is one of the most challenging steps of the calorimeter calibration procedure, and is therefore briefly discussed here. Only the hadronic machine case is considered, since at e^+e^- colliders the precise knowledge of the centre-of-mass energy provides useful constraints and renders this operation easier.

The electromagnetic absolute energy scale at hadronic machines is set mainly by using well known resonances, such as $\pi^0 \to \gamma\gamma$, $J/\Psi \to ee$, $\Upsilon \to ee$ in the low energy range and $Z \to ee$ at higher energies. Using an iterative procedure, the calibration factor of each calorimeter cell is varied until the reconstructed mass peak for the above particles is in agreement with the nominal value. As an example, Fig. 20 (left plot) shows the reconstructed invariant mass of the two electrons produced in Z decays from the D0 Run I data [31], before the final scale calibration.

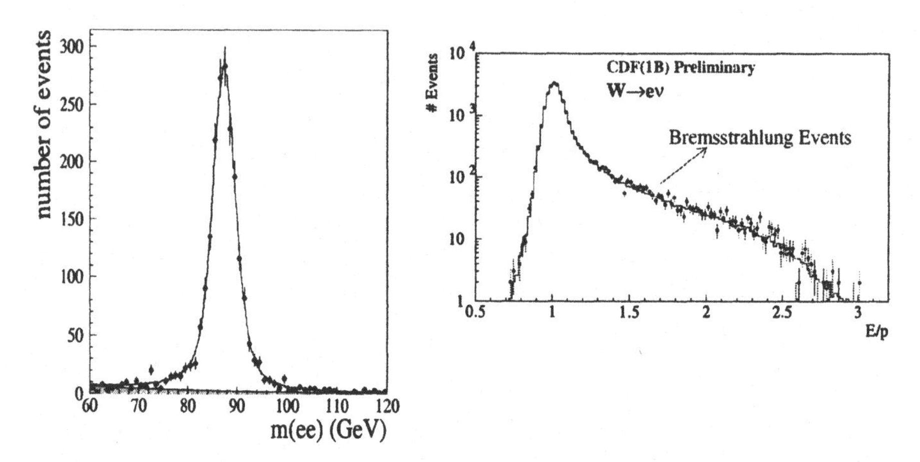

Figure 20 Left: the di-electron mass spectrum reconstructed in the D0 central calorimeter using the Run I Z data sample. The superimposed curve shows the fit and the shaded region the background. Right: the E/p ratio for isolated electrons from W decays as obtained from CDF Run 1B data. The superimposed curve shows the fit.

The calorimeter energy was calibrated by using the relation $E_{true} = \alpha E_{meas} + \delta$, where E_{meas} is the electron energy measured in the calorimeter and the parameters α and δ were varied until the reconstructed Z mass peak agreed with the nominal value. As it can be seen in Fig. 20, the mass peak before calibration is ~5% below the nominal Z mass value. This wrong initial scale has been attributed mainly to the fact that no module of the final D0 central calorimeter was calibrated with test beams (only prototypes), and indicates the importance of performing such test beam measurements to keep the energy correction factors (and therefore the related systematic uncertainties) to a small level.

An alternative (and complementary) method consists of transferring the energy scale from the tracker to the electromagnetic calorimeter by measuring the E/p ratio for isolated electrons, where E is the electron energy as measured in the calorimeter and p is the electron momentum as measured in the inner tracker. This procedure includes several steps. The tracker is first calibrated by using isolated muons. The momentum scale for electrons is not automatically available at this stage because electrons undergo Bremsstrahlung and therefore the original tracks lose part of their energy in Bremsstrahlung photons. A detailed Monte Carlo simulation of the tracker, which contains all the details of the material distribution, is used to compute the electron energy losses and therefore obtain the initial momentum. Finally the momentum scale is transferred to the electromagnetic calorimeter by asking that the E/p distribution for electrons peaks at one. An example from CDF [32] is shown in Fig. 20 (right plot).

By using these methods, a precision on the absolute electron energy scale of $\sim$0.1% has been achieved by the CDF and D0 experiments, limited by the statistics of the above-mentioned physics samples. The dominant sources of systematic uncertainties come from the knowledge of the tracker material, from calorimeter response non-linearities and from radiative Z decays. The uncertainty on the electron energy scale is the dominant systematic error on the W mass as measured at the Tevatron in the electron-neutrino channel [31, 32]. One of the LHC physics goals is to measure the W mass to about 15 MeV, which requires a calibration of the electron scale to the very challenging precision of 0.02% [13]. The CP-violation experiments KTeV and NA48 have a similar requirement, since an excellent knowledge of the electromagnetic energy scale is needed to reconstruct with precision the $K^0_{S,L}$ decay vertices and therefore their relative production rate (which enters directly in the ϵ'/ϵ measurement).

Setting the jet energy scale, that is inferring the original parton energy from the measured jet debris, is more complex than setting the electron scale, since there are more numerous (and more difficult to control) sources of uncertainties. The calorimeters are calibrated with test beams of single particles (electrons, pions) and not with jets; part of the jet energy can be carried away from neutrinos produced for instance in pion decays; part of the energy can be lost outside the cone which is used to collect the jet energy (this is usually the case if the original parton has radiated one or more gluons); the rest of the event can contribute some energy inside the cone used to reconstruct the jet energy, which needs to be subtracted; the calorimeter response to hadrons is usually non-compensated; etc.. Therefore, even more than in the electron case, physics samples are necessary to set the final jet scale. The main samples used at hadron colliders are the associated production of a single jet with a photon or a $Z \to \ell\ell$. If there are only one jet and one boson in the event, then the boson and the jet must have equal and opposite momenta in the plane transverse to the beam axis (i.e. $\vec{p}_T(\gamma, Z) = -\vec{p}_T(jet)$). This is because the momenta of the interacting partons have negligible components transverse to the beam axis. The transverse momentum of the photon or Z can be determined with precisions of $\sim$ 0.1% by using the electromagnetic calorimeter and the tracker as explained above. Therefore the jet scale can be obtained by requiring $|\vec{p}_T(jet)| = |\vec{p}_T(\gamma, Z)|$, that is from the electromagnetic and tracker scales.

Figure 21 shows the inverse of the correction factor to the measured jet energy obtained in D0 as a function of energy by using a sample of γ+1 jet events from the Run I data [33]. The experimental points can

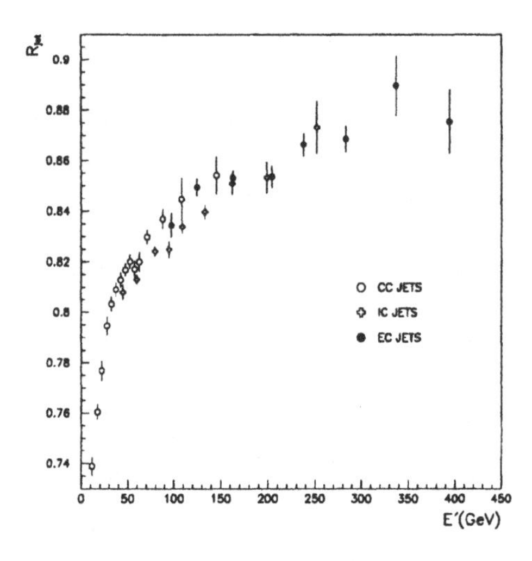

Figure 21 Inverse of the correction factor to the jet energy as a function of energy, as obtained with a sample of γ+1 jet events in the D0 experiment (Run I data). The different symbols indicate different calorimeter regions (central, end-cap, inter-cryostat).

be fitted with the function $R_{jet}(E) = a + b\ln(E) + c\ln(E)^2$, where the logarithmic energy dependence reflects the logarithmic increase of the electromagnetic component of the shower with energy (see Eq. 1.21). The correction decreases with energy mainly because the calorimeter response to a jet becomes more compensating at high energy and because the energy losses in the dead material become less important. The residual uncertainties on the jet scale after correction are at the level of 3% both in CDF and D0. They come mainly from the limited statistics of the physics samples, from the subtraction of the underlying event and of the background, and from the corrections for the energy lost by longitudinal leakage, in the dead material and out of the jet cone. The knowledge of the jet energy scale is the dominant systematic uncertainty on the top mass measurement at the Tevatron [34], contributing about 4 GeV out of a total systematic error of about 5.5 GeV in both CDF and D0.

The goal of the LHC experiments is to calibrate the jet energy scale with a precision of ~1%, in order to measure the top mass to ~ 1.5 GeV. In addition to the above-mentioned physics samples, also $W \rightarrow jj$ produced from top decays can be used at the LHC to achieve this goal. Events due to $t\bar{t}$ production, where one top decays as $t \rightarrow bW \rightarrow bjj$ and one top as $t \rightarrow bW \rightarrow b\ell\nu$, are expected to be collected at the rate of one million per year of LHC operation and to have negligible backgrounds. The calorimeter jet scale can then be determined by requiring that the invariant mass of the two jets from $W \rightarrow jj$ decays in $t\bar{t}$ events be compatible with the W mass. The latter should be known to better than 30 MeV at the time of the LHC start-up.

9. CALORIMETER INTEGRATION IN AN EXPERIMENT

Unlike in nuclear physics, for most high energy physics applications calorimeters are integrated in often very complex experiments, consisting of several sub-detectors of which the calorimeter is only one component. This has both negative and positive consequences. The drawback is that calorimeters have to satisfy overall constraints, such as space limitations, and work in an environment which can deteriorate their response (material, magnetic field, etc.). In many cases additional devices or special software techniques are needed in order to recover, at least in part, the loss in performance. The advantage is that the calorimeter task can be made more effective by combining the calorimeter measurements with the information from other sub-detectors. For instance, the use of an energy-flow algorithm, where the momentum of charged particles is measured in the inner detector and the energy of neutral particles in the calorimeters, has allowed the ALEPH experiment to improve the energy resolution for $Z \rightarrow q\bar{q}$ events by almost a factor of two compared to a purely calorimetric measurement of the event energy [35]. Similar results have been obtained by the other LEP experiments.

Two issues related to the calorimeter integration and performance in large-scale experiments are discussed here as examples: material effects and particle identification.

9.1. IMPACT OF MATERIAL

Energy losses in the material (e.g. from tracking devices) that particles have to traverse in an experiment before reaching the active part of the calorimeter are particularly important in the case of electrons and photons and therefore mostly detrimental for the performance of electromagnetic calorimeters. In running experiments (e.g. at LEP and Tevatron) the material in the inner detectors is typically a few percent of a radiation length, but the inner detectors of future LHC experiments will be more massive. Furthermore, often the coil providing the magnetic field in the inner cavity sits in front of the electromagnetic calorimeter (e.g. in the OPAL experiment), and the calorimeter support structures and cables provide additional dead layers (not to mention the contribution of the cryostat in the case of noble liquid calorimeters).

Although the average energy lost by electrons and photons in the upstream material can be determined and corrected for, the event-by-event fluctuations can not, unless dedicated devices are used. These fluctuations provide an additional contribution to the energy resolution, and if the upstream material is large ($> 1\ X_0$) they can spoil the calorimeter intrinsic performance. This is illustrated in Fig. 22, which shows the electron energy resolution of the lead-glass end-cap electromagnetic calorimeter of the OPAL experiment as obtained in a beam test performed in various conditions [36]. The open circles and open squares give the calorimeter energy resolutions without and with a slab of material (1.6 X_0 of aluminum) in front, respectively. It can be seen that over the energy range 10-50 GeV the calorimeter resolution is deteriorated by a factor of 1.7-1.3 by the presence of additional material.

Both simulation and test beam measurements performed with various calorimeters have shown that an acceptable recovery of the energy resolution is possible by using dedicated devices, provided that the upstream material does not exceed 2.5–3 X_0. Examples of such devices are massless gaps and presampler detectors. These are thin layers of active medium, placed in front of electromagnetic calorimeters or in the cracks between calorimeter parts (e.g. in the D0 and ATLAS calorimeters). Massless gaps are usually integrated in the calorimeter structure, whereas presamplers are separate devices read out independently of the calorimeter. In both cases, the principle of work is based on the fact that the energy released by the incident particles in these devices is proportional to the energy lost upstream. Therefore, by measuring the energy in these detectors, and by adding it (suitably weighted) to the energy measured in the calorimeter, it is possible to recover for energy losses

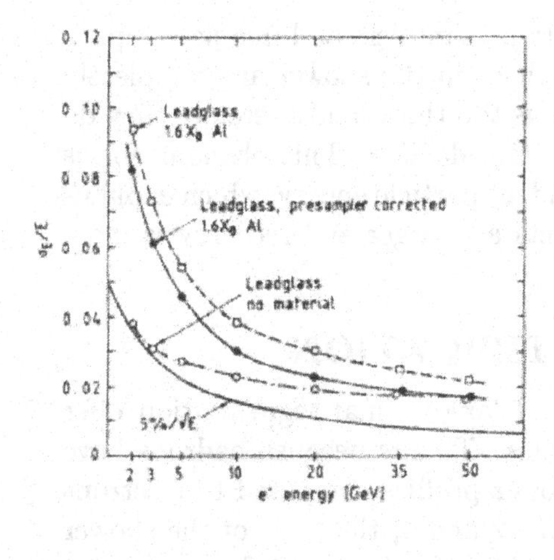

Figure 22 Fractional energy resolution of the OPAL lead-glass end-cap electromagnetic calorimeter as a function of the energy of the incident electron beam. The open circles show the calorimeter resolution without radiator, the open squares show the calorimeter resolution with a 1.6 X_0 Al radiator in front and no presampler corrections, the closed circles show the resolution obtained after presampler corrections (see text).

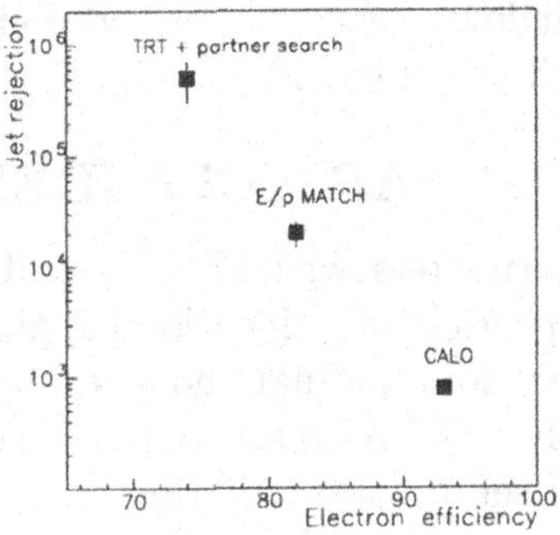

Figure 23 Jet rejection as a function of the electron identification efficiency in the p_T range 20–50 GeV, as obtained from a GEANT simulation of the ATLAS detector. The improvement obtained by using the information of the various sub-detectors is shown (see text).

event by event, and thus take into account fluctuations. In most cases massless gaps and presamplers have a coarse granularity, since they are used for energy measurements and not for position measurements

The OPAL end-cap presampler, which is made of thin multiwire gas chambers operated in saturated mode, is installed in front of the end-cap lead glass calorimeter. The expected contribution of this device to the electromagnetic energy measurement is shown by the closed circles in Fig. 22. By adding the presampler energy, suitably weighted, to the lead glass energy, about half of the deterioration in resolution due to the material is gained back at 10 GeV. At lower energies the improvement is smaller, at higher energies larger. If the material in front becomes too large (2.5-3 X_0), then the correlation between the energy deposited upstream and the energy deposited in presamplers and massless gaps is lost. This is because low-energy particles in the shower are completely absorbed by the dead material if this is too thick, and therefore they do not contribute to the signal in the active devices. This phenomenon is more pronounced the smaller the incident particle energy, which explains why at low energy the material effects are larger and recovery is more difficult (see Fig. 22).

9.2. PARTICLE IDENTIFICATION

Calorimeters with good lateral and longitudinal segmentation offer good particle identification capabilities. This is because hadrons have different longitudinal and lateral shower profiles compared to electrons and photons. As discussed in Sections 2 and 4, the scale of the shower size is set by the radiation length (which is of order 1-2 cm in most calorimeters) for electromagnetic showers and by the interaction length (which is larger than 10 cm in most calorimeters) for hadronic showers. As a consequence, hadronic showers are usually longer and broader than electromagnetic showers. Therefore, by measuring the energy fractions deposited in the cells of a segmented calorimeter it is possible to distinguish the two types of showers.

When calorimeters are operated inside high energy physics experiments, their particle identification capability can be made more powerful by combining the calorimeter information with that coming from other sub-detectors. An example is presented in Fig. 23, which shows the expected electron/jet separation capability of the ATLAS detector, as obtained from detailed GEANT simulations. It can be seen that the electromagnetic and hadronic calorimeters alone provide a jet rejection of 1000, for more than 90% electron efficiency. Another factor of ten in rejection is obtained by combining the electromagnetic calorimeter information with the inner detector, that is by requiring a track in the inner detector pointing to the shower in the calorimeter, with momentum matching the shower energy. A final rejection of more than 10^5 is achieved by using the Transition Radiation Tracker [37] to recognise charged hadrons faking electrons, and by rejecting converted photons.

10. CONCLUSIONS

Calorimeters are extremely versatile and flexible detectors. They are able to perform a large variety of measurements, including energy, position, angle and time measurements, they provide particle identification and triggering capabilities, and they are sensitive to essentially all types of particles. These features, together with the fact that their performance improves with energy, make them very suitable detectors for operation in present and future high energy physics experiments.

Big progresses have been made over the last thirty years in the field of calorimetry. Many techniques with different merits and drawbacks have been developed, and have been used in a large variety of applications (nuclear physics, accelerator and non-accelerator particle physics, space experiments, etc.). Large-scale calorimeters have been built and integrated in often very complex systems. Remarkable progress also in electronics (e.g. in terms of speed, noise optimisation, dynamic range) and signal treatment methods have allowed the intrinsic performance of these detectors to be exploited in the best way.

However, and despite these achievements, calorimeter operation at the LHC will represent yet another challenge for these detectors.

Acknowledgments

I wish to thank Michael Danilov, Tom Ferbel and Harrison Prosper for a very nice and special school.

References

[1] D.E. Groom et al., European Physical Journal **C15**, 1 (2000).

[2] ATLAS Collaboration, *Liquid argon calorimeter Technical Design Report*, CERN/LHCC/96-41.

[3] CMS Collaboration, *The Electromagnetic Calorimeter Project Technical Design Report*, CERN/LHCC/97-33.

[4] F. Melot, Rapport de stage, LAPP-Annecy (France), 1999; M. Maire, *Physics processes in electromagnetic showers*, talk given at the IX International Conference on Calorimetry in High Energy Physics (CALOR2000), Annecy, October 2000.

[5] U. Amaldi, Physica Scripta **23**, 409 (1981).

[6] P.B. Cushman, *Electromagnetic and hadronic calorimeters*, in *Instrumentation in high energy physics*, editor F. Sauli, World Scientific Singapore, 1993.

[7] NA48 Collaboration, Nucl. Instr. and Meth. **A360**, 224 (1995).

[8] R. Wigmans, Nucl. Instr. and Meth. **A259**, 389 (1987).

[9] ATLAS Collaboration, *Tile calorimeter Technical Design Report*, CERN/LHCC/96-42.

[10] S. Abachi et al., Nucl. Instr. and Meth. **A338**, 185 (1994).

[11] U. Behrens et al., Nucl. Instr. and Meth. **A289**, 115 (1990).

[12] ATLAS Collaboration, *Calorimeter performance Technical Design Report*, CERN/LHCC/96-40.

[13] ATLAS Collaboration, *Detector and Physics performance Technical Design Report*, CERN/LHCC/99-14.

[14] U. Fano, Phys. Rev. **72**, 26 (1947).

[15] G.F. Knoll, *Radiation detection and measurements*, ed. J.Wiley and Sons.

[16] M. Akrawy et al., Nucl. Instr. and Meth. **A290**, 76 (1990).

[17] Super-Kamiokande Collaboration, Nucl. Instr. and Meth. **A421**, 113 (1999).

[18] J. Conrad, *Neutrino Oscillations*, these Proceedings.

[19] KTeV Collaboration, Phys. Rev. Lett. **83** 22 (1999).

[20] D. Boutigny et al., *The BaBar Technical Design Report*, SLAC-R-457.

[21] BABAR Collaboration, *The first year of the BaBar experiment at PEP-II*, SLAC-PUB-8539.

[22] J.A. Bakken et al., Nucl. Instr. and Meth. **A228**, 296 (1985).

[23] Y. Karyotakis, *The L3 electromagnetic calorimeter*, LAPP-EXP-95-02, paper contributed to the 1994 Beijing Calorimetry Symposium, China.

[24] H.J. Crawford et al., Nucl. Instr. and Meth. **A256**, 47 (1987).

[25] G.D. Barr et al., *Proposal for a precision measurement of ϵ'/ϵ in CP-violating $K^0 \rightarrow 2\pi$ decays*, CERN/SPSC/90-22, SPSC/P253; V. Fanti et al., Phys. Lett. **B465** 335 (1999).

[26] G. Unal (for the NA48 Collaboration), hep-ex/0012011.

[27] M. Martini, *Performance of the NA48 liquid krypton calorimeter for the ϵ'/ϵ measurement at the CERN SPS*, in Proceedings of the VII International Conference on Calorimetry in High Energy Physics, Tucson, Arizona, 1997.

[28] L. Balka et al., Nucl. Instr. and Meth. **A267**, 272 (1988); S. Bertolucci et al., Nucl. Instr. and Meth. **A267**, 301 (1988).

[29] KLOE Collaboration, *The KLOE detector Technical Proposal*, LNF-93/003.

[30] B. Aubert et al., *Liquid argon calorimetry with LHC performance specifications*, CERN/DRDC/90-31.

[31] B. Abbott et al., Phys. Review D **58** (1998).

[32] F. Abe et al., Phys. Review D **52** (1995); Y-K. Kim, *W mass measurement at CDF*, talk given at the XIII Rencontres de Physique de la Vallée d'Aoste, La Thuile, Italy, March 1999.

[33] B. Abbott et al., Nucl. Instr. and Meth. **A424**, 352 (1999).

[34] F. Abe et al., Phys. Rev. Lett. **82** 271 (1999); B. Abbott et al., Phys. Rev. **D58** 052001 (1998).

[35] D. Buskulic et al., Nucl. Instr. and Meth. **A360** 481 (1995).

[36] C. Beard et al., Nucl. Instr. and Meth. **A286**, 117 (1990).

[37] ATLAS Collaboration, *Inner Detector Technical Design Report*, CERN/LHCC/97-17.

AN UPDATE ON THE PROPERTIES OF THE TOP QUARK

T. Ferbel
University of Rochester
Rochester, NY 14627 USA
ferbel@pas.rochester.edu

Abstract Properties of the top quark such as its mass, its decay characteristics, and the $t\bar{t}$ production cross section, have been studied by the CDF and DØ experiments at the Tevatron Collider. Thus far, all the observations conform to expectations from the standard model. Nevertheless, most conclusions are currently limited by statistical uncertainty, and it is hoped that with the anticipated improvements in the accelerator complex and in quality of the detectors, the consequent increase by over a factor of 100 in data before the turn-on of the LHC will provide the enhanced sensitivity needed to reveal the presence of new particle interactions and phenomena.

1. INTRODUCTION

It has been almost five years since the definitive observation of the top quark by the CDF and DØ experiments [1, 2]. The first hints of a possible signal were gleaned somewhat before then: (i) by DØ in their famous Event 417 [3], and (ii) by CDF in the large excess of events found in their initial data sample, and published in 1994 as "evidence for" top [4]. Event 417 survived the passage of time and withstood greater scrutiny, and is still regarded as one of the best examples of top-antitop production, but the first cross section reported by CDF for top production turned out to be more than a factor of two larger than the currently accepted value. For the early measurements of the mass of the top quark, DØ obtained a rather large value, but CDF got pretty much what is now accepted as the mass of top [1, 2].

Everything we know about top has been learned from studies of $t\bar{t}$ production, which, at the energy of the Tevatron, is dominated by the $q\bar{q}$ incident channel [5]. With top decaying into $W + b$ in the standard model (SM), the final states with least background arise from events that have $W \to \ell + \nu_\ell$ decays (with branching fractions of $\frac{1}{9}$ for each ℓ). When

H.B. Prosper and M. Danilov (eds.), Techniques and Concepts of High-Energy Physics, 369–379.

both W bosons decay leptonically (either $e\nu_\ell$ or $\mu\nu_\ell$), the events contain two (isolated) leptons of large transverse momentum (p_T). Such events, with their accompanying jets, correspond to "dilepton" channels. When one W decays leptonically and the other one via a quark and antiquark pair (because of the three quark colors, the $u\bar{d}$ and $c\bar{s}$ modes each have branching fractions of $3 \times \frac{1}{9} = \frac{1}{3}$) the events comprise the single-lepton channels, and when both W bosons decay via quarks the final state is called the all-jets channel. In addition to the nominal six objects, events can have extra jets arising from gluon emission in the initial or final state. The all-jets channel has the largest yield, but an enormous background from QCD jet production, and is therefore the most difficult to analyze.

About two years ago, CDF and DØ joined forces to produce an averaged top mass (M_t) and cross section that would best summarize the results from the analyses at the Tevatron. The averaging of the mass yielded $M_t = 174.3 \pm 5.1$ GeV [6], but the results on the $t\bar{t}$ cross section are still not ready (an unofficial value is 6.2 ± 1.2 pb). A summary of the latest measurements of the mass and cross section for all available final states is given in Fig. 1 [7, 8].

In principle, it is relatively straightforward to measure the cross section for some expected signal. What is required is a count of the number of events observed beyond the number expected from all known background processes. When backgrounds are far larger than signal, as is often the case for rare events, specific selection criteria, based on distributions in well-chosen variables, can be defined to minimize background but not remove much signal. This relies on having an excellent understanding of the production processes for signal, and especially so for background. That is, using selections that do not have much effect on signal, implies that the calculated differential distributions do not have to be exceedingly accurate. However, for large backgrounds that are reduced using severe selection criteria, the tails of the contributions become important, and any background subtraction becomes highly sensitive to the detailed understanding of such sources.

The two largest sources of background to $t\bar{t}$ events arise from QCD multi-jet production, where jets mimic isolated muon or electron signals, and W+jets events, a more insidious but somewhat better understood source of QCD background. Although, in the case of multijets, the probability that a jet is mistaken for an electron is rather small ($< 10^{-3}$), QCD calculations are not sufficently precise to provide a reliable model for the remaining level of contamination. Consequently, data obtained using QCD triggers are used to estimate the contribution of such sources of background to the signal of interest.

It is also straightforward, in principle, to measure the mass of the top quark. All the characteristics of a top event, e.g., the transverse momentum (or E_T) of any jet from the decay of top, the E_T of any lepton from the decay of the W, the invariant mass of any lepton and jet system, etc., reflect the mass M_t. Consequently, templates that depend on the specific value of M_t can be devised for such variables via Monte Carlo, and these can be fitted to data, with an appropriate admixture of background, and used to extract the value of M_t from the best fit. The key issue is, of course, to find variables that are especially sensitive to M_t, and, in this respect, the analyses have been remarkably clever [9].

Considering the few events and the subtleties of the mass extraction, the 3% precision achieved on M_t is quite astonishing. Cross sections observed in separate channels are consistent with branching fractions expected for $t \to W + b$ decay. It would be good to establish the electric charge of the top quark, but the events certainly look like top, feel like top, and, undoubtedly, are top. Although the uncertainties are still quite large, the superb agreement between theory [10] and observed cross section is one of the great triumphs of the SM and perturbative QCD. The value of the mass of the top quark is very large, and as a result its Yukawa coupling is close to unity, suggesting that top may hold an especially fundamental position in the SM. Nevertheless, the mass is completely consistent with expectations from electroweak theory. In fact, the top mass, taken with the well measured mass of the W obtained at the Tevatron and at LEP [11], has provided additional constraint on the mass of the Higgs in the standard model, which is now favored to be well below 200 GeV.

With the small sample of top events available from previous runs of the Tevatron, one might wonder whether there are any other important properties of the top quark that could be extracted from the data. Several studies carried out by CDF and D0, although not terribly sensitive, have nevertheless provided some interesting limits and tests of the SM. Recently completed searches and some of the still ongoing analyses are itemized below:

- Spin correlations in $t\bar{t}$ decays.
- Helicity of the W in $t\bar{t}$ final states.
- Extraction of the branching ratio of $t \to W + b$, and thereby the value of the Cabibbo-Kobayashi-Maskawa matrix element $|V_{tb}|$.
- Production of single-top events.
- Flavor-changing decays of the top quark via neutral currents (FCNC).
- The decay of top into a charged Higgs boson: $t \to H^+ + b$.
- Anomalous contributions to $t\bar{t}$ production from possible $t\bar{t}$ resonances

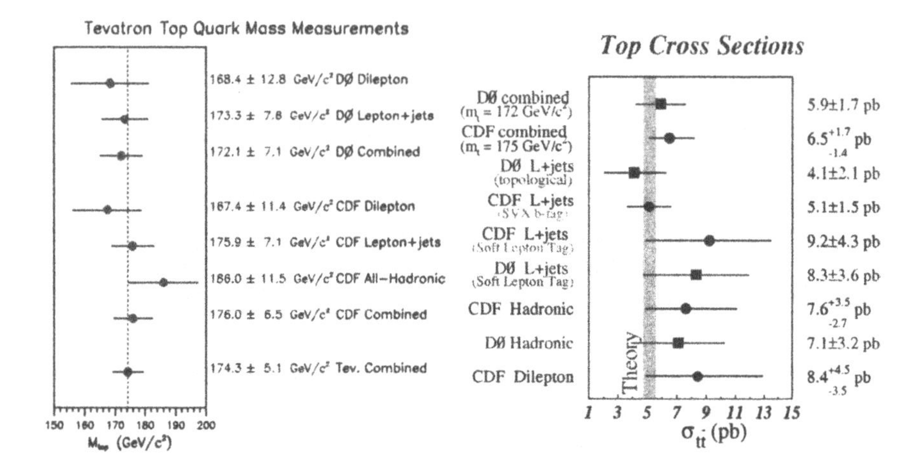

Figure 1 **In the left pane, are the measured values of the mass of the top quark, and, in the right pane, the cross sections obtained for individual $t\bar{t}$ channels at CDF and at DØ .**

We will discuss only several of the above analyses, some of which were intended primarily as vehicles for assessing the eventual sensitivity expected for such studies once data from future runs of the Tevatron become available. The next run is now scheduled to commence in Spring 2001 at a center of mass energy $\sqrt{s}$= 2 TeV, and the first goal is to reach an integrated luminosity of 2 events/fb. With the 10% increase in $\sqrt{s}$ and improvements in both detectors, the 20-fold increase in luminosity will correspond to a far greater increase in signal, especially for the more rare dilepton events and for events that will have b jets tagged either via displaced vertices based on silicon microstrip detectors or through "soft" (not isolated) leptons that often accompany b jets. It has been estimated [12] that an extra factor of at least four in the yield of $t\bar{t}$ events, and an extra factor of more than ten for the more difficult single-top events, will be obtained just from the upgrading of the detectors and increase in $\sqrt{s}$.

2. MORE ON MASS AND CROSS SECTION

Although DØ is still working on the extraction of the mass in the all-jets channel, and the latest values of cross sections from CDF have yet to appear in the journals, most of the results on top mass and cross sections are now relatively well known. In their independent approaches, each group has used both ingenuity and the strengths of their detectors to great advantage. CDF has concentrated on their excellent silicon system, and DØ has relied on its calorimetry and muon coverage, and

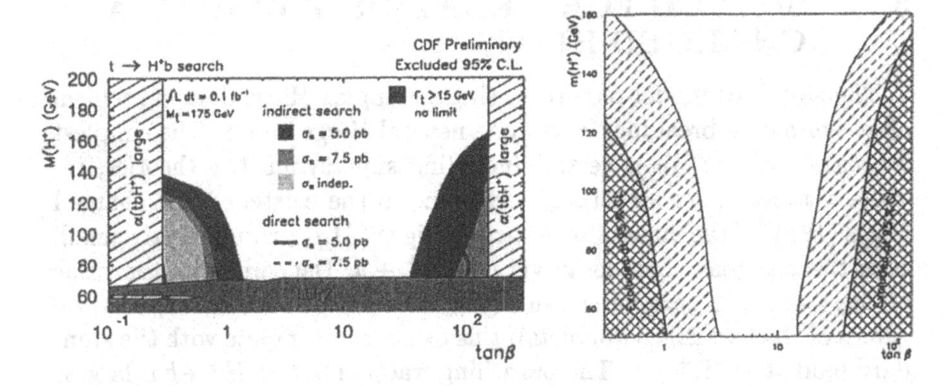

Figure 2 **The left pane displays regions of parameter space for charged Higgs bosons excluded by CDF. In right pane, are regions excluded by DØ, and expectations for sensitivity in the next run of the Tevatron.**

has pioneered novel approaches in analysis through bold application of neural networks.

For example, DØ has recently re-examined the yield of $e\mu$ dilepton events using a neural network approach rather than more classical means (e.g., random grid search) of implementing cutoffs on variables used to maximize separation between signal and background [13]. A modest improvement has been achieved in the yield of signal, with a simultaneous reduction in background. The net gain corresponds to $\sim$ 18% in statistics or $\sim$ 40% in running time. These kinds of approaches will be used more often in the next run, and will help reduce uncertainties in many analyses.

In the future, limitations on the accuracy of the top mass will be dominated mainly by the uncertainty in the energy scale used for reconstructing jets, and by ambiguities in the model for production and decay of top quarks. These are expected to improve by about a factor of two, and bring the total uncertainty down to 2 – 3 GeV. The major improvement in measurements of cross sections will be from an increase in statistics for the individual channels, which will also provide better checks of branching fractions into different final states. The absolute uncertainty will be limited by comparable contributions ($\sim$ 5%) from absolute luminosity, b–tagging efficiency, statistics, energy scale, and the model used for $t\bar{t}$ production. Thus, about a 10% uncertainty on the cross section should be within reach [14].

3. SEARCH FOR DECAY OF TOP INTO A CHARGED HIGGS

The standard model requires a single complex Higgs doublet, which, after symmetry breaking, leaves one neutral Higgs boson. The simplest extensions of the Higgs sector, including supersymmetric theories, involve a two-doublet structure, and point to the existence of a charged Higgs ($H^{\pm}$). If the mass of the charged Higgs ($M_{H^{\pm}}$) is sufficiently small, then the top quark can decay via $t \rightarrow H^{+} + b$. Depending on the value of $M_{H^{\pm}}$ and the parameter tanβ (the ratio of the vacuum expectation values of the two Higgs doublets), this decay can compete with the standard mode $t \rightarrow W^{+} + b$. The branching fraction of $t \rightarrow H^{+} + b$ is largest for both very small and very large tanβ (tan$\beta < 1$ and tan$\beta > 50$), and smallest when tan$\beta = \sqrt{M_t/M_b} \sim 6$, where the decay is dominated by $t \rightarrow W^{+} + b$. The decay of the $H^{\pm}$ also depends strongly on tanβ, with the branching to $c\bar{s}$ and $t^{*}\bar{b} \rightarrow Wb\bar{b}$ dominating for tan$\beta < 1$, and into $\tau\nu_{\tau}$ for tan$\beta > 1$. The relative decay rates into the two hadronic modes are quite sensitive to $M_{H^{\pm}}$, especially near the upper edge of allowed kinematics.

The search for $t \rightarrow H^{+} + b$ relies on a violation of lepton universality in Higgs decay, and has proceeded along two lines. First, the more direct approach is based on the appearance of excess $t\bar{t}$ signal in the $\tau + X$ channels, where the analyses rely on the specific decay of $H^{\pm} \rightarrow \tau^{\pm}\nu_{\tau}$ (and $\tau \rightarrow$ hadrons $+\nu_{\tau}$), which is dominant at large tanβ. The other route involves an indirect search, and is based on the disappearance of top signal. Because the standard analysis of $t\bar{t} \rightarrow$ lepton+jets has selection criteria optimized for the SM modes, it therefore ignores the possibility of a contribution from $H^{\pm}$. Consequently, if a large fraction of top quarks decay via a $H^{\pm}$, then, assuming that there are no additional sources of $t\bar{t}$ signal from mechanisms beyond the SM, there will be fewer events observed than expected in final states based purely on $t \rightarrow W^{+} + b$ decays. A less model-dependent approach, but one that is not very sensitive at current level of statistics, is used by CDF in searches for an anomaly in the ratio of lepton+jets and dilepton+jets $t\bar{t}$ final states. This indirect method is not affected by uncertainties in the $t\bar{t}$ production cross section [15].

Lower limits on $M_{H^{\pm}}$ of about 77 GeV, essentially independent of tanβ, have been obtained at LEP from searches for direct coupling of $Z \rightarrow H^{+}H^{-}$ [16], and a more model-dependent limit of $M_{H^{\pm}} > 244$ GeV has been extracted from the $b \rightarrow s\gamma$ transition studied at CLEO [17].

The results from CDF and DØ are given in Fig. 2 [15, 18] as a function of $M_{H^\pm}$ and tanβ, and are observed to exclude much of the phase space for tan$\beta < 1$ and tan$\beta > 30$. From the connection between tanβ and the branching fraction of $t \to Hb$, we can exclude the existence of a charged Higgs with $M_H < 120$ GeV, for $B(t \to Hb > 0.4)$, at $\sim 95\%$ confidence. The next run of the Tevatron is expected to reduce the unexcluded region of phase space by about a factor of two (as shown in Fig. 2), or, possibly, find the $H^\pm$.

4. HELICITY OF THE W AND SPIN CORRELATIONS IN TOP DECAYS

Spin provides another window for viewing the predictions of possible departures from the standard model. Two areas that have been studied at CDF and DØ involve the helicity of the W boson in top decay, and correlations among the decay products of the two top quarks in $t\bar{t}$ events. Given the $V-A$ form of the weak interaction, a top quark should decay into either a left handed or a longitudinally polarized W^+. This implies that charged leptons in $W \to \ell\nu_\ell$ decay tend to be emitted in a direction opposite to the line of flight of the W. The angular distribution of the lepton in the rest frame of the W, with the axis of quantization defined by the line of flight of the W, will therefore be asymmetric, and characterized by the fraction of left-handed W^+ in top decay (with helicities of -1), $f_{left} = 2M_W^2/(M_t^2 + 2M_W^2) = 1 - f_{long} \sim 0.3$. DØ has made preliminary studies to ascertain prospects for the next run, and CDF has already presented analyses of lepton p_T spectra for W decays in $t\bar{t}$ events in lepton and dilepton channels [19], yielding $f_{long} = 0.91 \pm 0.37 \pm 0.13$ (statistical and systematic uncertainties, respectively), in full agreement with the SM. The fits to lepton and dilepton top events are shown in Fig. 3.

The dominance of the $q\bar{q}$ incident channel for $t\bar{t}$ production, guarantees that the two top spins will tend to point along the same direction in the center of mass of the parton-parton collision. Because the lifetime of the top quark is $\sim 4 \times 10^{-25}$sec, and far shorter than hadronization time, the spin information carried by the top quarks is transmitted to their decay products. In fact, any depolarization could provide limits on the lifetime of top, and consequently on $\Gamma(t \to W + b)$ and $|V_{tb}|$.

For a polarized top quark, the angular distribution of the decay products in the top rest frame is given by $(1 + \alpha\cos\theta)/2$, where $\alpha = 1$ for the charged lepton or the d quark from W decay, and $|\alpha| \leq 0.41$ for the other decay products (W, ν, b or the up quark). (The α parameters for $\bar{t}$ have opposite sign to those for t.) Because of the difficulty of reconstructing down quarks from W decay, charged leptons would seem to offer the best means for extracting values of α. However, for interactions of unpolarized $p\bar{p}$, α cannot be measured in top decay. Nevertheless, α can be determined from the correlated distribution in the decay angles θ_+ and θ_- of the t and $\bar{t}$:

$$\frac{1}{\sigma}\frac{d\sigma}{d\cos\theta_+ d\cos\theta_-} = \frac{1 - \kappa\alpha_+\alpha_-\cos\theta_-\cos\theta_+}{4}$$

The value of κ depends on the axis of quantization chosen for analyzing the decays. The more standard axes of the incident beam ("Gottfried-Jackson" frame) or the line of flight of the top quark ("helicity" frame) are not the preferred ones, but instead there is an optimal axis, or "off diagonal" basis, as defined by the transformation [20]:

$$\tan\psi = \frac{\beta^{*2}\sin\theta^*\cos\theta^*}{1 - \beta^{*2}\sin^2\theta^*}$$

where ψ and θ^* are, respectively, the angle of the optimal axis and the angle for the line of flight of the top quarks, defined relative to the incident direction of the $\bar{p}$ in the parton-parton rest frame, and β^* refers to the velocity of the top quarks in that frame. In the off diagonal basis, the impact from opposite spin orientations of the top quarks (e.g., from gluon-gluon production) vanishes to leading order in α_{strong}, providing an expected value of $\kappa \sim 0.9$. To measure the decay angles, requires the full kinematic reconstruction of $t\bar{t}$ events. Unfortunately, dilepton events are kinematically underconstrained, and a special procedure was therefore developed at DØ [21] to handle the ambiguities and poor resolution brought about by the two missing neutrinos in these channels. Using its 6 dilepton events, DØ calculated all possible neutrino solutions, with smeared resolutions, and obtained a likelihood for each event permutation. These were added for all events, and are shown in the density plot in Fig. 3. A likelihood fit was then performed to signal (based on a spin-correlated $t\bar{t}$ Monte Carlo) and small sources of background, with κ as arbitrary parameter, which established that $\kappa > -0.25$ at 68% confidence [22], consistent with production through an intermediary gluon. A value of $\kappa \sim -1.0$ would correspond to an intermediary Higgs-like $J = 0$ boson.

Clearly, the results of spin studies to date have not been electrifying, however, with the statistics expected from the next run of the Tevatron, such data should provide delicate and sensitive tests of the SM.

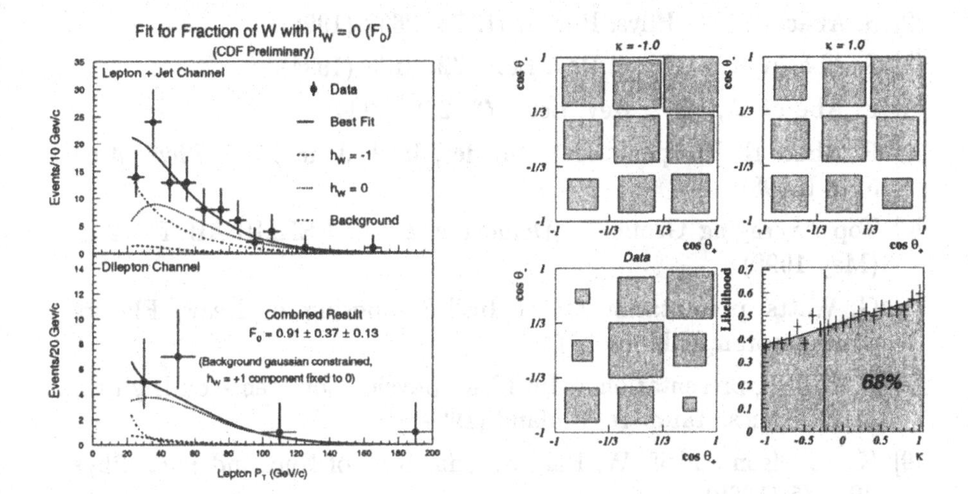

Figure 3 In left pane are results of a CDF fit to W helicities in $t\bar{t}$ events. The right pane is from a study of spin correlations in dilepton events reported by DØ.

5. CONCLUSION

Considering the small number of events collected thus far, the properties of the top quark are known to remarkable precision. The mass is 174.3 ± 5.1 GeV, the $t\bar{t}$ cross section (unofficial) is 6.2 ± 1.2 pb, the branching modes of the top quark are in line with expectation from $t \to W + b$ decay, and all observations are consistent with the SM. The upcoming enormous increase in statistical accuracy will hopefully reveal new interactions and the shortcomings of current theory.

Acknowledgments

I wish to thank my colleagues on CDF and DØ for their many contributions that have provided the basis for this review. I am especially grateful to Dhiman Chakraborty, Suyong Choi, John Conway, Mark Kruse, Ann Heinson, Andrew Robinson, Harpreet Singh, Eric Smith, Kirsten Tollefson, Gordon Watts and John Womersley for their input and helpful comments. Finally, I thank Misha Danilov and Harrison Prosper for relieving me of responsibilities as Director of the Advanced Study Institute on Techniques and Concepts of High Energy Physics, but allowing me to participate in the School.

References

[1] F. Abe *et al.*, Phys. Rev. Lett. **74**, 2626 (1995)

[2] S. Abachi *et al.*, Phys. Rev. Lett. **74**, 2632 (1995).

[3] S. Abachi *et al.*, Phys. Rev. Lett. **72**, 2138 (1994).

[4] F. Abe *et al.*, Phys. Rev. Lett. **73**, 225 (1994)

[5] P. Bhat, H. Prosper and S. Snyder, Int'l. Jour. Mod. Phys. **A13**, 5113 (1998)

[6] Top Averaging Group, L. Demortier *et al.*, FERMILAB TM 2084 (May 1999)

[7] G. Watts, presentation at 8th Int'l Symposium on Heavy Flavors, Southampton, U.K. (1999).

[8] F. Ptohos, presentation at Int'l Europhysics Conference on High Energy Physics, Tampere, Finland (1999).

[9] K. Tollefson and E. W. Barnes, Ann. Rev. of Nuc. and Part. Phys. **49**, 435 (1999).

[10] E. Laenen, J. Smith, and W. L. Van Neerven, Phys. Lett. **B321**, 254 (1994); E. L. Berger and H. Contopanagos, Phys. Rev. **D54**, 3085 (1996); S. Catani *et al.*, Phys. Lett. **B378**, 329 (1996); R. Bonciani *et al.*, Nuc. Phys. **B529**, 424 (1998).

[11] B. Abbott et al., Phys. Rev. Lett. **84**, 222(2000); LEP EW Working Group, CERN-EP/99-15; and for CDF's measurement see the report Fermilab-conf-99-173e.

[12] Report at DØ Workshop, Seattle (1999), and private communication from A. Heinson.

[13] H. Singh, presented at conference on Neural Computing in High Energy Physics, Israel (1999).

[14] Thinkshop: Meeting on top-quark physics for Run II (can be accessed at URL http://lutece.fnal.gov/thinkshop/default.html)

[15] F. Abe *et al.*, Phys. Rev. Lett. **79**, 357 (1997). Some of these results are about to be updated because of the recent change in the $\bar{t}t$ cross section measured by CDF.

[16] G Abbiendi *et al.*, Eur. Phys. J. **C7**, 407 (1999); M. Acciari *et al.*, Phys. Lett. **B446**, 368 (1999); R. Barate, *et al.*, Phys. Lett. **B450**, 467 (1999).

[17] M.S. Alam *et al.*, Phys. Rev. Lett. **74**, 2885 (1995).

[18] B. Abbott *et al.*, Phys. Rev. Lett. **82**, 4974 (1999), and D. Chakaborty in Proc. of Hadron Collider Physics Meeting at Mumbai, India (1999), World Scientific Pub., B. S. Acharya *et al.*, eds.

[19] For CDF measurements, see their public web page http://www-cdf.fnal.gov/physics/new/top/latest_results.html.

[20] G. Mahlon and S. Parke, Phys. Lett. **B411**, 173 (1997).

[21] B. Abbott *et al.*, Phys. Rev. **D60**, 052001 (1999).

[22] B. Abbott *et al.*, Phys. Rev. Lett **85**, 256 (2000).

ACCELERATOR PHYSICS AND CIRCULAR COLLIDERS

A reading guide

John M. Jowett
Accelerator Physics Group,
SL Division,
CERN
CH-1211 Geneva 23
John.Jowett@cern.ch

Keywords: Accelerator, beam-beam, betatron, synchrotron

Abstract These lectures provided a tutorial introduction to the physics of particle accelerators illustrated with applications to circular colliders of the present and near future. Unfortunately, it was not possible to write them up completely. Instead, we provide a guide to introductory literature at a suitably advanced level.

1. ACCELERATOR PHYSICS CONCEPTS

The first lecture introduced some physical principles, key concepts and terms of accelerator physics. It concentrated on ideas that are generally useful in present day circular colliders for high energy physics. Although the basic periodicity of a circular machine is fundamental to its beam dynamics, many of the concepts introduced here carry over to linear colliders too.

Luminosity, L, is the quantity which, together with the relevant cross-section, σ determines the event rate σL in a collider. It is introduced, in varying ways, in many texts and papers. A practical compendium of formulas is given in [1], pp. 247–251. In fact, this handbook is an extremely useful reference in itself for almost all topics in modern accelerator physics. It provides many further references as a starting point for further study.

H.B. Prosper and M. Danilov (eds.), Techniques and Concepts of High-Energy Physics, 381–387.

Trivial it may be but it is worth noting the relationships between some possible units of luminosity, in order of decreasing frequency of use:

$$10^{30}\,\mathrm{cm^{-2}s^{-1}} = 0.0864\,\mathrm{pb^{-1}day^{-1}} = 1\,\mu\mathrm{b^{-1}s^{-1}} = 10^{34}\,\mathrm{m^{-2}s^{-1}}. \quad (1.1)$$

Much of this lecture was devoted to the analysis of single-particle motion in a circular accelerator with alternating-gradient focusing. It can fairly be said that the concepts of closed orbit, transverse betatron oscillations, tune, transfer matrices, Twiss functions (most importantly the ubiquitous β-function), linear perturbations, beam size and emittances form the theoretical, and terminological foundation for most of accelerator physics. They help understand the main features of colliders and provide the basic intuitions needed for qualitative understanding of the more complicated phenomena that limit performance.

A thorough treatment of these topics at a mathematical level appropriate for participants in this institute can be found in a recent textbook [2]. Other modern texts with somewhat different approaches are [3, 4, 5]. The early lectures [6] remain very popular as an elementary introduction. Other comprehensive accounts by several authors can be found in proceedings of the CERN and US Particle Accelerator Schools [7, 8].

Highly individual theoretical discussions going far beyond the basic theory can be found in [9, 10].

A general reference for recent developments in the accelerator world, including status reports on the various projects is [11].

Finally, a tour of the optical design of the LEP collider (for reviews, see [13, 14]) was used as a concrete illustration.

2. PRESENT DAY CIRCULAR COLLIDERS

This lecture started was a survey of the main features of some currently operating colliders interspersed with further accelerator physics topics.

DAΦNE, a double ring collider that recently started operation at the very low beam energy of 0.51 GeV, is one of the most challenging colliders ever constructed. A general project description can be found in [15] and details of the interaction region design in [16, 17]. Progress to the date of this school is summarised in [18].

The introduction of basic concepts continued with the extension of the transfer matrix formalism to treat off-momentum orbits and the introduction of the (badly-named) dispersion function, momentum compaction and transition energy. This led on to the treatment of the third degree of freedom, longitudinal or "synchrotron" motion and the role of the radio-frequency accelerating system. These topics are well covered in the textbooks [2, 3, 4, 5, 7, 8].

Normally, in the space of a short course like this, there is no time to discuss the many-particle collective phenomena that arise with high intensity beams. Numerous notions such as wake-fields, impedances, beam distributions, collective modes, etc. have to be introduced and there are many types of instability to discuss. Nevertheless, in order to give a flavour of the subject, a single instability mechanism, the so-called Transverse Mode-Coupling Instability was chosen for treatment by means of a two-macro-particle model of a particle bunch. This example is both non-trivial, in that it involves coupling of longitudinal and transverse motions and important, in that it is often the ultimate limit on bunch intensity in large rings such as LEP. Furthermore the model provides a respectable approximation to the full theory; a similar treatment and full references can be found in [26], a comprehensive text on this subject. Other introductions can be found in [2, 7, 8].

In e^+e^- rings (and, to some extent, the largest hadron colliders), synchrotron radiation [25] introduces dissipative terms into the particle equations of motion, resulting in *radiation damping.* Furthermore the quantum fluctuations of this radiation lead to the finite beam size and energy spreads of the beams–these can be regarded as macroscopic quantum effects. For treatments of this topic consult [6, 3, 2, 1]; [27] derives the fundamentals from a statistical-mechanical viewpoint.

The vertical emittance of the beam arises from several sources, including these radiation effects and is usually an important performance factor determining the luminosity. Monte-Carlo of machine imperfections and operational correction procedures obtained good statistical agreement with measured vertical emittances in LEP [19].

Wiggler magnets are an integral part of the design of DAΦNE, LEP, PEP-II [20], KEK-B [21] and other machines. Their functions in e^+e^- colliders are quite different different from those in synchrotron light sources. They can help to fight instabilities by increasing the damping effect of synchrotron radiation or lengthening the particle bunches, modify emittances to maximise luminosity or enhance the spin-polarization of the beams [1, 3, 2].

The electromagnetic interactions of colliding beams, the *beam-beam effect* (really many effects) is a fundamental limit to the luminosity. The concepts introduced in the first lecture were sufficient to motivate the introduction of the famous beam-beam tune-shift parameter in the framework of linear perturbation theory and to show its role in beam-beam phenomenology. Particularly clear presentations of the incoherent and coherent aspects of beam-beam effects can be found in [22, 23].

More recent discussions of beam-beam effects can be found in numerous articles in [7, 8, 12, 11]. Extremely high values of this parameter have recently been obtained at LEP [24].

These concepts were illustrated by the examples of modern-day "factory" colliders such as PEP-II, KEK-B, and DAΦNE. Interaction region design is particularly important and relevant to the experimenters so some attention was paid to this.

At the time of these lectures, the performance of these machines is rising rapidly; the latest status information is presented at the particle accelerator conferences [11]. Comparison with expectations (see, e.g., the workshop [12] that took place shortly before the current generation of e^+e^- colliders came on line) is generally very favourable, showing that the e^+e^- circular collider technology is now mature; performance is quite predictable even when the beam parameters push hard against fundamental physical limits.

So far, there have been just a handful of hadron colliders built in the world and it is fair to ask whether the same can be said for them, especially given the significant differences among the colliders built to date.

3. FUTURE CIRCULAR COLLIDERS

Circular colliders will dominate experimental particle physics at both the energy and luminosity frontiers for many years to come. In a few years' time, the LHC will become the main energy frontier machine for the whole world and we discussed its design and parameters. The latest information, including many detailed reports, on this large project can be found at [28].

The design of superconducting hadron colliders such as HERA, the Tevatron or LHC is dominated by issues connected with superconducting magnet technology; [29] is a text on this subject. We briefly discussed the latest design and the field quality results of the LHC magnet prototypes [30, 11].

Field quality is important because higher multipole components in the magnetic fields lead to non-linear terms in the equations of transverse betatron motion. Single-particle dynamics becomes much more complex and there is a need to understand the stability of particle motion over very large numbers of turns around the ring. The zone of stable phase space around the closed orbit is called the "dynamic aperture" and considerable effort has been deployed to compute it for hadron colliders. A brief discussion of non-linear resonances was given in the lectures and this and further physics of non-linear particle dynamics is discussed in [2, 31, 9, 10, 7, 8].

As the performance of colliders is pushed further, new classes of instability have arisen. One example, illustrated qualitatively in the lectures is the so=-called electron cloud instability [32, 33], currently an active field of research. This class of instabilities is relevant both to high-intensity double-ring e^+e^- factories and to hadron colliders like the LHC.

The lectures closed with a short discussion of a new type of "circular" collider, for muons. Although this is also an rapidly evolving field, a good expansive introduction can be found in [35]. Muon colliders and the related neutrino factories throw up many formidable accelerator physics challenges some of which are inciting the direct involvement of experimental particle physicists. This is a point I would like to stress to this audience—we should not be shy of coming out of our compartments.

Despite doubts occasionally expressed in the past, we have not come to the end of the line with circular colliders for particle physics at either the energy or luminosity frontier. They will continue to challenge the ingenuity of physicists for many years yet.

References

[1] A.W. Chao, M. Tigner (Eds.), *Handbook of Accelerator Physics and Engineering*, World Scientific, 1999. Comprehensive reference (not a text).

[2] S.Y. Lee, *Accelerator Physics*, World Scientific, Singapore 1999.

[3] H. Wiedemann, *Particle Accelerator Physics: Basic Principles and Linear Beam Dynamics,* Springer-Verlag, Berlin, 1993.

[4] D. Edwards & M. Syphers, An Introduction to the Physics of High Energy Accelerators, Wiley, New York, 1993.

[5] M. Conte, W.M. MacKay, *An Introduction to the Physics of Particle Accelerators.* World Scientific, Singapore, 1995.

[6] M. Sands, The Physics of Electron Storage Rings, SLAC-121 (1970).

[7] CERN Accelerator School, e.g., Fifth General Accelerator Physics Course, Ed. S. Turner, CERN Report 94-01 (1994). `http://www.cern.ch/Schools/CAS/`

[8] US Particle Accelerator School, AIP Conference Proceedings, Nos. 87, 153, 184, ... `http://fnalpubs.fnal.gov/uspas/`

[9] L. Michelotti, Intermediate Classical Dynamics with Applications to Beam Physics, Wiley, New York 1995. Mathematically-oriented modern dynamics text with accelerator applications.

[10] E. Forest, Beam Dynamics - A new attitude and framework, Harwood Academic, Amsterdam 1998. Modern approach to single particle dynamics via maps, etc.

[11] Joint Accelerator Conference Website for the Asian, European and American Particle Accelerator Conferences, ICALEPCS and LINAC conferences. Accelerator Conferences, `http://accelconf.web.cern.ch/accelconf/`

[12] *Proceedings of the Advanced ICFA Beam Dynamics Workshop on Beam Dynamics Issues for* e^+e^- *Factories*, Eds. L. Palumbo and G. Vignola, , Frascati Physics Series No. 10, Frascati, 1998.

[13] J.M. Jowett, in [12]; also CERN SL/98-029 (AP).

[14] D. Brandt *et al*, Rep. Prog. Phys. 63 (2000) 939.

[15] G. Vignola, in [12].

[16] S. Guiducci, in [12].

[17] H. Koiso, in [12].

[18] M. Zobov, *Status Report on DAΦNE Performance*, Proceedings of EPAC 2000, Vienna, available via [11].

[19] J.M. Jowett, *Realistic Prediction of Dynamic Aperture and Optics Performance for LEP*, Proceedings of US PAC 1999, New York, available via [11].

[20] J.T. Seeman, in [12].

[21] K. Oide, in [12].

[22] A.A. Zholents, *Beam-beam Effects in Electron-Positron Storage Rings*, Proc. Joint US-CERN School on Particle Accelerators, Lecture Notes in Physics 400, M. Dienes, M. Month, S. Turner (Eds.), Springer-Verlag, 1990.

[23] A.W. Chao, *Coherent Beam-beam Effects* , same proceedings as [22].

[24] R. Assmann, K. Cornelis, *The Beam-beam Interaction in the Presence of Strong Radiation Damping*, Proceedings of EPAC 2000, Vienna, available via [11].

[25] J.D. Jackson, Classical Electrodynamics, Wiley, New York 1975.

[26] A.W. Chao, Physics of Collective Beam Instabilities in High Energy Accelerators, Wiley, New York 1993.

[27] J.M. Jowett, Introductory Statistical Mechanics for Electron Storage Rings, M. Month and M. Dienes (Eds.), "Physics of Particle Accelerators", AIP Conf. Proc. 153 (1987) 864.

[28] LHC, The Large Hadron Collider Project, `http://lhc.web.cern.ch/lhc/`.

[29] K.-H. Mess, S. Wolff, P. Schmüser *Superconducting Accelerator Magnets*, World Scientific, Singapore, 1996.

[30] M. Modena *et al*, *Manufacture and Performance of the LHC Main Dipole Final Prototypes*, Proceedings of EPAC 2000, Vienna, available via [11].

[31] H. Wiedemann, *Particle Accelerator Physics II: Nonlinear and Higher-Order Beam Dynamics,* Springer-Verlag, Berlin, 1995.

[32] http://www.aps.anl.gov/conferences/icfa/two-stream.html

[33] Electron Cloud in the LHC, http://wwwslap.cern.ch/collective/electron-cloud/electron-cloud.html

[34] ICFA 8th Advanced Beam Dynamics Mini-Workshop on Two-Stream Instabilities in Particle Accelerators and Storage Rings, http://wwwslap.cern.ch/~jowett/accelerator.html

[35] D. Neuffer, μ^+-μ^- Colliders, CERN Report 99-12 (1999).

WORKSHOP ON CONFIDENCE LIMITS

A Report

Harrison B. Prosper
Department of Physics, Florida State University
Tallahassee, FL 32306, USA *
harry@hep.fsu.edu

Abstract In early 2000, workshops were held at CERN and at Fermilab to discuss the topic of confidence limits. I report on some of the key issues that were addressed at these meetings.

1. INTRODUCTION

In January 2000, Fred James and Louis Lyons convened a meeting at CERN to discuss confidence limits. The meeting attracted about 140 physicists, many of whom were happy to voice strongly held opinions about this topic. A disinterested observer might well puzzle over why such an arcane topic should elicit such intense interest. The reason of course is that we high energy physicists insist on looking for things that most probably do not exist. Having found nothing interesting we feel obliged, nonetheless, to conclude something scientifically useful from our null result. Our conclusions are invariably stated in terms of confidence limits.

The meeting was unusual not only because of the topic of discussion, and the fact the meeting was the first of its kind in our field, but also because of the presence and active participation of professional statisticians. Mike Woodroofe and Peter Clifford attended the CERN workshop while Jim Berger attended the follow up meeting at Fermilab. These colleagues did a splendid job of politely noting and correcting some of our misconceptions. Both meetings were extremely helpful in clarifying ideas and points of view about this troublesome topic. One point became abundantly clear to the participants: much has happened in statistics since the 1930s about which we physicists are, for the most

*Work supported in part by the U.S. Dept. of Energy under Grant No. DE-FG02-97ER41022

H.B. Prosper and M. Danilov (eds.), Techniques and Concepts of High-Energy Physics, 389–399.

part, completely ignorant. It was generally agreed that we would do well to update our understanding of the concepts and terminology of statistics, if only so that we might converse more fruitfully with our statistician friends about problems of mutual interest.

A complete record of the deliberations of the CERN workshop can be found in a CERN Yellow book[1]. In this lecture I shall touch upon the key issues that arose from the numerous formal and informal discussions.

Notation

Unless stated otherwise, I shall use the following notation.

N	Number of observed events
s	Mean number of signal events
b	Mean number of background events
$\hat{s}$	Estimate of s
$\hat{b}$	Estimate of b
$\overline{s}$	Upper limit ob s
$\underline{s}$	Lower limit on s
$P(A\|B)$	Probability of A given B

2. GOAL OF WORKSHOP

The problem addressed by the workshop can be stated as follows: Given N observed events (with N, typically near zero) and an estimate $\hat{b}$ of the mean background count, deduce a statement of the form

$$\text{Prob}\left[\underline{s}(N,\hat{b}) < s < \overline{s}(N,\hat{b})\right] \geq \text{CL}, \tag{1.1}$$

where CL is the confidence level (e.g., CL $= 0.95$). The results of the KARMEN[2] experiment, $N = 0$, $\hat{b} = 2.8$, is an example of a null result from which one may wish to deduce such a statement. The goal of the workshop was to improve general understanding of the meaning of Eq. (1.1).

The organizers of the workshop clearly understood the need to demand clarity from the speakers if the workshop was to be successful. To that end, Fred James stated some ground rules that speakers were asked to follow:

- Give the mathematical basis of the proposed method for constructing confidence limits.
- Methods should be based on accepted principles of statistics.
- *Frequentists* should

  - Give the coverage properties of their limits.
  - Give the ensemble used to define the coverage.
- *Bayesians* should
  - Define probability.
  - Discuss issue of priors.

Fred also asked Bayesians to make some comments about probability and quantum mechanics. This topic is extremely interesting, but is beyond the scope of this lecture.

3. MAIN ISSUES

Three issues dominated the discussions:

- The nature of probability.
- The meaning of confidence limits.
- The handling of nuisance parameters.

At the end of the CERN workshop, Bob Cousins summarized the points on which there was thought to be general agreement. Below I report on each of these issues in turn.

3.1. WHAT IS PROBABILITY?

In the context of the workshops, this question can be re-stated more concretely as: What do *we* wish Eq. (1.1) to mean? That is, what meaning do we wish to ascribe to the symbol "Prob?" It may not be obvious at first that we have a choice. In fact, we do because the theory of probabilityis a purely mathematical theory, which, being mere mathematics, does not provide an *interpretation* of the symbols for which theorems have been derived. Probability theory is not a physical theory. Unlike a law of physics, we cannot ask Nature for her opinion regarding the meaning we have ascribed to probability.

It was possible, however, to gauge the opinion of the workshop participants, the overwhelming majority of whom took probability to be synonymous with *relative frequency.* Probability, according to this interpretation, is simply the relative frequency with which a given outcome is realized in an unlimited number of trials, in each of which different outcomes are possible a priori. This interpretation of probability became popular in the latter half of the 19th century and is associated with the likes of Venn and Boole and later von Mises, Fisher and Neyman to name but a few of the chief proponents.

A century earlier, probability was interpreted by Bayes and his contemporaries as a *degree of belief*, an interpretation that was later vigorously championed by the mathematical physicist Pierre Simon de Laplace. Probability, according to Laplace—the famed author of the *Mechanique Celeste*, is merely a formal way to quantify how strongly one should believe a given statement. People who subscribe to Laplace's point of view are generally referred to as *Bayesians*, in honor of the Reverend Thomas Bayes' seminal paper of 1763[3] which introduced a special case of a theorem that now bears his name. People of the other persuasion are called *Frequentists*.

The meaning of Eq. (1.1) therefore depends on which interpretation of probability one chooses to adopt. I summarize the essential differences in the table below.

Degree of Belief	Relative Frequency
This *singular* statement $\underline{s} < s < \overline{s}$ should be accepted as true with a degree of belief of $100 \times$ CL%.	Statements of this type $\underline{s} < s < \overline{s}$ are true $100 \times$ CL%. of the time.
The probability CL does not need an ensemble to define it	The probability CL is defined only with respect to an ensemble of statements.

There was much debate and discussion about the essential difference between the statements Prob(*Theory*|*Data*) and Prob(*Data*|*Theory*), a point that was emphasized in talks by Bob Cousins. In particular, Cousins noted the necessity of specifying a prior probability if one wished to go from one probability to another via Bayes' Theorem

$$\text{Prob}(Theory|Data) \propto \text{Prob}(Data|Theory)\,\text{Prior}(Theory). \qquad (1.2)$$

The probability Prob(*Data*|*Theory*) is proportional to the likelihood, and of the three probabilities that appear in Bayes' Theorem, it is the only one that is used in Frequentist analysis. Consequently, no statement of the form "the probability that the theory is true given the data" can ever be made. Yet this is precisely the kind of statement that most physicists wish to make, or make unwittingly, in spite of their claim to be champions of Frequentist methods. The purported advantage of these methods is that they avoid the need to agonize over prior probabilities.

Agony over priors, however, is replaced by disagreements about which (ficticious) ensemble is to be used to define the frequencies.

On the other hand, the need for a Prior(*Theory*) (the prior probability for the theory or hypothesis being advanced) is unavoidable in a Bayesian analysis. Unfortunately, its specification, especially for multi-parameter models, can be extremely difficult and is often open to the charge of arbitrariness. But the use of a prior allows one to compute Prob(*Theory*|*Data*) and thus make the kind of statements we wish to make.

In statistics, it would seem, it is not possible to have your cake and eat it!

3.2. WHAT ARE CONFIDENCE LIMITS?

In view of the two rather different interpretations of probability it should come as no surprise that the interpretations of confidence limits differ for Frequentists and Bayesians. In this section, I shall summarize three of the Frequentist methods for finding confidence limits that were discussed at the CERN workshop, and end with a brief statement about what a Bayesian would do.

3.2.1 Neyman. Neyman introduced the concept of confidence intervals in his classic 1937 paper[4]. At the Fermilab workshop, Jim Berger alerted us to the fact that many statisticians today who consider themselves to be card carrying Frequentists do not necessarily subscribe to all things Neyman. However, most high energy physicists, when pressed, would admit to being thoroughgoing "Neymanists!" Therefore, it is important to be clear about what Neyman said in his 1937 paper.

For clarity, I'll describe Neyman's idea in the context of the kind of problem that arises in experimental high energy physics: one searches for something, finds few or no events and constructs an interval that is designed to bracket the mean signal count with some probability, in a sense that I shall now make clear.

Suppose, by whatever procedure, one has constructed a set of intervals $[\underline{s}(N), \overline{s}(N)]$, indexed by the observed count N. In general, for each value of N the interval will be different. Now imagine an ensemble in which each value of N occurs in proportion to its relative frequency, that is, its probability if you're a Frequentist. That ensemble is characterized by a *fixed* but unknown value for the mean signal count s. Although one does not know which intervals contain the mean signal count, one knows a priori that some *fraction* of the intervals in this ensemble will contain the mean signal count s. That fraction is called the *coverage probability*, or coverage, for the ensemble. Neyman required that the coverage probability be greater than or equal to a specified fraction CL, the confidence level. But, the crucial thing is that he required the coverage *never* to fall below CL for *all* a priori possible ensembles, each characterized by a different value of s. Neyman went further. Not only must the coverage match or exceed CL for all possible fixed values of s but it must do so for *all* possible fixed values of the other parameters on which the problem depends. For example, if the experimental results depend on two parameters, say the mean signal s and mean background b, then the coverage criterion should hold for all possible fixed values of s and b.

Neyman's idea is simple and elegant: whatever the actual value of the mean signal count in your experiment, provided that you construct the set of intervals $[\underline{s}(N), \overline{s}(N)]$ appropriately, you are guaranteed that your particular experiment will have a $100 \times$ CL % chance of yielding an interval that will bracket the mean signal count s. Moreover, as Bob Cousins and Gary Feldman noted[1, 6], this property holds over a collection of actual experiments, each of which may be measuring a different physical quantity! The claim is that if you consider the set of high energy physics experiments that have been performed over the past half-century and that have published, say, 90% limits, about 90% of them should have yielded limits that bracket whatever quantity was being measured. However, that claim can be challenged[5].

Neyman showed that it is possible to construct intervals having this remarkable property. For a discrete distribution $p(n|s)$, an infinite number of sets of limits can be computed as follows

$$\sum_{n=0}^{N} p(n|\overline{s}) = \alpha_L, \tag{1.3}$$

$$\sum_{n=N}^{\infty} p(n|\underline{s}) = \alpha_R, \tag{1.4}$$

where $\alpha_L + \alpha_R = 1 - \text{CL}$. To render the set of limits unique one often takes $\alpha_L = \alpha_R$ and thus arrive at *central limits*. Before 1998, the Particle Data Group advocated this method of constructing Poisson upper limits. For example, when one takes $\alpha_R = 0$, in order to get upper limits, one obtains the well-known result $\overline{s} = 2.3$, for $N = 0$ and $\text{CL} = 0.90$.

However, a careful reading of Neyman's paper reveals that there is considerable freedom in how one might construct confidence limits, an observation that was cleverly exploited by Feldman and Cousins in their "unified approach," described next.

3.2.2 Feldman and Cousins: The Unified Approach.

In an interesting paper[6], Gary Feldman and Bob Cousins solved two problems: 1) to construct non-negative intervals in the presence of background b even when the latter exceeds the observed count N and 2) to switch smoothly between one-sided and two-sided limits in a way that satisfies the Neyman criterion. For one-parameter problems, this method is optimal; therefore, every other Frequentist method must necessarily give larger limits and or greater over-coverage.

Moreover, the method is reasonably simple to implement. Here is the algorithm described for the case of precisely known background $\hat{b} = b$:

Feldman-Cousins algorithm

for each s in $[0, s_{\max}]$

{

for each N in $[0, N_{\max}]$

{

compute $\lambda(N) = \frac{p(N|s,\hat{b})}{p(N|\hat{s},\hat{b})}$,

with

$$\hat{s} = \begin{cases} N - \hat{b} & \text{if } N > \hat{b} \\ 0 & \text{otherwise} \end{cases}$$

}

sort sequence $\{\lambda_i\}$ in decreasing order

$sum = 0$

$i = 1$

while $sum <$ CL

{

$sum = sum + p(N_i|s, \hat{b})$

with

$N_i = \arg_N \lambda_i(N)$

(i.e., the N associated with λ)

$i = i + 1$

}

}

(Note that each N is associated with a *set* of s values)

for each N in $[0, N_{\max}]$

{

$\underline{s}(N) = \min s$

$\overline{s}(N) = \max s$

}

In this example the likelihood $p(n|s,b)$ depends on the single parameter s, since b is assumed to be perfectly known. Consequently, by construction, the Feldman-Cousins limits satisfy the Neyman criterion. When applied to the KARMEN data, $N = 0$ and $\hat{b} = 2.8$, assuming the background uncertainty to be negligible, one obtains $[\underline{s}, \overline{s}] = [0.0, 1.71]$. Note, however, the following property, which is shared by other methods: if the KARMEN collaboration worked hard to reduce the background so that $\hat{b} = 0$, they would find that the interval $[\underline{s}, \overline{s}]$ broadens to $[0.0, 3.09]$, that is, they would obtain a worse limit.

There was an inconclusive debate at the CERN workshop about whether or not this behaviour is pathological. Feldman and Cousins[6] noted that the better experiment is better *on the average* than the worse experiment. This can be quantified by inventing a measure of robustness that tells us how much the limit would change on average were we to repeat the experiment many times. Feldman and Cousins suggested that, in addition to the limits, one publish the mean upper limit $< \overline{s} >$ in the absence of signal. That number would be larger, that is, worse, for the experiment with the larger background. This is a sensible proposal. However, it is not clear how that information could actually be used to construct, for example, an exclusion plot.

3.2.3 Alex Read: The $\mathbf{CL_S}$ Method. In the course of their systematic efforts to find the Higgs boson the LEP physicists have devised a novel way to characterize a null result that (unfortunately[1]) they refer to as the CL_S method. They compute two p-values (that is, tail probabilities)

$$CL_{s+b} \equiv \text{Prob}(q \leq Q|s + b)$$

and

$$CL_b \equiv \text{Prob}(q \leq Q|b),$$

where Q is the observed value of some observable q, such as the number of events N. They then define CL_S as

$$CL_s \equiv \frac{CL_{s+b}}{CL_b}.$$

From this quantity an upper limit can be extracted by equating CL_s to an agreed upon value.

The problem with this method is that, from the point of view of a Frequentist, it is not clear what CL_s means. However, one could adopt a charitable attitude that sidesteps the interpretation of CL_s and merely demand that the limits so obtained satisfy Neyman's criterion. Alex Read reported that the LEP limits do indeed possess reasonable

[1]Warning: CL_S is not a confidence level, in spite of its name!

Frequentist properties and therefore, in that regard, the precise meaning of CL_s is unimportant.

From a Bayesian viewpoint, the CL_s method can be viewed as a Frequentist mock-up of a Bayesian procedure in which the ratio of the probability of two hypotheses, signal plus background and background, is computed. A Bayesian would compute

$$\frac{\text{Prob}(s+b|Q)}{\text{Prob}(b|Q)} = \frac{\text{Prob}(Q|s+b)}{\text{Prob}(Q|b)} \times \frac{\text{Prior}(s+b)}{\text{Prior}(b)} \tag{1.5}$$

and report how much more likely one hypothesis is than the other.

3.2.4 Bayesian. For a Bayesian the calculation of a Bayesian confidence interval, or *credible interval*, is straightforward in principle. One merely solves the following

$$\int_{\underline{s}(N)}^{\overline{s}(N)} \text{Prob}(s, \hat{b}|N) = CL, \tag{1.6}$$

where

$$\text{Prob}(s|N, \hat{b}) = \frac{\text{Prob}(N|s, \hat{b})\text{Prior}(s)}{\int_s \text{Prob}(N|s, \hat{b})\text{Prior}(s)}, \tag{1.7}$$

and $\text{Prior}(s) \equiv f(s)ds$ is the prior probability. Normally, one finds the values of $\underline{s}$ and $\overline{s}$ that minimizes the length of the interval; that is, one computes what Bayesians call the *highest posterior density* (HPD) intervals. Like the Feldman-Cousins intervals, these intervals provide a smooth transition from one to two-sided intervals.

The practical problem of the Bayesian approach is deciding how to choose $f(s)$. Some physicists[1, 6] argue that $f(s)$ should always be subjectively chosen, because herein lies the power of the Bayesian method. Of course, the inferences arrived at would depend necessarily on the opinion of the physicist, whose opinion you may not share. This is clearly unsatisfactory as a way to report results "objectively"; and for this reason, the Bayesian approach is declared unfit for scientific reporting. Other physicists (see, for example, Ref. [5]) argue that it is perfectly reasonable to adopt a convention or to arrive at $f(s)$ by applying some sensible principles, such as Jeffreys[7] idea that inferences should be invariant with respect to how the problem has been parameterized. For a contrary viewpoint, see D'Agostini[8].

The problem is that many plausible arguments have been suggested that yield, unfortunately, different functional forms for $f(s)$. However, I believe this is not a good reason to reject the Bayesian approach. If we did so on the grounds that it is "subjective" we would be forced to reject *all* approaches, since each one contains irreducible elements of subjectivity, and or arbitrariness and or convention.

3.3. HOW SHOULD ONE HANDLE NUISANCE PARAMETERS?

A nuisance parameter is any parameter which must, somehow, be removed from the problem in order to make inferences on the parameters of interest. For example, in most experiments the backgrounds and acceptances are usually known imprecisely. Nonetheless, the true values of these parameters must somehow be excised from the problem in order to make a statement about the cross section only. Apart from possible computational difficulties this is not an issue for Bayesians: one simply integrates the nuisance parameters out of the problem.

The Frequentist, however, does not have that option. Consequently, the problem of nuisance parameters is rendered difficult. In practice, what is usually done is to replace all nuisance parameters by estimates thereof and thereby convert the exact likelihood function into an approximation called the *profile* likelihood. The profile likelihood is a one-parameter object to which any of the usual Frequentist methods can be applied. In principle, one should study the frequency properties of the procedure to verify that approximate coverage is obtained over the whole parameter space. I would hazard a guess, however, that this is seldom done, the main excuse being the large the computational burden involved.

3.4. WHAT CAN WE AGREE ON?

At the end of the workshop Bob Cousins tried to summarize the points on which there was thought to be broad agreement. Happily, most of the points listed below were uncontroversial.

- Bayesian limits depend on priors.
- Bayesian limits do not need ensembles to define them.
- The likelihood function should be published.
- The usual goodnes-of-fit tests do not exist in Bayesian statistics [2].
- Frequentist intervals only make sense with respect to an ensemble and they are calculated from Prob($Data|Theory$) only.

[2] Jim Berger has, however, made considerable progress recently on this question[9].

- To compute Prob($Theory|Data$) requires the use of a prior probability.
- The coverage of Frequentist intervals should always be checked.
- Sensitivity to priors should always be checked.

4. CONCLUSIONS

The workshops at CERN and Fermilab succeeded in clarifying several issues regarding confidence limits. However, much disagreement remains about which method or methods should be recommended. One thing that became clear is that the Feldman-Cousins limits have rapidly gained acceptance in certain experiments. Most objections to their method I found unconvincing, except my own: it is not Bayesian!

It was also clear that Bayesian-inspired methods are widely used, by high energy physicists, in conjunction with Frequentist ideas. Unfortunately, all too often, the consequence has been a lack of clarity and some confusion. Our relative ignorance of the advances made in statistics since the 1930s prompted many of us to resolve to do the following: read more modern books on statistics and learn the jargon!

References

[1] James, F., Lyons, L. and Perrin, Y. (2000) *Workshop On Confidence Limits, Proceedings*, CERN, Geneva, Switzerland.

[2] Drexlin, G. et al. (1990) *Nucl. Instrum. Methods* A **289**, 490.

[3] See for example, Jeffreys, H. (1961) *Theory of Probability*, Clarendon Press, Oxford.

[4] Neyman, J. (1937) *Phil. Trans. R. Soc.* A **236**, 333.

[5] Prosper, H.B. (2000) Bayesian Analysis, CERN Report **2000-005**, 29.

[6] Feldman, G.J. and Cousins, R.D. (1998) Unified approach to the classical statistical analysis of small signals, *Phys. Rev.* D **57**, 3873.

[7] Jeffreys, H. (1961) *Theory of Probability*, Clarendon Press, Oxford.

[8] D'Agostini, G. (2000) Confidence Limits: What is the Problem? Is There *the* Solution, CERN Report **2000-005**, 3.

[9] Berger, J., http://www.stat.duke.edu/~berger/papers.html.

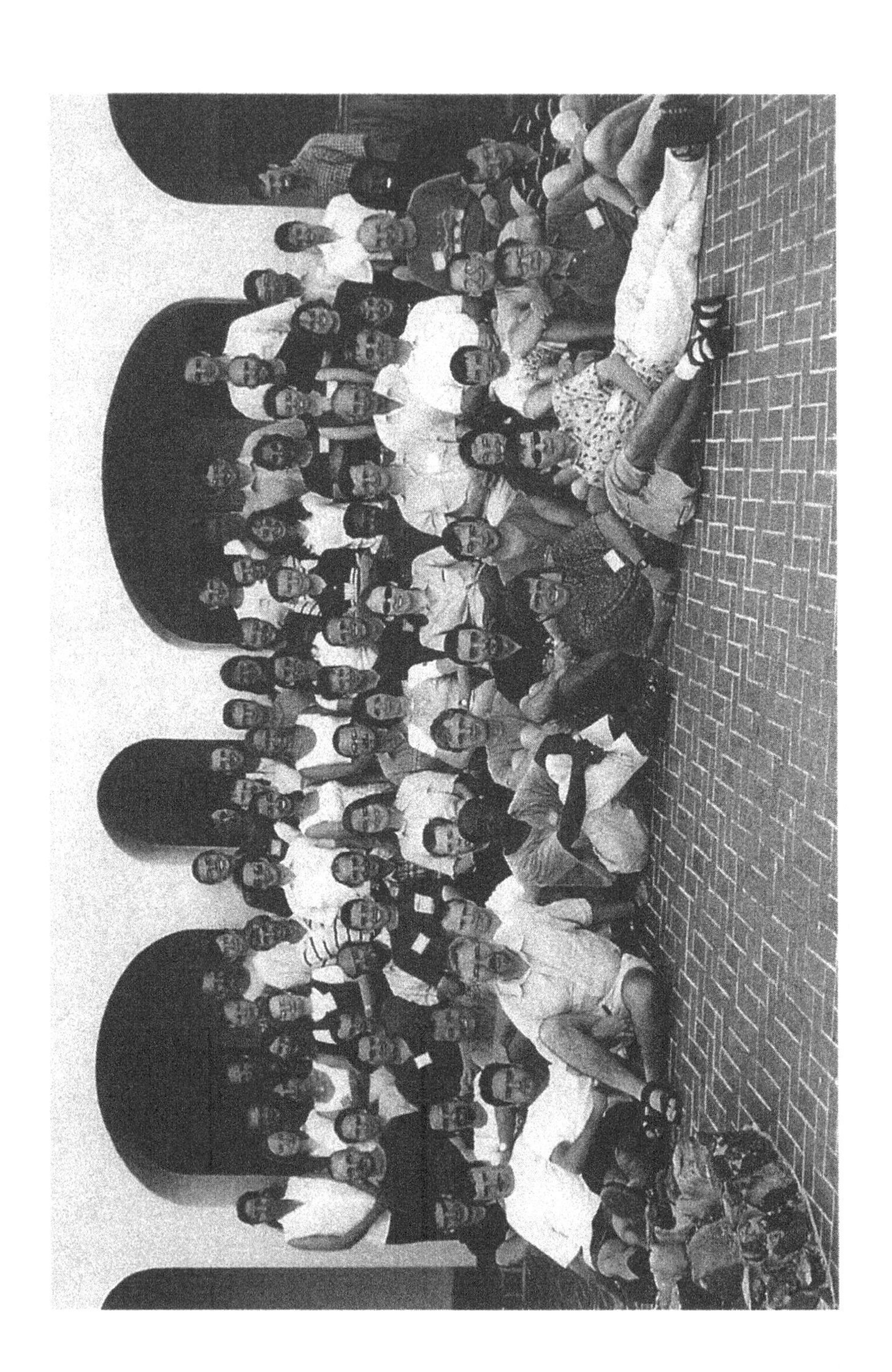

Participants at the ASI (scanning from left to right): Jamie Boyd, Awatif Belymam, Roman Mizuk, Carsten Hensel, Elaine McLeod, Ben West, Ricardo Goncalo, Karel Soustruznik, Maria de Jesus Varanda, Joao Afonso Bastos, Elliot Lipeles, Krzysztof Cieslik, Robyn Madrak, Kamil Sedlak, Marian Zdrazil, Ilya Narsky, Natalia Kuznetsova, Andreas Schwarz, Ron Ruth, Yuri Dokshitzer, Aron Soha, Misha Danilov, Bruce Knuteson, Konstantin Anikeev, Borislav Pavlov, Piotr Niezurawski, Valery Rubakov, Sven Schaller, Marcela Carena, Louise Aspinwall, Harrison Prosper, Tuba Conka-Nurdan, Torsten Jagla, Stephen Asztalos, Anne Oppelt-Pohl, Peter Walsham, Alexander Kupco, Bruno Serfass, Elisa Bernardini, Sean Carroll, Janet Conrad, Magda Pedace, Massimo Casarsa, Alessandro Cerri, Daniel Jeans, Victor Riabov, Boris Leissner, Dennis Shpakov, Szymon Gadomski, John Strologas, Ali Anjomshoaa, Margherita Obertino, Marco Maggiora, Ajit Kurup, Roger Moore, Manfred Lindner, Manuela Cirilli, Marcella Bona, Evgeny Popkov, Anna Burrage, Jacques Lafrancois, Daniel Wicke, Levan Babuhkhadia, Mikolai Cwiok, John Jowett, Jodi Whittlin, Fabiola Gianotti, Alexander Rakitine, Sergey Avvakumov, Tom Ferbel, Barbara Smalska, John Schwarz, Emmanuel Olaiya, Mat Dobbs, Dmitriy Karmanov. (Missing Pavel Zarzhitsky.)

Index

LECTURERS

Janet Conrad	Columbia University, New York, NY
Y. Dokshitzer	LPT, Paris, France
T. Ferbel	University of Rochester, Rochester, NY
F. Gianotti	CERN, Geneva, Switzerland
J. Jowett	CERN, Geneva, Switzerland
J. Lefrancois	Lab de l'Accelerateur Lineaire, Orsay, France
H. Prosper	Florida State University, Tallahassee, FL
V. Rubakov	Institute for Nuclear Research, Moscow, Russia
A. Schwarz	Deutsches Elektronen Synchrotron, Hamburg, Germany
J. Schwarz	California Institute of Technology, Pasadena, CA

ADVISORY COMMITTEE

R. Cashmore,	University of Oxford, Oxford, England
T. Ferbel	University of Rochester, Rochester, NY
D. Gross	University of California, Santa Barbara, CA
C. Jarlskog	University of Lund, Lund, Sweden
J. Lefrancois	Lab de l'Accelerateur Lineaire, Orsay, France
C. Quigg	Fermi National Accelerator Laboratory, Batavia, IL
F. Sauli	CERN, Geneva, Switzerland
A. Skrinsky	BINP-Novosibirsk, Russia
M. Spiro	Saclay, Paris, France
G. Wolf	University of Hamburg, Hamburg, Germany

CO-DIRECTOR

M. Danilov	Institute for Theoretical and Experimental Physics Moscow, Russia
H. Prosper	Florida State University, Tallahassee, FL